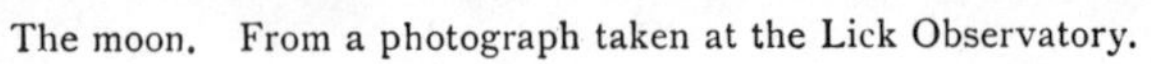

The moon. From a photograph taken at the Lick Observatory.

A Short History of Astronomy

From Earliest Times Through the Nineteenth Century

By ARTHUR BERRY, M.A.
LATE FELLOW AND ASSISTANT TUTOR OF KING'S COLLEGE, CAMBRIDGE;
FELLOW OF UNIVERSITY COLLEGE, LONDON

Wagner. Verzeiht! es ist ein gross Ergetzen
Sich in den Geist der Zeiten zu versetzen,
Zu schauen wie vor uns ein weiser Mann gedacht,
Und wie wir's dann zuletzt so herrlich weit gebracht.
Faust. O ja, bis an die Sterne weit!

GOETHE'S *Faust.*

DOVER PUBLICATIONS, INC.
NEW YORK

Published in Canada by General Publishing Company, Ltd., 30 Lesmill Road, Don Mills, Toronto, Ontario.

Published in the United Kingdom by Constable and Company, Ltd., 10 Orange Street, London WC 2.

This Dover edition, first published in 1961, is an unabridged and unaltered republication, except for the substitution of new portraits for those in the original edition. The work was originally published by John Murray in 1898.

Standard Book Number: 486-20210-0
Library of Congress Catalog Card Number: 61-3503

Manufactured in the United States of America
Dover Publications, Inc.
180 Varick Street
New York, N. Y. 10014

PREFACE.

I HAVE tried to give in this book an outline of the history of astronomy from the earliest historical times to the present day, and to present it in a form which shall be intelligible to a reader who has no special knowledge of either astronomy or mathematics, and has only an ordinary educated person's power of following scientific reasoning.

In order to accomplish my object within the limits of one small volume it has been necessary to pay the strictest attention to compression; this has been effected to some extent by the omission of all but the scantiest treatment of several branches of the subject which would figure prominently in a book written on a different plan or on a different scale. I have deliberately abstained from giving any connected account of the astronomy of the Egyptians, Chaldaeans, Chinese, and others to whom the early development of astronomy is usually attributed. On the one hand, it does not appear to me possible to form an independent opinion on the subject without a first-hand knowledge of the documents and inscriptions from which our information is derived; and on the other, the various Oriental scholars who have this knowledge still differ so widely from one another in the interpretations that they give that it appears premature to embody their results in the dogmatic form of a text-book. It has also seemed advisable to lighten the book by omitting—except in a very few simple and important cases—all accounts of astronomical instruments; I do not remember ever to have derived any pleasure or profit from a written description of a scientific instrument before seeing the instrument itself, or one very similar to it, and I have abstained from attempting to give to my readers what I have never succeeded in obtaining myself. The aim of the book has also necessitated the omission of a number of important astronomical discoveries, which find their natural expression in the technical language of mathematics. I have on this account only been able to describe in the briefest and most general way the wonderful and beautiful superstructure which several generations of mathematicians have erected on the foundations laid by Newton. For the same reason I have been compelled occasionally to occupy a good deal of space in stating in ordinary English what might have been expressed much more briefly, as well as more clearly, by an algebraical formula: for the benefit of such mathematicians as may happen to read the book I have added a few mathematical footnotes; otherwise I have tried to abstain scrupulously from the use of any mathematics beyond simple arithmetic and a few technical terms which are explained in the text. A good deal of space has also been saved by the total omission of, or the briefest possible reference to, a very large number of astronomical facts which do not bear on any well-established general theory; and for similar reasons I have generally abstained from noticing speculative theories which have not yet been established or refuted. In particular, for these and for other reasons (stated more fully at the beginning of chapter XIII.), I have dealt in the briefest possible way with the immense mass of observations which modern astronomy has accumulated; it would, for example, have been easy to have filled one or more volumes with an account of observations of sun-spots made during the last half-century, and of theories based on them, but I have in fact only given a page or two to the subject.

I have given short biographical sketches of leading astronomers (other than living ones), whenever the material existed, and have attempted in this way to make their personalities and surroundings tolerably vivid; but I have tried to resist the temptation of filling up space with merely picturesque details having no real bearing on scientific progress. The trial of Kepler's mother for witchcraft is probably quite as interesting as that of Galilei before the Inquisition, but I have entirely omitted the first and given a good deal of space to the second, because, while the former appeared to be chiefly of curious interest, the latter appeared to me to be not merely a striking incident in the life of a great astronomer, but a part of the history of astronomical thought. I have also inserted a large number of dates, as they occupy very little space, and may be found useful by some readers, while they can be ignored with great ease by others; to facilitate reference the dates of birth and death (when known) of every astronomer of note mentioned in the book (other than living ones) have been put into the Index of Names.

I have not scrupled to give a good deal of space to descriptions of such obsolete theories as appeared to me to form an integral part of astronomical progress. One of the reasons why the history of a science is worth studying is that it sheds light on the processes whereby a scientific theory is formed in order to account for certain facts, and then undergoes successive modifications as new facts are gradually brought to bear on it, and is perhaps finally abandoned when its discrepancies with facts can no longer be explained or concealed. For example, no modern astronomer as such need be concerned with the Greek scheme of epicycles, but the history of its invention, of its gradual perfection as fresh observations were obtained, of its subsequent failure to stand more stringent tests, and of its final abandonment in favour of a more satisfactory theory, is, I think, a valuable and interesting object-lesson in scientific method. I have at any rate written this book with that conviction, and have decided very largely from that point of view what to omit and what to include.

The book makes no claim to be an original contribution to the subject; it is written largely from second-hand sources, of which, however, many are not very accessible to the general reader. Particulars of the authorities which have been used are given in an appendix.

It remains gratefully to acknowledge the help that I have received in my work. Mr. W. W. Rouse Ball, Tutor of Trinity College, whose great knowledge of the history of mathematics—a subject very closely connected with astronomy—has made his criticisms of special value, has been kind enough to read the proofs, and has thereby saved me from several errors; he has also given me valuable information with regard to portraits of astronomers. Miss H. M. Johnson has undertaken the laborious and tedious task of reading the whole book in manuscript as well as in proof, and of verifying the cross-references. Miss F. Hardcastle, of Girton College, has also read the proofs, and verified most of the numerical calculations, as well as the cross-references. To both I am indebted for the detection of a large number of obscurities in expression, as well as of clerical and other errors and of misprints. Miss Johnson has also saved me much time by making the Index of Names, and Miss Hardcastle has rendered me a further service of great value by drawing a considerable number of the diagrams. I am also indebted to Mr. C. E. Inglis, of this College, for fig 81; and I have to thank Mr. W. H. Wesley, of the Royal Astronomical Society, for various references to the literature of the subject, and in particular for help in obtaining access to various illustrations.

I am further indebted to the following bodies and individual astronomers for permission to reproduce photographs and drawings, and in some cases also for the gift of copies of the originals : the Council of the Royal Society, the Council of the Royal Astronomical Society, the Director of the Lick Observatory, the Director of the Instituto Geographico-Militare of Florence, Professor Barnard, Major Darwin, Dr. Gill, M. Janssen, M. Loewy, Mr. E. W. Maunder, Mr. H. Pain, Professor E. C. Pickering, Dr Schuster, Dr. Max Wolf.

ARTHUR BERRY.

KING'S COLLEGE, CAMBRIDGE,
September 1898.

CONTENTS.

CHAPTER II.

CHAPTER III.

CHAPTER IV.

CHAPTER V.

CHAPTER VI.

CHAPTER VII.

CHAPTER VIII.

CHAPTER IX.

CHAPTER X.

CHAPTER XI.

CHAPTER XII.

CHAPTER XIII.

LIST OF ILLUSTRATIONS.

A SHORT HISTORY OF ASTRONOMY.

CHAPTER I.

PRIMITIVE ASTRONOMY.

"The never-wearied Sun, the Moon exactly round,
And all those Stars with which the brows of ample heaven are
crowned,
Orion, all the Pleiades, and those seven Atlas got,
The close beamed Hyades, the Bear, surnam'd the Chariot,
That turns about heaven's axle tree, holds ope a constant eye
Upon Orion, and of all the cressets in the sky
His golden forehead never bows to th' Ocean empery."

The Iliad (Chapman's translation).

1. ASTRONOMY is the science which treats of the sun, the moon, the stars, and other objects such as comets which are seen in the sky. It deals to some extent also with the earth, but only in so far as it has properties in common with the heavenly bodies. In early times astronomy was concerned almost entirely with the observed motions of the heavenly bodies. At a later stage astronomers were able to discover the distances and sizes of many of the heavenly bodies, and to weigh some of them; and more recently they have acquired a considerable amount of knowledge as to their nature and the material of which they are made.

2. We know nothing of the beginnings of astronomy, and can only conjecture how certain of the simpler facts of the science—particularly those with a direct influence on human life and comfort—gradually became familiar to early mankind, very much as they are familiar to modern savages.

With these facts it is convenient to begin, taking them in the order in which they most readily present themselves to any ordinary observer.

3. The sun is daily seen to rise in the eastern part of the sky, to travel across the sky, to reach its highest position in the south in the middle of the day, then to sink, and finally to set in the western part of the sky. But its daily path across the sky is not always the same: the points of the horizon at which it rises and sets, its height in the sky at midday, and the time from sunrise to sunset, all go through a series of changes, which are accompanied by changes in the weather, in vegetation, etc.; and we are thus able to recognise the existence of the seasons, and their recurrence after a certain interval of time which is known as a year.

4. But while the sun always appears as a bright circular disc, the next most conspicuous of the heavenly bodies, the moon, undergoes changes of form which readily strike the observer, and are at once seen to take place in a regular order and at about the same intervals of time. A little more care, however, is necessary in order to observe the connection between the form of the moon and her position in the sky with respect to the sun. Thus when the moon is first visible soon after sunset near the place where the sun has set, her form is a thin crescent (cf. fig. 11 on p. 31), the hollow side being turned away from the sun, and she sets soon after the sun. Next night the moon is farther from the sun, the crescent is thicker, and she sets later; and so on, until after rather less than a week from the first appearance of the crescent, she appears as a semicircular disc, with the flat side turned away from the sun. The semicircle enlarges, and after another week has grown into a complete disc; the moon is now nearly in the opposite direction to the sun, and therefore rises about at sunset and sets about at sunrise. She then begins to approach the sun on the other side, rising before it and setting in the daytime; her size again diminishes, until after another week she is again semicircular, the flat side being still turned away from the sun, but being now turned towards the west instead of towards the east. The semicircle then becomes a gradually diminishing crescent, and the time of rising

approaches the time of sunrise, until the moon becomes altogether invisible. After two or three nights the new moon reappears, and the whole series of changes is repeated. The different forms thus assumed by the moon are now known as her **phases**; the time occupied by this series of changes, the month, would naturally suggest itself as a convenient measure of time; and the day, month, and year would thus form the basis of a rough system of time-measurement.

5. From a few observations of the stars it could also clearly be seen that they too, like the sun and moon, changed their positions in the sky, those towards the east being seen to rise, and those towards the west to sink and finally set, while others moved across the sky from east to west, and those in a certain northern part of the sky, though also in motion, were never seen either to rise or set. Although anything like a complete classification of the stars belongs to a more advanced stage of the subject, a few star groups could easily be recognised, and their position in the sky could be used as a rough means of measuring time at night, just as the position of the sun to indicate the time of day.

6. To these rudimentary notions important additions were made when rather more careful and prolonged observations became possible, and some little thought was devoted to their interpretation.

Several peoples who reached a high stage of civilisation at an early period claim to have made important progress in astronomy. Greek traditions assign considerable astronomical knowledge to Egyptian priests who lived some thousands of years B.C., and some of the peculiarities of the pyramids which were built at some such period are at any rate plausibly interpreted as evidence of pretty accurate astronomical observations; Chinese records describe observations supposed to have been made in the 25th century B.C.; some of the Indian sacred books refer to astronomical knowledge acquired several centuries before this time; and the first observations of the Chaldaean priests of Babylon have been attributed to times not much later.

On the other hand, the earliest recorded astronomical observation the authenticity of which may be accepted without scruple belongs only to the 8th century B.C.

For the purposes of this book it is not worth while to make any attempt to disentangle from the mass of doubtful tradition and conjectural interpretation of inscriptions, bearing on this early astronomy, the few facts which lie embedded therein; and we may proceed at once to give some account of the astronomical knowledge, other than that already dealt with, which is discovered in the possession of the earliest really historical astronomers—the Greeks—at the beginning of their scientific history, leaving it an open question what portions of it were derived from Egyptians, Chaldaeans, their own ancestors, or other sources.

7. If an observer looks at the stars on any clear night he sees an apparently innumerable * host of them, which seem to lie on a portion of a spherical surface, of which he is the centre. This spherical surface is commonly spoken of as the sky, and is known to astronomy as the **celestial sphere.** The visible part of this sphere is bounded by the earth, so that only half can be seen at once; but only the slightest effort of the imagination is required to think of the other half as lying below the earth, and containing other stars, as well as the sun. This sphere appears to the observer to be very large, though he is incapable of forming any precise estimate of its size. †

Most of us at the present day have been taught in childhood that the stars are at different distances, and that this sphere has in consequence no real existence. The early peoples had no knowledge of this, and for them the celestial sphere really existed, and was often thought to be a solid sphere of crystal.

Moreover modern astronomers, as well as ancient, find it convenient for very many purposes to make use of this sphere, though it has no material existence, as a means of representing the directions in which the heavenly bodies are seen and their motions. For all that direct observation

* In our climate 2,000 is about the greatest number ever visible at once, even to a keen-sighted person.

† Owing to the greater brightness of the stars overhead they usually seem a little nearer than those near the horizon, and consequently the visible portion of the celestial sphere appears to be rather less than a half of a complete sphere. This is, however, of no importance, and will for the future be ignored.

can tell us about the position of such an object as a star is its *direction*; its distance can only be ascertained by indirect methods, if at all. If we draw a sphere, and suppose the observer's eye placed at its centre O (fig. 1), and then draw a straight line from O to a star S, meeting the surface of the sphere in the point *s*; then the star appears exactly in the same position as if it were at *s*, nor would its apparent position be changed if it were placed at any other point, such as S′ or S″, on this same

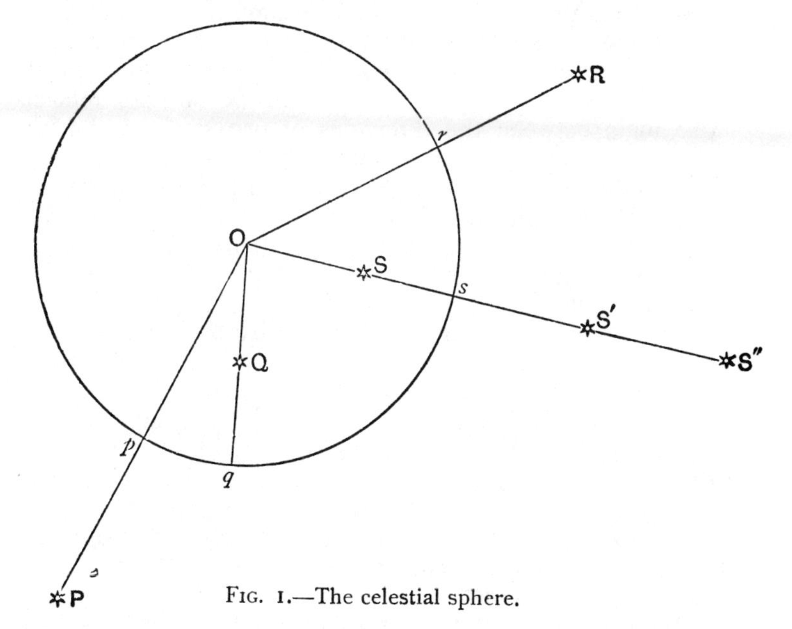

FIG. 1.—The celestial sphere.

line. When we speak, therefore, of a star as being at a point *s* on the celestial sphere, all that we mean is that it is in the same direction as the point *s*, or, in other words, that it is situated somewhere on the straight line through O and S. The advantages of this method of representing the position of a star become evident when we wish to compare the positions of several stars. The difference of direction of two stars is the angle between the lines drawn from the eye to the stars; *e.g.*, if the stars are R, S, it is the angle R O S. Similarly the difference of direction of another pair of stars, P, Q, is the angle P O Q. The two stars P and Q appear nearer together than do R and S, or farther apart, according as the angle P O Q is less or greater than the angle R O S. But if we represent the stars by the corresponding points *p*, *q*, *r*, *s* on the celestial sphere, then (by an obvious property of the sphere) the angle P O Q (which is the same as *p* O *q*) is less or greater than the angle R O S (or *r* O *s*) according as the arc joining *p q* on the sphere is less or greater than the arc joining *r s*, and in the same proportion; if, for example, the angle R O S is twice as great as the angle P O Q, so also is the arc *p q* twice as great as the arc *r s*. We may therefore, in all questions relating only to the directions of the stars, replace the angle between the directions of two stars by the arc joining the corresponding points on the celestial sphere, or, in other words, by the distance between these points on the celestial sphere. But such arcs on a sphere are easier both to estimate by eye and to treat geometrically than angles, and the use of the celestial sphere is therefore of great value, apart from its historical origin. It is important to note that this **apparent distance** of two stars, *i.e.* their distance from one another on the celestial sphere, is an entirely different thing from their actual distance from one another in space. In the figure, for example, Q is actually much nearer to S than it is to P, but the apparent distance measured by the arc *q s* is several times greater than *q p*. The apparent distance of two points on the celestial sphere is measured numerically by the angle between the lines joining the eye to the two points, expressed in **degrees, minutes,** and **seconds.***

We might of course agree to regard the celestial sphere as of a particular size, and then express the distance between two points on it in miles, feet, or inches; but it is practically very inconvenient to do so. To say, as some people occasionally do, that the distance between two stars is so many feet is meaningless, unless the supposed size of the celestial sphere is given at the same time.

It has already been pointed out that the observer is always at the centre of the celestial sphere; this remains

* A right angle is divided into ninety degrees (90°), a degree into sixty minutes (60′), and a minute into sixty seconds (60″).

true even if he moves to another place. A sphere has, however, only one centre, and therefore if the sphere remains fixed the observer cannot move about and yet always remain at the centre. The old astronomers met this difficulty by supposing that the celestial sphere was so large that any possible motion of the observer would be insignificant in comparison with the radius of the sphere and could be neglected. It is often more convenient—when we are using the sphere as a mere geometrical device for representing the position of the stars—to regard the sphere as moving with the observer, so that he always remains at the centre.

8. Although the stars all appear to move across the sky (§ 5), and their rates of motion differ, yet the distance between any two stars remains unchanged, and they were consequently regarded as being attached to the celestial sphere. Moreover a little careful observation would have shown that the motions of the stars in different parts of the sky, though at first sight very different, were just such as would have been produced by the celestial sphere—with the stars attached to it—turning about an axis passing through the centre and through a point in the northern sky close to the familiar pole-star. This point is called the **pole.** As, however, a straight line drawn through the centre of a sphere meets it in two points, the axis of the celestial sphere meets it again in a second point, opposite the first, lying in a part of the celestial sphere which is permanently below the horizon. This second point is also called a pole; and if the two poles have to be distinguished, the one mentioned first is called the **north pole,** and the other the **south pole.** The direction of the rotation of the celestial sphere about its axis is such that stars near the north pole are seen to move round it in circles in the direction opposite to that in which the hands of a clock move; the motion is uniform, and a complete revolution is performed in four minutes less than twenty-four hours; so that the position of any star in the sky at twelve o'clock to-night is the same as its position at four minutes to twelve to-morrow night.

The moon, like the stars, shares this motion of the celestial sphere, and so also does the sun, though this is more difficult to recognise owing to the fact that the sun and stars are not seen together.

As other motions of the celestial bodies have to be dealt with, the general motion just described may be conveniently referred to as the **daily motion** or **daily rotation** of the celestial sphere.

9. A further study of the daily motion would lead to the recognition of certain important circles of the celestial sphere.

Each star describes in its daily motion a circle, the size of which depends on its distance from the poles. Fig. 2 shews the paths described by a number of stars near the pole, recorded photographically, during part of a night. The pole-star describes so small a circle that its motion can only with difficulty be detected with the naked eye, stars a little farther off the pole describe larger circles, and so on, until we come to stars half-way between the two poles, which describe the largest circle which can be drawn on the celestial sphere. The circle on which these stars lie and which is described by any one of them daily is called the **equator.** By looking at a diagram such as fig. 3, or, better still, by looking at an actual globe, it can easily be seen that half the equator (E Q W) lies above and half (the dotted part, W R E) below the horizon, and that in consequence a star, such as *s*, lying on the equator, is in its daily motion as long a time above the horizon as below. If a star, such as S, lies on the north side of the equator, *i.e.* on the side on which the north pole P lies, more than half of its daily path lies above the horizon and less than half (as shewn by the dotted line) lies below; and if a star is near enough to the north pole (more precisely, if it is nearer to the north pole than the nearest point, K, of the horizon), as σ, it never sets, but remains continually above the horizon. Such a star is called a (**northern**) **circumpolar** star. On the other hand, less than half of the daily path of a star on the south side of the equator, as S′, is above the horizon, and a star, such as σ', the distance of which from the north pole is greater than the distance of the farthest point, H, of the horizon, or which is nearer than H to the south pole, remains continually below the horizon.

10. A slight familiarity with the stars is enough to shew any one that the same stars are not always visible at the

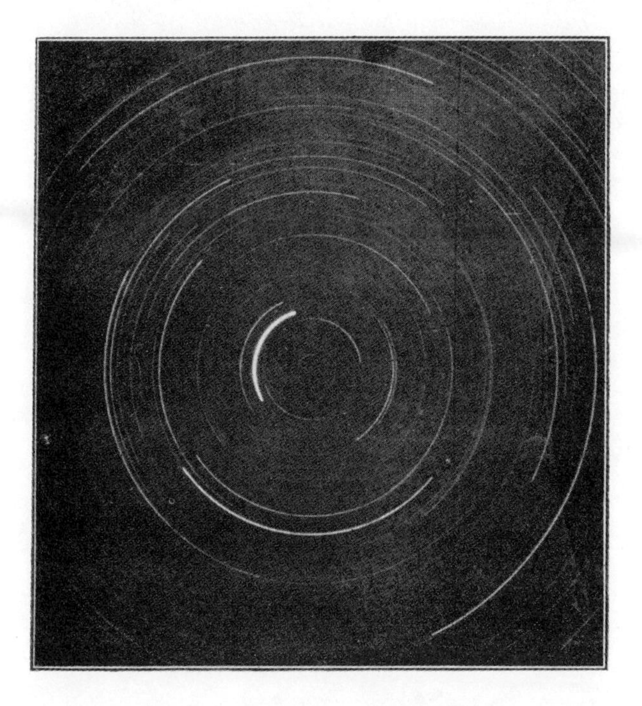

FIG. 2.—The paths of circumpolar stars, shewing their movement during seven hours. From a photograph by Mr. H. Pain. The thickest line is the path of the pole star.

[*To face p.* 8.

same time of night. Rather more careful observation, carried out for a considerable time, is necessary in order to see that the aspect of the sky changes in a regular way from night to night, and that after the lapse of a year the same stars become again visible at the same time. The explanation of these changes as due to the motion of the sun on the celestial sphere is more difficult, and the

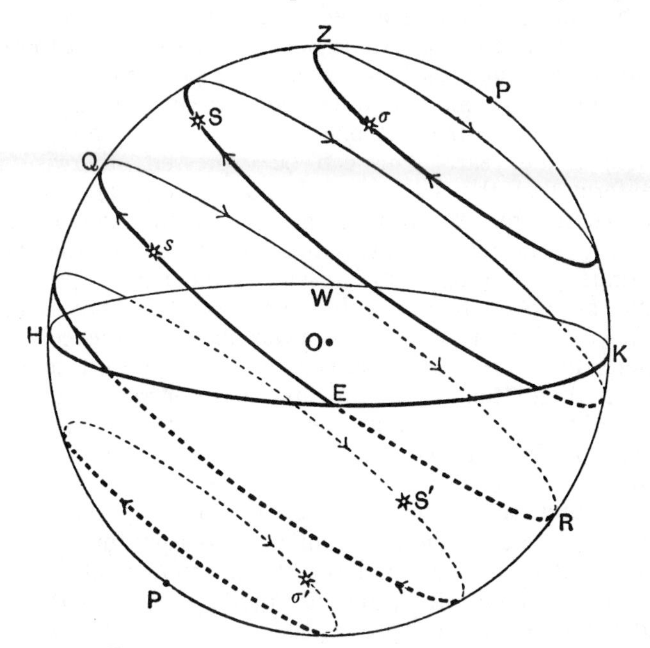

FIG. 3.—The circles of the celestial sphere.

unknown discoverer of this fact certainly made one of the most important steps in early astronomy.

If an observer notices soon after sunset a star somewhere in the west, and looks for it again a few evenings later at about the same time, he finds it lower down and nearer to the sun; a few evenings later still it is invisible, while its place has now been taken by some other star which was at first farther east in the sky. This star can in turn be observed to approach the sun evening by evening. Or if the stars visible after sunset low down in the east are noticed a few days later, they are found to be higher up in the sky, and their place is taken by other stars at first too low down to be seen. Such observations of stars rising or setting about sunrise or sunset shewed to early observers that the stars were gradually changing their position with respect to the sun, or that the sun was changing its position with respect to the stars.

The changes just described, coupled with the fact that the stars do not change their positions with respect to one another, shew that the stars as a whole perform their daily revolution rather more rapidly than the sun, and at such a rate that they gain on it one complete revolution in the course of the year. This can be expressed otherwise in the form that the stars are all moving westward on the celestial sphere, relatively to the sun, so that stars on the east are continually approaching and those on the west continually receding from the sun. But, again, the same facts can be expressed with equal accuracy and greater simplicity if we regard the stars as fixed on the celestial sphere, and the sun as moving on it from west to east among them (that is, in the direction *opposite* to that of the daily motion), and at such a rate as to complete a circuit of the celestial sphere and to return to the same position after a year.

This **annual motion** of the sun is, however, readily seen not to be merely a motion from west to east, for if so the sun would always rise and set at the same points of the horizon, as a star does, and its midday height in the sky and the time from sunrise to sunset would always be the same. We have already seen that if a star lies on the equator half of its daily path is above the horizon, if the star is north of the equator more than half, and if south of the equator less than half; and what is true of a star is true for the same reason of any body sharing the daily motion of the celestial sphere. During the summer months therefore (March to September), when the day is longer than the night, and more than half of the sun's daily path is above the horizon, the sun must be north of the equator, and during the winter months (September to March) the sun must be south of the equator. The change in the sun's distance from the pole is also evident from the fact that in the winter months the sun is on the whole lower down in the sky than in summer, and that in particular its midday height is less.

11. The sun's path on the celestial sphere is therefore oblique to the equator, lying partly on one side of it and partly on the other. A good deal of careful observation of the kind we have been describing must, however, have been necessary before it was ascertained that the sun's annual path on the celestial sphere (see fig. 4) is a **great circle** (that is, a circle having its centre at the centre of the sphere). This great circle is now called the **ecliptic** (because eclipses take place only when the moon is in or near it), and the angle at which it cuts the equator is called the **obliquity** of the ecliptic. The Chinese claim to have measured the obliquity in 1100 B.C., and to have found the remarkably accurate value 23° 52′ (cf. chapter II., § 35). The truth of this statement may reasonably be doubted, but on the other hand the statement of some late Greek writers that either Pythagoras or Anaximander (6th century B.C.) was the *first* to discover the obliquity of the ecliptic is almost certainly wrong. It must have been known with reasonable accuracy to both Chaldaeans and Egyptians long before.

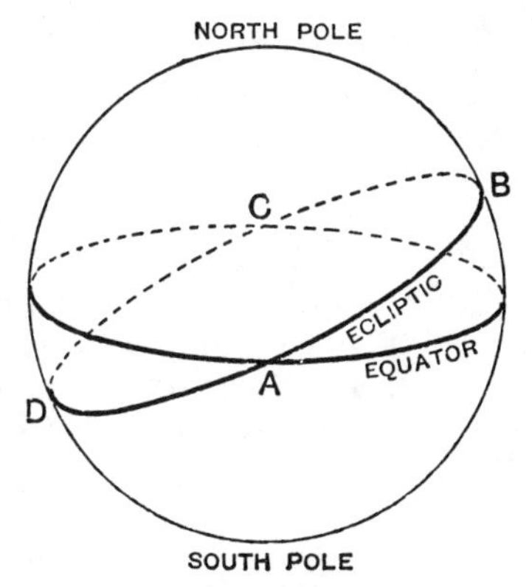

FIG. 4.—The equator and the ecliptic.

When the sun crosses the equator the day is equal to the night, and the times when this occurs are consequently known as the **equinoxes**, the **vernal equinox** occurring when the sun crosses the equator from south to north (about March 21st), and the **autumnal equinox** when it crosses back (about September 23rd). The points on the celestial sphere where the sun crosses the equator (A, C in fig. 4), *i.e.* where ecliptic and equator cross one another, are called the **equinoctial points,** occasionally also the equinoxes.

After the vernal equinox the sun in its path along the

ecliptic recedes from the equator towards the north, until it reaches, about three months afterwards, its greatest distance from the equator, and then approaches the equator again. The time when the sun is at its greatest distance from the equator on the north side is called the **summer solstice,** because then the northward motion of the sun is arrested and it temporarily appears to stand still. Similarly the sun is at its greatest distance from the equator towards the south at the **winter solstice**. The points on the ecliptic (B, D in fig. 4) where the sun is at the solstices are called the **solstitial points,** and are half-way between the equinoctial points.

12. The earliest observers probably noticed particular groups of stars remarkable for their form or for the presence of bright stars among them, and occupied their fancy by tracing resemblances between them and familiar objects, etc. We have thus at a very early period a rough attempt at dividing the stars into groups called **constellations** and at naming the latter.

In some cases the stars regarded as belonging to a constellation form a well-marked group on the sky, sufficiently separated from other stars to be conveniently classed together, although the resemblance which the group bears to the object after which it is named is often very slight. The seven bright stars of the Great Bear, for example, form a group which any observer would very soon notice and naturally make into a constellation, but the resemblance to a bear of these and the fainter stars of the constellation is sufficiently remote (see fig. 5), and as a matter of fact this part of the Bear has also been called a Waggon and is in America familiarly known as the Dipper; another constellation has sometimes been called the Lyre and sometimes also the Vulture. In very many cases the choice of stars seems to have been made in such an arbitrary manner, as to suggest that some fanciful figure was first imagined and that stars were then selected so as to represent it in some rough sort of way. In fact, as Sir John Herschel remarks, "The constellations seem to have been purposely named and delineated to cause as much confusion and inconvenience as possible. Innumerable snakes twine through long and contorted areas of the heavens where no

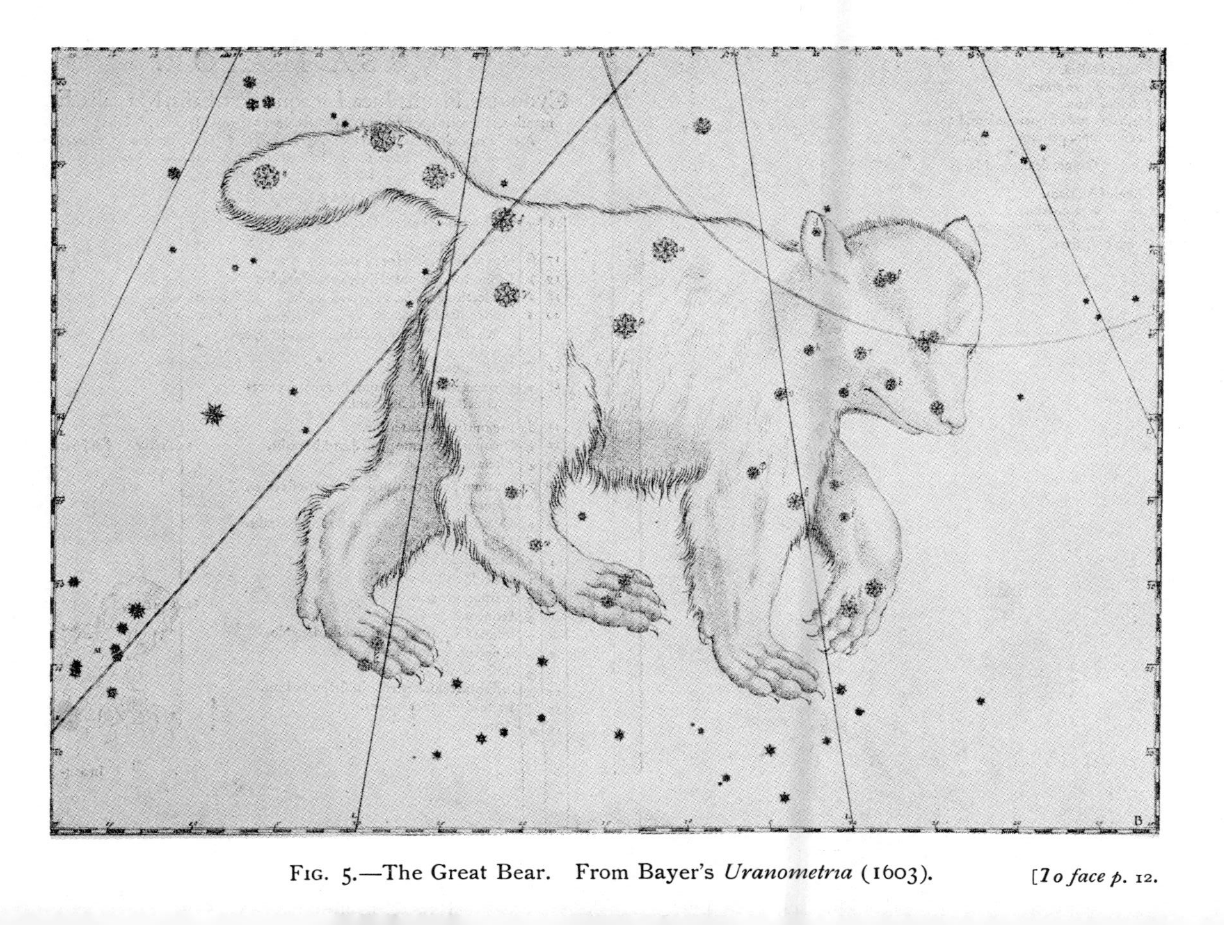

FIG. 5.—The Great Bear. From Bayer's *Uranometria* (1603).

[To face p. 12.

memory can follow them ; bears, lions, and fishes, large and small, confuse all nomenclature." (*Outlines of Astronomy*, § 301.)

The constellations as we now have them are, with the exception of a certain number (chiefly in the southern skies) which have been added in modern times, substantially those which existed in early Greek astronomy ; and such information as we possess of the Chaldaean and Egyptian constellations shews resemblances indicating that the Greeks borrowed some of them. The names, as far as they are not those of animals or common objects (Bear, Serpent, Lyre, etc.), are largely taken from characters in the Greek mythology (Hercules, Perseus, Orion, etc.). The constellation Berenice's Hair, named after an Egyptian queen of the 3rd century B.C., is one of the few which commemorate a historical personage.*

13. Among the constellations which first received names were those through which the sun passes in its annual circuit of the celestial sphere, that is those through which the ecliptic passes. The moon's monthly path is also a great circle, never differing very much from the ecliptic, and the paths of the planets (§ 14) are such that they also are never far from the ecliptic. Consequently the sun, the moon, and the five planets were always to be found within a region of the sky extending about 8° on each side of the ecliptic. This strip of the celestial sphere was called the **zodiac,** because the constellations in it were (with one exception) named after living things (Greek ζῷον, an animal) ; it was divided into twelve equal parts, the **signs of the zodiac,** through one of which the sun passed every month, so that the position of the sun at any time could be roughly described by stating in what "sign" it was. The stars in each "sign" were formed into a constellation, the "sign" and the constellation each receiving the same name. Thus

* I have made no attempt either here or elsewhere to describe the constellations and their positions, as I believe such verbal descriptions to be almost useless. For a beginner who wishes to become familiar with them the best plan is to get some better informed friend to point out a few of the more conspicuous ones, in different parts of the sky. Others can then be readily added by means of a star-atlas, or of the star-maps given in many textbooks.

arose twelve **zodiacal constellations**, the names of which have come down to us with unimportant changes from early Greek times.* Owing, however, to an alteration of the position of the equator, and consequently of the equinoctial points, the sign Aries, which was defined by Hipparchus in the second century B.C. (see chapter II., § 42) as beginning at the vernal equinoctial point, no longer contains the constellation Aries, but the preceding one, Pisces; and there is a corresponding change throughout the zodiac. The more precise numerical methods of modern astronomy have, however, rendered the signs of the zodiac almost obsolete; but the **first point of Aries** (♈), and the **first point of Libra** (♎), are still the recognised names for the equinoctial points.

In some cases individual stars also received special names, or were called after the part of the constellation in which they were situated, *e.g.* Sirius, the Eye of the Bull, the Heart of the Lion, etc.; but the majority of the present names of single stars are of Arabic origin (chapter III., § 64).

14. We have seen that the stars, as a whole, retain invariable positions on the celestial sphere,† whereas the sun and moon change their positions. It was, however, discovered in prehistoric times that five bodies, at first sight barely distinguishable from the other stars, also changed their places. These five—Mercury, Venus, Mars, Jupiter, and Saturn—with the sun and moon, were called **planets**,‡ or wanderers, as distinguished from the **fixed stars**.

* The names, in the customary Latin forms, are: Aries, Taurus, Gemini, Cancer, Leo, Virgo, Libra, Scorpio, Sagittarius, Capricornus, Aquarius, and Pisces; they are easily remembered by the doggerel verses:—

> The Ram, the Bull, the Heavenly Twins,
> And next the Crab, the Lion shines,
> The Virgin and the Scales,
> The Scorpion, Archer, and He-Goat,
> The Man that bears the Watering-pot,
> And Fish with glittering tails.

† This statement leaves out of account small motions nearly or quite invisible to the naked eye, some of which are among the most interesting discoveries of telescopic astronomy; see, for example, chapter X., §§ 207-215.

‡ The custom of calling the sun and moon planets has now died out, and the modern usage will be adopted henceforward in this book.

Mercury is never seen except occasionally near the horizon just after sunset or before sunrise, and in a climate like ours requires a good deal of looking for; and it is rather remarkable that no record of its discovery should exist. Venus is conspicuous as the Evening Star or as the Morning Star. The discovery of the identity of the Evening and Morning Stars is attributed to Pythagoras (6th century B.C.), but must almost certainly have been made earlier, though the Homeric poems contain references to both, without any indication of their identity. Jupiter is at times as conspicuous as Venus at her brightest, while Mars and Saturn, when well situated, rank with the brightest of the fixed stars.

The paths of the planets on the celestial sphere are, as we have seen (§ 13), never very far from the ecliptic; but whereas the sun and moon move continuously along their paths from west to east, the motion of a planet is sometimes from west to east, or **direct,** and sometimes from east to west, or **retrograde.** If we begin to watch a planet when it is moving eastwards among the stars, we find that after a time the motion becomes slower and slower, until the planet hardly seems to move at all, and then begins to move with gradually increasing speed in the opposite direction; after a time this westward motion becomes slower and then ceases, and the planet then begins to move eastwards again, at first slowly and then faster, until it returns to its original condition, and the changes are repeated. When the planet is just reversing its motion it is said to be **stationary,** and its position then is called a **stationary point.** The time during which a planet's motion is retrograde is, however, always considerably less than that during which it is direct; Jupiter's motion, for example, is direct for about 39 weeks and retrograde for 17, while Mercury's direct motion lasts 13 or 14 weeks and the retrograde motion only about 3 weeks (see figs. 6, 7). On the whole the planets advance from west to east and describe circuits round the celestial sphere in periods which are different for each planet. The explanation of these irregularities in the planetary motions was long one of the great difficulties of astronomy.

15. The idea that some of the heavenly bodies are

nearer to the earth than others must have been suggested by eclipses (§ 17) and **occultations,** *i.e.* passages of the moon over a planet or fixed star. In this way the moon would be recognised as nearer than any of the other celestial bodies. No direct means being available for determining the distances, rapidity of motion was employed as a test of probable nearness. Now Saturn returns to the same place among the stars in about $29\frac{1}{2}$ years, Jupiter in 12 years, Mars in 2 years, the sun in one year, Venus in 225

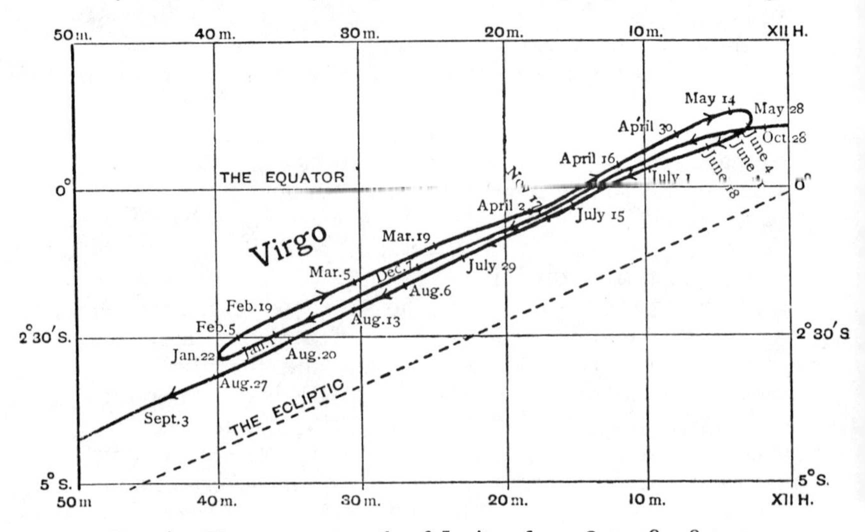

FIG. 6.—The apparent path of Jupiter from Oct. 28, 1897, to Sept. 3, 1898. The dates printed in the diagram shew the positions of Jupiter.

days, Mercury in 88 days, and the moon in 27 days; and this order was usually taken to be the order of distance, Saturn being the most distant, the moon the nearest. The stars being seen above us it was natural to think of the most distant celestial bodies as being the highest, and accordingly Saturn, Jupiter, and Mars being beyond the sun were called **superior planets,** as distinguished from the two **inferior planets** Venus and Mercury. This division corresponds also to a difference in the observed motions, as Venus and Mercury seem to accompany the sun in its annual journey, being never more than about 47° and 29° respectively distant from it, on either side; while the other planets are not thus restricted in their motions.

16. One of the purposes to which applications of astronomical knowledge was first applied was to the measurement of time. As the alternate appearance and disappearance of the sun, bringing with it light and heat, is the most obvious of astronomical facts, so the day is

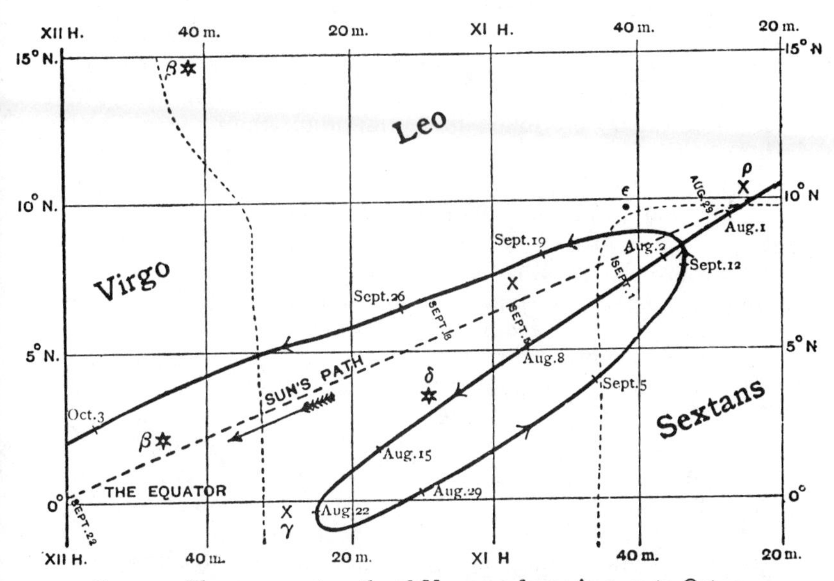

FIG. 7.—The apparent path of Mercury from Aug. 1 to Oct. 3, 1898. The dates printed in capital letters shew the positions of the sun; the other dates shew those of Mercury.

the simplest unit of time.* Some of the early civilised nations divided the time from sunrise to sunset and also the night each into 12 equal hours. According to this arrangement a day-hour was in summer longer than a

* It may be noted that our word "day" (and the corresponding word in other languages) is commonly used in two senses, either for the time between sunrise and sunset (day as distinguished from night), or for the whole period of 24 hours or day-and-night. The Greeks, however, used for the latter a special word, νυχθήμερον.

night-hour and in winter shorter, and the length of an hour varied during the year. At Babylon, for example, where this arrangement existed, the length of a day-hour was at midsummer about half as long again as in midwinter, and in London it would be about twice as long. It was therefore a great improvement when the Greeks, in comparatively late times, divided the whole day into 24 equal hours. Other early nations divided the same period into 12 double hours, and others again into 60 hours.

The next most obvious unit of time is the **lunar month,** or period during which the moon goes through her phases. A third independent unit is the year. Although the year is for ordinary life much more important than the month, yet as it is much longer and any one time of year is harder to recognise than a particular phase of the moon, the length of the year is more difficult to determine, and the earliest known systems of time-measurement were accordingly based on the month, not on the year. The month was found to be nearly equal to 29½ days, and as a period consisting of an exact number of days was obviously convenient for most ordinary purposes, months of 29 or 30 days were used, and subsequently the calendar was brought into closer accord with the moon by the use of months containing alternately 29 and 30 days (cf. chapter II., § 19).

Both Chaldaeans and Egyptians appear to have known that the year consisted of about 365¼ days; and the latter, for whom the importance of the year was emphasised by the rising and falling of the Nile, were probably the first nation to use the year in preference to the month as a measure of time. They chose a year of 365 days.

The origin of the week is quite different from that of the month or year, and rests on certain astrological ideas about the planets. To each hour of the day one of the seven planets (sun and moon included) was assigned as a "ruler," and each day named after the planet which ruled its first hour. The planets being taken in the order already given (§ 15), Saturn ruled the first hour of the first day, and therefore also the 8th, 15th, and 22nd hours of the first day, the 5th, 12th, and 19th of the second day, and so on; Jupiter ruled the 2nd, 9th, 16th, and 23rd hours of the first day, and subsequently the 1st hour of the 6th day. In this way the first hours of successive days fell respectively to Saturn, the Sun, the Moon, Mars, Mercury, Jupiter, and Venus. The first three are easily recognised in our Saturday, Sunday, and Monday; in the other days the names of the Roman gods have been replaced by their supposed Teutonic equivalents—Mercury by Wodan, Mars by Thues, Jupiter by Thor, Venus by Freia.*

17. **Eclipses** of the sun and moon must from very early times have excited great interest, mingled with superstitious terror, and the hope of acquiring some knowledge of them was probably an important stimulus to early astronomical work. That eclipses of the sun only take place at new moon, and those of the moon only at full moon, must have been noticed after very little observation; that eclipses of the sun are caused by the passage of the moon in front of it must have been only a little less obvious; but the discovery that eclipses of the moon are caused by the earth's shadow was probably made much later. In fact even in the time of Anaxagoras (5th century B.C.) the idea was so unfamiliar to the Athenian public as to be regarded as blasphemous.

One of the most remarkable of the Chaldaean contributions to astronomy was the discovery (made at any rate several centuries B.C.) of the recurrence of eclipses after a period, known as the **saros,** consisting of 6,585 days (or eighteen of our years and ten or eleven days, according as five or four leap-years are included). It is probable that the discovery was made, not by calculations based on knowledge of the motions of the sun and moon, but by mere study of the dates on which eclipses were recorded to have taken place. As, however, an eclipse of the sun (unlike an eclipse of the moon) is only visible over a small part of the surface of the earth, and eclipses of the sun occurring at intervals of eighteen years are not generally visible at the same place, it is not at all easy to see how the Chaldaeans could have established their cycle for this case, nor is it in fact clear that the saros was supposed to apply to solar as well as to lunar eclipses. The saros may

* Compare the French: Mardi, Mercredi, Jeudi, Vendredi; or better still the Italian: Martedi, Mercoledi, Giovedi, Venerdi.

be illustrated in modern times by the eclipses of the sun which took place on July 18th, 1860, on July 29th, 1878, and on August 9th, 1896; but the first was visible in Southern Europe, the second in North America, and the third in Northern Europe and Asia.

18. To the Chaldaeans may be assigned also the doubtful honour of having been among the first to develop **astrology**, the false science which has professed to ascertain the influence of the stars on human affairs, to predict by celestial observations wars, famines, and pestilences, and to discover the fate of individuals from the positions of the stars at their birth. A belief in some form of astrology has always prevailed in oriental countries; it flourished at times among the Greeks and the Romans; it formed an important part of the thought of the Middle Ages, and is not even quite extinct among ourselves at the present day.* It should, however, be remembered that if the history of astrology is a painful one, owing to the numerous illustrations which it affords of human credulity and knavery, the belief in it has undoubtedly been a powerful stimulus to genuine astronomical study (cf. chapter III., § 56, and chapter V., §§ 99, 100).

* See, for example, *Old Moore's* or *Zadkiel's Almanack*.

CHAPTER II.

GREEK ASTRONOMY.

"The astronomer discovers that geometry, a pure abstraction of the human mind, is the measure of planetary motion."

EMERSON.

19. In the earlier period of Greek history one of the chief functions expected of astronomers was the proper regulation of the calendar. The Greeks, like earlier nations, began with a calendar based on the moon. In the time of Hesiod a year consisting of 12 months of 30 days was in common use; at a later date a year made up of 6 **full** months of 30 days and 6 **empty** months of 29 days was introduced. To Solon is attributed the merit of having introduced at Athens, about 594 B.C., the practice of adding to every alternate year a "full" month. Thus a period of two years would contain 13 months of 30 days and 12 of 29 days, or 738 days in all, distributed among 25 months, giving, for the average length of the year and month, 369 days and about $29\frac{1}{2}$ days respectively. This arrangement was further improved by the introduction, probably during the 5th century B.C., of the **octaeteris**, or eight-year cycle, in three of the years of which an additional "full" month was introduced, while the remaining years consisted as before of 6 "full" and 6 "empty" months. By this arrangement the average length of the year was reduced to $365\frac{1}{4}$ days, that of the month remaining nearly unchanged. As, however, the Greeks laid some stress on beginning the month when the new moon was first visible, it was necessary to make from time to time arbitrary alterations in the calendar, and considerable confusion

resulted, of which Aristophanes makes the Moon complain in his play *The Clouds*, acted in 423 B.C.:

> "Yet you will not mark your days
> As she bids you, but confuse them, jumbling them all sorts of ways.
> And, she says, the Gods in chorus shower reproaches on her head,
> When, in bitter disappointment, they go supperless to bed,
> Not obtaining festal banquets, duly on the festal day."

20. A little later, the astronomer *Meton* (born about 460 B.C.) made the discovery that the length of 19 years is very nearly equal to that of 235 lunar months (the difference being in fact less than a day), and he devised accordingly an arrangement of 12 years of 12 months and 7 of 13 months, 125 of the months in the whole cycle being "full" and the others "empty." Nearly a century later *Callippus* made a slight improvement, by substituting in every fourth period of 19 years a "full" month for one of the "empty" ones. Whether **Meton's cycle,** as it is called, was introduced for the civil calendar or not is uncertain, but if not it was used as a standard by reference to which the actual calendar was from time to time adjusted. The use of this cycle seems to have soon spread to other parts of Greece, and it is the basis of the present ecclesiastical rule for fixing Easter. The difficulty of ensuring satisfactory correspondence between the civil calendar and the actual motions of the sun and moon led to the practice of publishing from time to time tables (παραπήγματα) not unlike our modern almanacks, giving for a series of years the dates of the phases of the moon, and the rising and setting of some of the fixed stars, together with predictions of the weather. Owing to the same cause the early writers on agriculture (*e.g.* Hesiod) fixed the dates for agricultural operations, not by the calendar, but by the times of the rising and setting of constellations, *i.e.* the times when they first became visible before sunrise or were last visible immediately after sunset—a practice which was continued long after the establishment of a fairly satisfactory calendar, and was apparently by no means extinct in the time of Galen (2nd century A.D.).

21. The Roman calendar was in early times even more confused than the Greek. There appears to have been

at one time a year of either 304 or 354 days; tradition assigned to Numa the introduction of a cycle of four years, which brought the calendar into fair agreement with the sun, but made the average length of the month considerably too short. Instead, however, of introducing further refinements the Romans cut the knot by entrusting to the ecclesiastical authorities the adjustment of the calendar from time to time, so as to make it agree with the sun and moon. According to one account, the first day of each month was proclaimed by a crier. Owing either to ignorance, or, as was alleged, to political and commercial favouritism, the priests allowed the calendar to fall into a state of great confusion, so that, as Voltaire remarked, "les généraux romains triomphaient toujours, mais ils ne savaient pas quel jour ils triomphaient."

A satisfactory reform of the calendar was finally effected by Julius Caesar during the short period of his supremacy at Rome, under the advice of an Alexandrine astronomer *Sosigenes*. The error in the calendar had mounted up to such an extent, that it was found necessary, in order to correct it, to interpolate three additional months in a single year (46 B.C.), bringing the total number of days in that year up to 445. For the future the year was to be independent of the moon; the ordinary year was to consist of 365 days, an extra day being added to February every fourth year (our leap-year), so that the average length of the year would be $365\frac{1}{4}$ days.

The new system began with the year 45 B.C., and soon spread, under the name of the **Julian Calendar**, over the civilised world.

22. To avoid returning to the subject, it may be convenient to deal here with the only later reform of any importance.

The difference between the average length of the year as fixed by Julius Caesar and the true year is so small as only to amount to about one day in 128 years. By the latter half of the 16th century the date of the vernal equinox was therefore about ten days earlier than it was at the time of the Council of Nice (A.D. 325), at which rules for the observance of Easter had been fixed. Pope

Gregory XIII. introduced therefore, in 1582, a slight change; ten days were omitted from that year, and it was arranged to omit for the future three leap-years in four centuries (viz. in 1700, 1800, 1900, 2100, etc., the years 1600, 2000, 2400, etc., remaining leap-years). The **Gregorian Calendar**, or **New Style**, as it was commonly called, was not adopted in England till 1752, when 11 days had to be omitted; and has not yet been adopted in Russia and Greece, the dates there being now 12 days behind those of Western Europe.

23. While their oriental predecessors had confined themselves chiefly to astronomical observations, the earlier Greek philosophers appear to have made next to no observations of importance, and to have been far more interested in inquiring into causes of phenomena. *Thales*, the founder of the Ionian school, was credited by later writers with the introduction of Egyptian astronomy into Greece, at about the end of the 7th century B.C.; but both Thales and the majority of his immediate successors appear to have added little or nothing to astronomy, except some rather vague speculations as to the form of the earth and its relation to the rest of the world. On the other hand, some real progress seems to have been made by *Pythagoras* * and his followers. Pythagoras taught that the earth, in common with the heavenly bodies, is a sphere, and that it rests without requiring support in the middle of the universe. Whether he had any real evidence in support of these views is doubtful, but it is at any rate a reasonable conjecture that he knew the moon to be bright because the sun shines on it, and the phases to be caused by the greater or less amount of the illuminated half turned towards us; and the curved form of the boundary between the bright and dark portions of the moon was correctly interpreted by him as evidence that the moon was spherical, and not a flat disc, as it appears at first sight. Analogy would then probably suggest that the earth also was spherical. However this may be, the belief in the spherical form of the earth never disappeared from

* We have little definite knowledge of his life. He was born in the earlier part of the 6th century B.C., and died at the end of the same century or beginning of the next.

Greek thought, and was in later times an established part of Greek systems, whence it has been handed down, almost unchanged, to modern times. This belief is thus 2,000 years older than the belief in the rotation of the earth and its revolution round the sun (chapter IV.), doctrines which we are sometimes inclined to couple with it as the foundations of modern astronomy.

In Pythagoras occurs also, perhaps for the first time, an idea which had an extremely important influence on ancient and mediaeval astronomy. Not only were the stars supposed to be attached to a crystal sphere, which revolved daily on an axis through the earth, but each of the seven planets (the sun and moon being included) moved on a sphere of its own. The distances of these spheres from the earth were fixed in accordance with certain speculative notions of Pythagoras as to numbers and music; hence the spheres as they revolved produced harmonious sounds which specially gifted persons might at times hear: this is the origin of the idea of the **music of the spheres** which recurs continually in mediaeval speculation and is found occasionally in modern literature. At a later stage these spheres of Pythagoras were developed into a scientific representation of the motions of the celestial bodies, which remained the basis of astronomy till the time of Kepler (chapter VII.).

24. The Pythagorean *Philolaus*, who lived about a century later than his master, introduced for the first time the idea of the motion of the earth: he appears to have regarded the earth, as well as the sun, moon, and five planets, as revolving round some central fire, the earth rotating on its own axis as it revolved, apparently in order to ensure that the central fire should always remain invisible to the inhabitants of the known parts of the earth. That the scheme was a purely fanciful one, and entirely different from the modern doctrine of the motion of the earth, with which later writers confused it, is sufficiently shewn by the invention as part of the scheme of a purely imaginary body, the counter-earth (ἀντιχθών), which brought the number of moving bodies up to ten, a sacred Pythagorean number. The suggestion of such an important idea as that of the motion of the earth, an idea so

repugnant to uninstructed common sense, although presented in such a crude form, without any of the evidence required to win general assent, was, however, undoubtedly a valuable contribution to astronomical thought. It is well worth notice that Coppernicus in the great book which is the foundation of modern astronomy (chapter IV., § 75) especially quotes Philolaus and other Pythagoreans as authorities for his doctrine of the motion of the earth.

Three other Pythagoreans, belonging to the end of the 6th century and to the 5th century B.C., *Hicetas* of Syracuse, *Heraclitus*, and *Ecphantus*, are explicitly mentioned by later writers as having believed in the rotation of the earth.

An obscure passage in one of Plato's dialogues (the *Timaeus*) has been interpreted by many ancient and modern commentators as implying a belief in the rotation of the earth, and Plutarch also tells us, partly on the authority of Theophrastus, that Plato in old age adopted the belief that the centre of the universe was not occupied by the earth but by some better body.*

Almost the only scientific Greek astronomer who believed in the motion of the earth was *Aristarchus* of Samos, who lived in the first half of the 3rd century B.C., and is best known by his measurements of the distances of the sun and moon (§ 32). He held that the sun and fixed stars were motionless, the sun being in the centre of the sphere on which the latter lay, and that the earth not only rotated on its axis, but also described an orbit round the sun. *Seleucus* of Seleucia, who belonged to the middle of the 2nd century B.C., also held a similar opinion. Unfortunately we know nothing of the grounds of this belief in either case, and their views appear to have found little favour among their contemporaries or successors.

It may also be mentioned in this connection that Aristotle (§ 27) clearly realised that the apparent daily motion of the stars could be explained by a motion either of the stars or of the earth, but that he rejected the latter explanation.

25. *Plato* (about 428–347 B.C.) devoted no dialogue especially to astronomy, but made a good many references

* Theophrastus was born about half a century, Plutarch nearly five centuries, later than Plato.

to the subject in various places. He condemned any careful study of the actual celestial motions as degrading rather than elevating, and apparently regarded the subject as worthy of attention chiefly on account of its connection with geometry, and because the actual celestial motions suggested ideal motions of greater beauty and interest. This view of astronomy he contrasts with the popular conception, according to which the subject was useful chiefly for giving to the agriculturist, the navigator, and others a knowledge of times and seasons.* At the end of the same dialogue he gives a short account of the celestial bodies, according to which the sun, moon, planets, and fixed stars revolve on eight concentric and closely fitting wheels or circles round an axis passing through the earth. Beginning with the body nearest to the earth, the order is Moon, Sun, Mercury, Venus, Mars, Jupiter, Saturn, stars. The Sun, Mercury, and Venus are said to perform their revolutions in the same time, while the other planets move more slowly, statements which shew that Plato was at any rate aware that the motions of Venus and Mercury are different from those of the other planets. He also states that the moon shines by reflected light received from the sun.

Plato is said to have suggested to his pupils as a worthy problem the explanation of the celestial motions by means of a combination of uniform circular or spherical motions. Anything like an accurate theory of the celestial motions, agreeing with actual observation, such as Hipparchus and Ptolemy afterwards constructed with fair success, would hardly seem to be in accordance with Plato's ideas of the true astronomy, but he may well have wished to see established some simple and harmonious geometrical scheme which would not be altogether at variance with known facts.

26. Acting to some extent on this idea of Plato's, *Eudoxus* of Cnidus (about 409–356 B.C.) attempted to explain the most obvious peculiarities of the celestial motions by means of a combination of uniform circular motions. He may be regarded as representative of the transition from speculative

* *Republic*, VII. 529, 530.

to scientific Greek astronomy. As in the schemes of several of his predecessors, the fixed stars lie on a sphere which revolves daily about an axis through the earth; the motion of each of the other bodies is produced by a combination of other spheres, the centre of each sphere lying on the surface of the preceding one. For the sun and moon three spheres were in each case necessary: one to produce the daily motion, shared by all the celestial bodies; one to produce the annual or monthly motion in the opposite direction along the ecliptic; and a third, with its axis inclined to the axis of the preceding, to produce the smaller motion to and from the ecliptic. Eudoxus evidently was well aware that the moon's path is not coincident with the ecliptic, and even that its path is not always the same, but changes continuously, so that the third sphere was in this case necessary; on the other hand, he could not possibly have been acquainted with the minute deviations of the sun from the ecliptic with which modern astronomy deals. Either therefore he used erroneous observations, or, as is more probable, the sun's third sphere was introduced to explain a purely imaginary motion conjectured to exist by "analogy" with the known motion of the moon. For each of the five planets four spheres were necessary, the additional one serving to produce the variations in the speed of the motion and the reversal of the direction of motion along the ecliptic (chapter I., § 14, and below, § 51). Thus the celestial motions were to some extent explained by means of a system of 27 spheres, 1 for the stars, 6 for the sun and moon, 20 for the planets. There is no clear evidence that Eudoxus made any serious attempt to arrange either the size or the time of revolution of the spheres so as to produce any precise agreement with the observed motions of the celestial bodies, though he knew with considerable accuracy the time required by each planet to return to the same position with respect to the sun; in other words, his scheme represented the celestial motions qualitatively but not quantitatively. On the other hand, there is no reason to suppose that Eudoxus regarded his spheres (with the possible exception of the sphere of the fixed stars) as material; his known devotion to mathematics renders it probable that in his eyes (as in those of most of the scientific Greek astronomers who succeeded him) the spheres were mere geometrical figures, useful as a means of resolving highly complicated motions into simpler elements. Eudoxus was also the first Greek recorded to have had an observatory, which was at Cnidus, but we have few details as to the instruments used or as to the observations made. We owe, however, to him the first systematic description of the constellations (see below, § 42), though it was probably based, to a large extent, on rough observations borrowed from his Greek predecessors or from the Egyptians. He was also an accomplished mathematician, and skilled in various other branches of learning.

Shortly afterwards Callippus (§ 20) further developed Eudoxus's scheme of revolving spheres by adding, for reasons not known to us, two spheres each for the sun and moon and one each for Venus, Mercury, and Mars, thus bringing the total number up to 34.

27. We have a tolerably full account of the astronomical views of *Aristotle* (384–322 B.C.), both by means of incidental references, and by two treatises—the *Meteorologica* and the *De Coelo*—though another book of his, dealing specially with the subject, has unfortunately been lost. He adopted the planetary scheme of Eudoxus and Callippus, but imagined on "metaphysical grounds" that the spheres would have certain disturbing effects on one another, and to counteract these found it necessary to add 22 fresh spheres, making 56 in all. At the same time he treated the spheres as material bodies, thus converting an ingenious and beautiful geometrical scheme into a confused mechanism.* Aristotle's spheres were, however, not adopted by the leading Greek astronomers who succeeded him, the systems of Hipparchus and Ptolemy being geometrical schemes based on ideas more like those of Eudoxus.

28. Aristotle, in common with other philosophers of his time, believed the heavens and the heavenly bodies to be spherical. In the case of the moon he supports this belief by the argument attributed to Pythagoras (§ 23), namely that the observed appearances of the moon in its several

* Confused, because the mechanical knowledge of the time was quite unequal to giving any explanation of the way in which these spheres acted on one another.

phases are those which would be assumed by a spherical body of which one half only is illuminated by the sun. Thus the visible portion of the moon is bounded by two planes passing nearly through its centre, perpendicular respectively to the lines joining the centre of the moon to those of the sun and earth. In the accompanying diagram, which represents a section through the centres of the sun

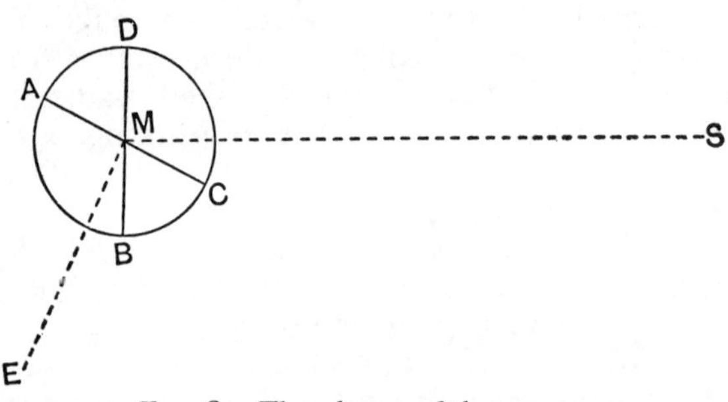

FIG. 8.—The phases of the moon.

(S), earth (E), and moon (M), A B C D representing on a much enlarged scale a section of the moon itself, the portion D A B which is turned away from the sun is dark, while the portion A D C, being turned away from the observer on the earth, is in any case invisible to him. The part of the moon which appears bright is therefore that of which B C is a section, or the portion represented by F B G C in fig. 9 (which represents the complete moon), which consequently appears to the eye as bounded by a semicircle F C G, and a portion F B G of an oval curve (actually an ellipse). The breadth of this bright surface clearly varies with the relative positions of sun, moon, and earth ; so that in the course of a month, during which the moon assumes successively the positions relative to sun and earth represented by 1, 2, 3, 4, 5, 6, 7, 8 in fig. 10, its appearances are those represented by the corresponding numbers in fig. 11, the moon thus passing

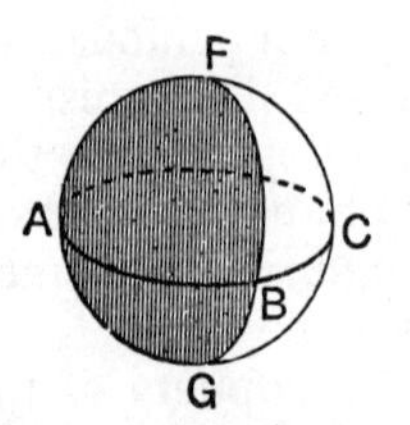

FIG. 9.—The phases of the moon.

through the familiar phases of crescent, half full, gibbous, full moon, and gibbous, half full, crescent again.*

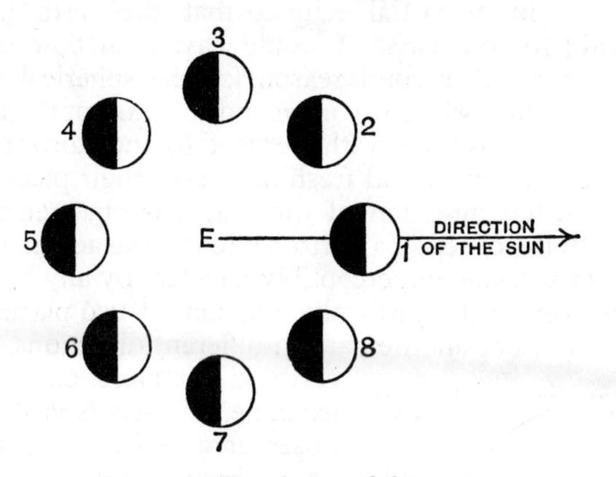

FIG. 10.—The phases of the moon.

Aristotle then argues that as one heavenly body is spherical, the others must be so also, and supports this conclusion by another argument, equally inconclusive to

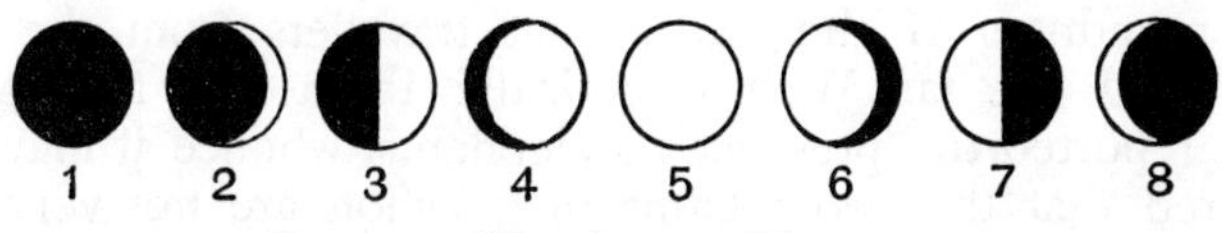
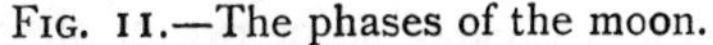

FIG. 11.—The phases of the moon.

us, that a spherical form is appropriate to bodies moving as the heavenly bodies appear to do.

29. His proofs that the earth is spherical are more interesting. After discussing and rejecting various other suggested forms, he points out that an eclipse of the moon is caused by the shadow of the earth cast by the sun, and

* I have introduced here the familiar explanation of the phases of the moon, and the argument based on it for the spherical shape of the moon, because, although probably known before Aristotle, there is, as far as I know, no clear and definite statement of the matter in any earlier writer, and after his time it becomes an accepted part of Greek elementary astronomy. It may be noticed that the explanation is unaffected either by the question of the rotation of the earth or by that of its motion round the sun.

argues from the circular form of the boundary of the shadow as seen on the face of the moon during the progress of the eclipse, or in a partial eclipse, that the earth must be spherical; for otherwise it would cast a shadow of a different shape. A second reason for the spherical form of the earth is that when we move north and south the stars change their positions with respect to the horizon, while some even disappear and fresh ones take their place. This shows that the direction of the stars has changed as compared with the observer's horizon; hence, the actual direction of the stars being imperceptibly affected by any motion of the observer on the earth, the horizons at two places, north and south of one another, are in different directions, and the earth is therefore curved. For example, if a star is visible to an observer at A (fig. 12), while to an observer at B it is at the same time invisible, *i.e.* hidden by the earth, the surface of the earth at A must be in a different direction from that at B. Aristotle quotes further, in confirmation of the roundness of the earth, that travellers from the far East and the far West (practically India and Morocco) alike reported the presence of elephants, whence it may be inferred that the two regions in question are not very far apart. He also makes use of some rather obscure arguments of an *a priori* character.

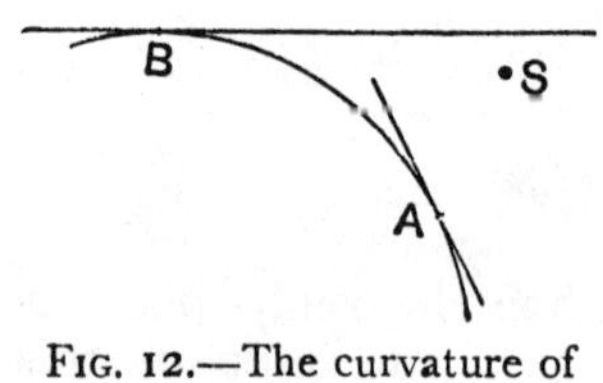

FIG. 12.—The curvature of the earth.

There can be but little doubt that the readiness with which Aristotle, as well as other Greeks, admitted the spherical form of the earth and of the heavenly bodies, was due to the affection which the Greeks always seem to have had for the circle and sphere as being "perfect," *i.e.* perfectly symmetrical figures.

30. Aristotle argues against the possibility of the revolution of the earth round the sun, on the ground that this motion, if it existed, ought to produce a corresponding apparent motion of the stars. We have here the first appearance of one of the most serious of the many objections ever brought against the belief in the motion of the earth, an objection really only finally disposed of during the

present century by the discovery that such a motion of the stars can be seen in a few cases, though owing to the almost inconceivably great distance of the stars the motion is imperceptible except by extremely refined methods of observation (cf. chapter XIII., §§ 278, 279). The question of the distances of the several celestial bodies is also discussed, and Aristotle arrives at the conclusion that the planets are farther off than the sun and moon, supporting his view by his observation of an occultation of Mars by the moon (*i.e.* a passage of the moon in front of Mars), and by the fact that similar observations had been made in the case of other planets by Egyptians and Babylonians. It is, however, difficult to see why he placed the planets beyond the sun, as he must have known that the intense brilliancy of the sun renders planets invisible in its neighbourhood, and that no occultations of planets by the sun could really have been seen even if they had been reported to have taken place. He quotes also, as an opinion of "the mathematicians," that the stars must be at least nine times as far off as the sun.

There are also in Aristotle's writings a number of astronomical speculations, founded on no solid evidence and of little value; thus among other questions he discusses the nature of comets, of the Milky Way, and of the stars, why the stars twinkle, and the causes which produce the various celestial motions.

In astronomy, as in other subjects, Aristotle appears to have collected and systematised the best knowledge of the time; but his original contributions are not only not comparable with his contributions to the mental and moral sciences, but are inferior in value to his work in other natural sciences, *e.g.* Natural History. Unfortunately the Greek astronomy of his time, still in an undeveloped state, was as it were crystallised in his writings, and his great authority was invoked, centuries afterwards, by comparatively unintelligent or ignorant disciples in support of doctrines which were plausible enough in his time, but which subsequent research was shewing to be untenable. The advice which he gives to his readers at the beginning of his exposition of the planetary motions, to compare his views with those which they arrived at themselves or met with elsewhere, might with advantage have been noted and followed by many of the so-called Aristotelians of the Middle Ages and of the Renaissance.*

31. After the time of Aristotle the centre of Greek scientific thought moved to Alexandria. Founded by Alexander the Great (who was for a time a pupil of Aristotle) in 332 B.C., Alexandria was the capital of Egypt during the reigns of the successive Ptolemies. These kings, especially the second of them, surnamed Philadelphos, were patrons of learning; they founded the famous Museum, which contained a magnificent library as well as an observatory, and Alexandria soon became the home of a distinguished body of mathematicians and astronomers. During the next five centuries the only astronomers of importance, with the great exception of Hipparchus (§ 37), were Alexandrines.

32. Among the earlier members of the Alexandrine school were *Aristarchus* of Samos, *Aristyllus*, and *Timocharis*, three nearly contemporary astronomers belonging

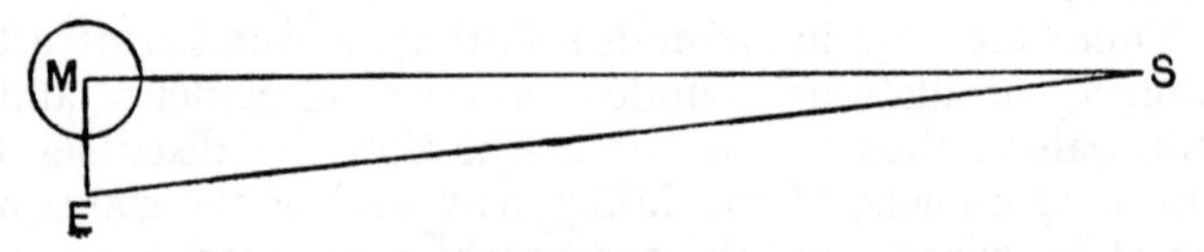

Fig. 13.—The method of Aristarchus for comparing the distances of the sun and moon.

to the first half of the 3rd century B.C. The views of Aristarchus on the motion of the earth have already been mentioned (§ 24). A treatise of his *On the Magnitudes and Distances of the Sun and Moon* is still extant: he there gives an extremely ingenious method for ascertaining the comparative distances of the sun and moon. If, in the figure, E, S, and M denote respectively the centres of the earth, sun, and moon, the moon evidently appears to an observer at E half full when the angle E M S is a right angle. If when this is the case the angular distance between the centres of the sun and moon, *i.e.* the angle M E S, is measured, two angles of the triangle M E S are

* See, for example, the account of Galilei's controversies, in chapter VI.

known; its shape is therefore completely determined, and the ratio of its sides E M, E S can be calculated without much difficulty. In fact, it being known (by a well-known result in elementary geometry) that the angles at E and S are together equal to a right angle, the angle at S is obtained by subtracting the angle S E M from a right angle. Aristarchus made the angle at S about 3°, and hence calculated that the distance of the sun was from 18 to 20 times that of the moon, whereas, in fact, the sun is about 400 times as distant as the moon. The enormous error is due to the difficulty of determining with sufficient accuracy the moment when the moon is half full: the boundary separating the bright and dark parts of the moon's face is in reality (owing to the irregularities on the surface of the moon) an ill-defined and broken line (cf. fig. 53 and the frontispiece), so that the observation on which Aristarchus based his work could not have been made with any accuracy even with our modern instruments, much less with those available in his time. Aristarchus further estimated the apparent sizes of the sun and moon to be about equal (as is shewn, for example, at an eclipse of the sun, when the moon sometimes rather more than hides the surface of the sun and sometimes does not quite cover it), and inferred correctly that the real diameters of the sun and moon were in proportion to their distances. By a method based on eclipse observations which was afterwards developed by Hipparchus (§ 41), he also found that the diameter of the moon was about $\frac{1}{3}$ that of the earth, a result very near to the truth; and the same method supplied data from which the distance of the moon could at once have been expressed in terms of the radius of the earth, but his work was spoilt at this point by a grossly inaccurate estimate of the apparent size of the moon (2° instead of $\frac{1}{2}$°), and his conclusions seem to contradict one another. He appears also to have believed the distance of the fixed stars to be immeasurably great as compared with that of the sun. Both his speculative opinions and his actual results mark therefore a decided advance in astronomy.

Timocharis and Aristyllus were the first to ascertain and to record the positions of the chief stars, by means of numerical measurements of their distances from fixed positions on the sky; they may thus be regarded as the authors of the first real star catalogue, earlier astronomers having only attempted to fix the position of the stars by more or less vague verbal descriptions. They also made a number of valuable observations of the planets, the sun, etc., of which succeeding astronomers, notably Hipparchus and Ptolemy, were able to make good use.

33. Among the important contributions of the Greeks to astronomy must be placed the development, chiefly from the mathematical point of view, of the consequences of the rotation of the celestial sphere and of some of the simpler motions of the celestial bodies, a development the individual steps of which it is difficult to trace. We have,

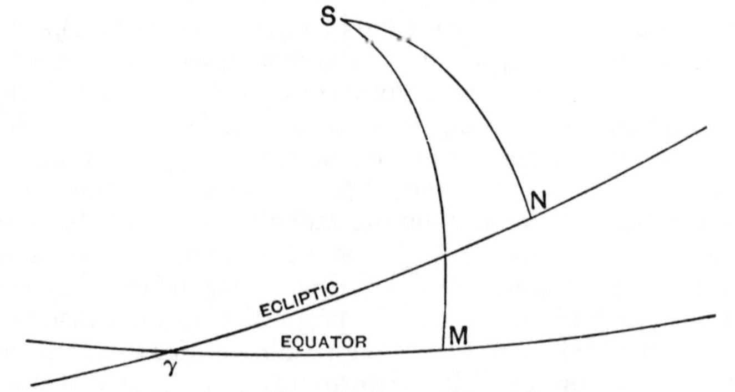

FIG. 14.—The equator and the ecliptic.

however, a series of minor treatises or textbooks, written for the most part during the Alexandrine period, dealing with this branch of the subject (known generally as **Spherics**, or the Doctrine of the Sphere), of which the *Phenomena* of the famous geometer *Euclid* (about 300 B.C.) is a good example. In addition to the points and circles of the sphere already mentioned (chapter I., §§ 8–11), we now find explicitly recognised the **horizon**, or the great circle in which a horizontal plane through the observer meets the celestial sphere, and its **pole**,* the **zenith**,† or

* The **poles** of a great circle on a sphere are the ends of a diameter perpendicular to the plane of the great circle. Every point on the great circle is at the same distance, 90°, from each pole.

† The *word* "zenith" is Arabic, not Greek: cf. chapter III., § 64.

point on the celestial sphere vertically above the observer; the **verticals**, or great circles through the zenith, meeting the horizon at right angles; and the **declination circles**, which pass through the north and south poles and cut the equator at right angles. Another important great circle was the **meridian**, passing through the zenith and the poles. The well-known Milky Way had been noticed, and was regarded as forming another great circle. There are also traces of the two chief methods in common use at the present day of indicating the position of a star on the celestial sphere, namely, by reference either to the equator or to the ecliptic. If through a star S we draw on the sphere a portion of a great circle S N, cutting the ecliptic ♈ N at right angles in N, and another great circle (a declination circle) cutting the equator at M, and if ♈ be the first point of Aries (§ 13), where the ecliptic crosses the equator, then the position of the star is completely defined *either* by the lengths of the arcs ♈ N, N S, which are called the **celestial longitude** and **latitude** respectively, *or* by the arcs ♈ M, M S, called respectively the **right ascension** and **declination.*** For some purposes it is more convenient to find the position of the star by the first method, *i.e.* by reference to the ecliptic; for other purposes in the second way, by making use of the equator.

34. One of the applications of Spherics was to the construction of sun-dials, which were supposed to have been originally introduced into Greece from Babylon, but which were much improved by the Greeks, and extensively used both in Greek and in mediaeval times. The proper graduation of sun-dials placed in various positions, horizontal, vertical, and oblique, required considerable mathematical skill. Much attention was also given to the time of the rising and setting of the various constellations, and to similar questions.

35. The discovery of the spherical form of the earth led to a scientific treatment of the differences between the seasons in different parts of the earth, and to a corresponding division of the earth into zones. We have already seen that the height of the pole above the horizon varies in

* Most of these names are not Greek, but of later origin.

different places, and that it was recognised that, if a traveller were to go far enough north, he would find the pole to coincide with the zenith, whereas by going south he would reach a region (not very far beyond the limits of actual Greek travel) where the pole would be on the horizon and the equator consequently pass through the zenith; in regions still farther south the north pole would be permanently invisible, and the south pole would appear above the horizon.

Further, if in the figure H E K W represents the horizon, meeting the equator Q E R W in the east and west points E W, and the meridian H Q Z P K in the south and north points H and K, Z being the zenith and P the pole, then it is easily seen that Q Z is equal to P K, the height of the pole above the horizon. Any celestial body, therefore, the distance of which from the equator towards the north (declination) is less than P K, will cross the meridian to the south of the zenith, whereas if its declination be greater than P K, it will cross to the north of the zenith. Now the greatest distance of the sun from the equator is equal to the angle between the ecliptic and the equator, or about $23\frac{1}{2}°$. Consequently at places at which the height of the pole is less than $23\frac{1}{2}°$ the sun will, during part of the year, cast shadows at midday towards the south. This was known actually to be the case not very far south of Alexandria. It was similarly recognised that on the other side of the equator there must be a region in which the sun ordinarily cast shadows towards the south, but occasionally towards the north. These two regions are the torrid zones of modern geographers.

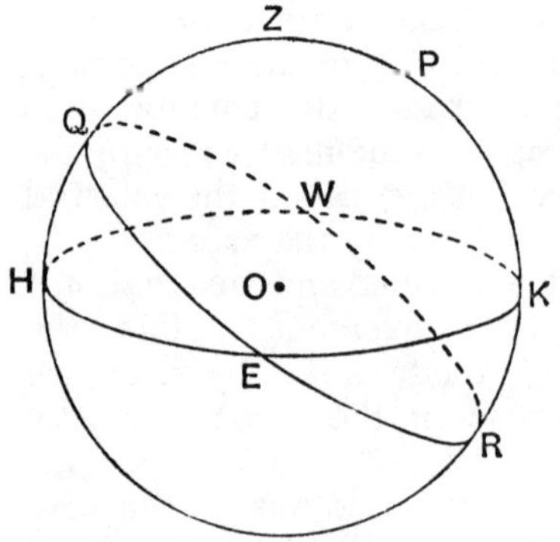

FIG. 15.—The equator, the horizon, and the meridian.

Again, if the distance of the sun from the equator is $23\frac{1}{2}°$, its distance from the pole is $66\frac{1}{2}°$; therefore in regions so far north that the height P K of the north pole is more than $66\frac{1}{2}°$, the sun passes in summer into the region of the circumpolar stars which never set (chapter I., § 9), and therefore during a portion of the summer the sun remains continuously above the horizon. Similarly in the same regions the sun is in winter so near the south pole that for a time it remains continuously below the horizon. Regions in which this occurs (our Arctic regions) were unknown to Greek travellers, but their existence was clearly indicated by the astronomers.

36. To *Eratosthenes* (276 B.C. to 195 or 196 B.C.), another member of the Alexandrine school, we owe one of the first scientific estimates of the size of the earth. He found

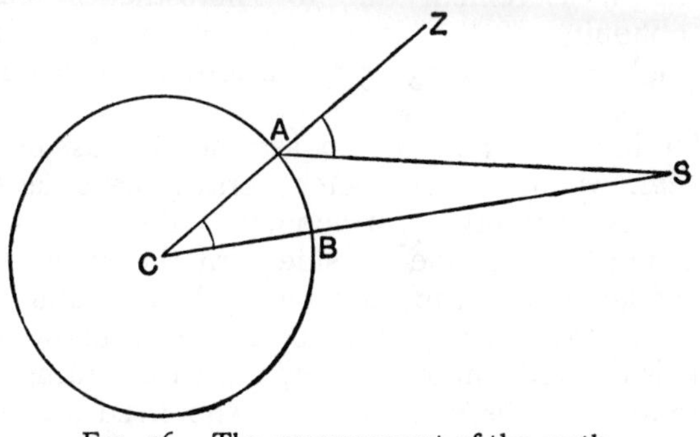

FIG. 16.—The measurement of the earth.

that at the summer solstice the angular distance of the sun from the zenith at Alexandria was at midday $\frac{1}{50}$th of a complete circumference, or about 7°, whereas at Syene in Upper Egypt the sun was known to be vertical at the same time. From this he inferred, assuming Syene to be due south of Alexandria, that the distance from Syene to Alexandria was also $\frac{1}{50}$th of the circumference of the earth. Thus if in the figure S denotes the sun, A and B Alexandria and Syene respectively, C the centre of the earth, and A Z the direction of the zenith at Alexandria, Eratosthenes estimated the angle S A Z, which, owing to the great distance of S, is sensibly equal to the angle S C A, to be 7°, and hence inferred that the arc A B was to the circumference of the earth in the proportion of 7° to 360° or 1 to 50. The distance between Alexandria and Syene being known to be 5,000 *stadia*, Eratosthenes thus arrived at 250,000 stadia as an estimate of the circumference of the earth, a number altered into 252,000 in order to give an exact number of stadia (700) for each degree on the earth. It is evident that the data employed were rough, though the principle of the method is perfectly sound; it is, however, difficult to estimate the correctness of the result on account of the uncertainty as to the value of the *stadium* used. If, as seems probable, it was the common Olympic stadium, the result is about 20 per cent. too great, but according to another interpretation * the result is less than 1 per cent. in error (cf. chapter x., § 221).

Another measurement due to Eratosthenes was that of the obliquity of the ecliptic, which he estimated at $\frac{22}{83}$ of a right angle, or $23^\circ\ 51'$, the error in which is only about $7'$.

37. An immense advance in astronomy was made by *Hipparchus*, whom all competent critics have agreed to rank far above any other astronomer of the ancient world, and who must stand side by side with the greatest astronomers of all time. Unfortunately only one unimportant book of his has been preserved, and our knowledge of his work is derived almost entirely from the writings of his great admirer and disciple Ptolemy, who lived nearly three centuries later (§§ 46 *seqq.*). We have also scarcely any information about his life. He was born either at Nicaea in Bithynia or in Rhodes, in which island he erected an observatory and did most of his work. There is no evidence that he belonged to the Alexandrine school, though he probably visited Alexandria and may have made some observations there. Ptolemy mentions observations made by him in 146 B.C., 126 B.C., and at many intermediate dates, as well as a rather doubtful one of 161 B.C. The period of his greatest activity must therefore have been about the middle of the 2nd century B.C.

Apart from individual astronomical discoveries, his chief services to astronomy may be put under four heads. He invented or greatly developed a special branch of mathe-

* That of M. Paul Tannery: *Recherches sur l'Histoire de l'Astronomie Ancienne*, chap. v.

matics,* which enabled processes of numerical calculation to be applied to geometrical figures, whether in a plane or on a sphere. He made an extensive series of observations, taken with all the accuracy that his instruments would permit. He systematically and critically made use of old observations for comparison with later ones so as to discover astronomical changes too slow to be detected within a single lifetime. Finally, he systematically employed a particular geometrical scheme (that of eccentrics, and to a less extent that of epicycles) for the representation of the motions of the sun and moon.

38. The merit of suggesting that the motions of the heavenly bodies could be represented more simply by combinations of uniform *circular* motions than by the revolving *spheres* of Eudoxus and his school (§ 26) is generally attributed to the great Alexandrine mathematician *Apollonius* of Perga, who lived in the latter half of the 3rd century B.C., but there is no clear evidence that he worked out a system in any detail.

On account of the important part that this idea played in astronomy for nearly 2,000 years, it may be worth while to examine in some detail Hipparchus's theory of the sun, the simplest and most successful application of the idea.

We have already seen (chapter I., § 10) that, in addition to the daily motion (from east to west) which it shares with the rest of the celestial bodies, and of which we need here take no further account, the sun has also an annual motion on the celestial sphere in the reverse direction (from west to east) in a path oblique to the equator, which was early recognised as a great circle, called the ecliptic. It must be remembered further that the celestial sphere, on which the sun appears to lie, is a mere geometrical fiction introduced for convenience; all that direct observation gives is the change in the sun's direction, and therefore the sun may consistently be supposed to move in such a way as to vary its distance from the earth in any arbitrary manner, provided only that the alterations in the apparent size of the sun, caused by the variations in its distance, agree with those observed, or that at any rate the differences

* Trigonometry.

are not great enough to be perceptible. It was, moreover, known (probably long before the time of Hipparchus) that the sun's apparent motion in the ecliptic is not quite uniform, the motion at some times of the year being slightly more rapid than at others.

Supposing that we had such a complete set of observations of the motion of the sun, that we knew its position from day to day, how should we set to work to record and describe its motion? For practical purposes nothing could be more satisfactory than the method adopted in our almanacks, of giving from day to day the position of the sun; after observations extending over a few years it would not be difficult to verify that the motion of the sun is (after allowing for the irregularities of our calendar) from year to year the same, and to predict in this way the place of the sun from day to day in future years.

But it is clear that such a description would not only be long, but would be felt as unsatisfactory by any one who approached the question from the point of view of intellectual curiosity or scientific interest. Such a person would feel that these detailed facts ought to be capable of being exhibited as consequences of some simpler general statement.

A modern astronomer would effect this by expressing the motion of the sun by means of an algebraical formula, *i.e.* he would represent the velocity of the sun or its distance from some fixed point in its path by some symbolic expression representing a quantity undergoing changes with the time in a certain definite way, and enabling an expert to compute with ease the required position of the sun at any assigned instant.*

The Greeks, however, had not the requisite algebraical knowledge for such a method of representation, and Hipparchus, like his predecessors, made use of a geometrical

* The process may be worth illustrating by means of a simpler problem. A heavy body, falling freely under gravity, is found (the resistance of the air being allowed for) to fall about 16 feet in 1 second, 64 feet in 2 seconds, 144 feet in 3 seconds, 256 feet in 4 seconds, 400 feet in 5 seconds, and so on. This series of figures carried on as far as may be required would satisfy practical requirements, supplemented if desired by the corresponding figures for fractions of seconds; but the mathematician represents the same

representation of the required variations in the sun's motion in the ecliptic, a method of representation which is in some respects more intelligible and vivid than the use of algebra, but which becomes unmanageable in complicated cases. It runs moreover the risk of being taken for a mechanism. The circle, being the simplest curve known, would naturally be thought of, and as any motion other than a uniform motion would itself require a special representation, the idea of Apollonius, adopted by Hipparchus, was to devise a proper combination of uniform circular motions.

39. The simplest device that was found to be satisfactory in the case of the sun was the use of the **eccentric**, *i.e.* a circle the centre of which (C) does not coincide with the position of the observer on the earth (E). If in fig. 17 a point, S, describes the eccentric circle A F G B uniformly, so that it always passes over equal arcs of the circle in equal times and the angle A C S increases uniformly, then it is evident that the angle A E S, or the apparent distance of S from A, does not increase uniformly. When S is near the point A, which is farthest from the earth and hence called the **apogee**, it appears on account of its greater distance from the observer to move more slowly than when near F or G; and it appears to move fastest when near B, the point nearest to E, hence called the **perigee**. Thus the motion of S varies in the same sort of way as the motion of the sun as actually observed. Before, however, the eccentric could be considered as satisfactory, it was necessary to show that it was possible to choose the direction of the line B E C A (the **line of apses**) which determines the positions of the sun when moving fastest and when moving most slowly, and the magnitude of the ratio of E C to the radius C A of the circle (the **eccentricity**), so as to make the calculated positions of the sun in various parts of its path differ from the observed positions at the corresponding

facts more simply and in a way more satisfactory to the mind by the formula $s = 16\ t^2$, where s denotes the number of feet fallen, and t the number of seconds. By giving t any assigned value, the corresponding space fallen through is at once obtained. Similarly the motion of the sun can be represented approximately by the more complicated formula $l = nt + 2\ e\ sin\ nt$, where l is the distance from a fixed point in the orbit, t the time, and n, e certain numerical quantities.

times of year by quantities so small that they might fairly be attributed to errors of observation.

This problem was much more difficult than might at first sight appear, on account of the great difficulty experienced in Greek times and long afterwards in getting satisfactory observations of the sun. As the sun and stars are not visible at the same time, it is not possible to measure directly the distance of the sun from neighbouring stars and so to fix its place on the celestial sphere. But it

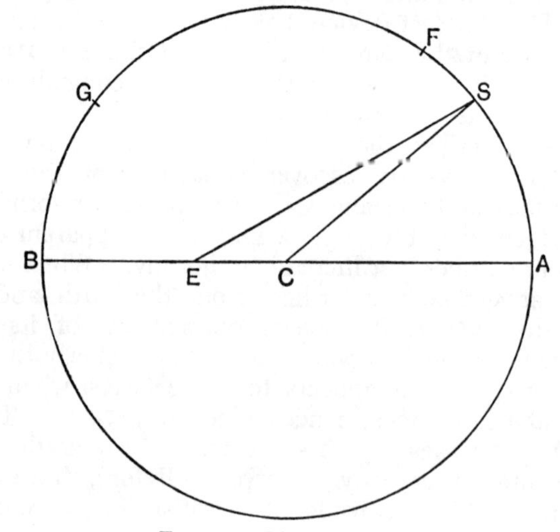

FIG. 17.—The eccentric.

is possible, by measuring the length of the shadow cast by a rod at midday, to ascertain with fair accuracy the height of the sun above the horizon, and hence to deduce its distance from the equator, or the declination (figs. 3, 14). This one quantity does not suffice to fix the sun's position, but if also the sun's right ascension (§ 33), or its distance east and west from the stars, can be accurately ascertained, its place on the celestial sphere is completely determined. The methods available for determining this second quantity were, however, very imperfect. One method was to note the time between the passage of the sun across some fixed position in the sky (*e.g.* the meridian), and the passage of a star across the same place, and thus to ascertain the angular distance between them (the celestial sphere being known to turn through 15° in an hour), a method which with modern clocks is extremely accurate, but with the rough water-clocks or sand-glasses of former times was very uncertain. In another method the moon was used as a connecting link between sun and stars, her position relative

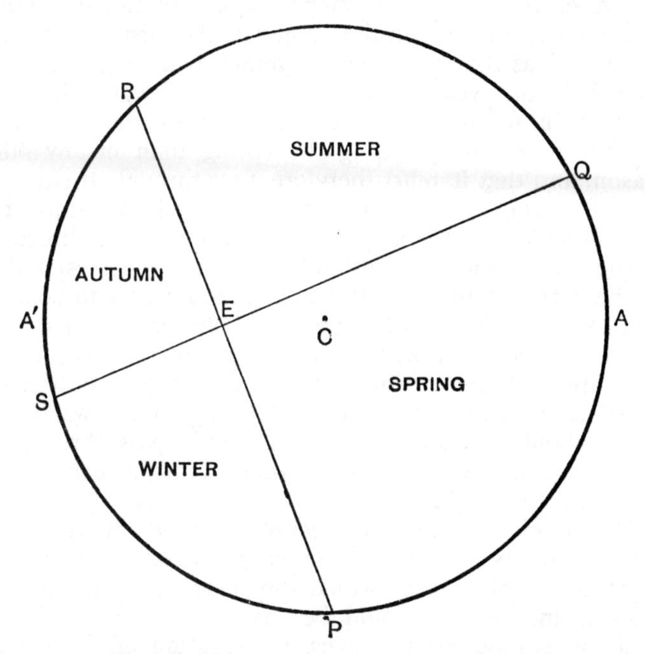

FIG. 18.—The position of the sun's apogee.

to the latter being observed by night, and with respect to the former by day; but owing to the rapid motion of the moon in the interval between the two observations, this method also was not susceptible of much accuracy.

In the case of the particular problem of the determination of the line of apses, Hipparchus made use of another method, and his skill is shewn in a striking manner by his recognition that both the eccentricity and position of the apse line could be determined from a knowledge of the lengths of two of the seasons of the year, *i.e.* of the intervals into which the year is divided by the solstices and the equinoxes (§ 11). By means of his own observations, and of others made by his predecessors, he ascertained the length of the spring (from the vernal equinox to the summer solstice) to be 94 days, and that of the summer (summer solstice to autumnal equinox) to be $92\frac{1}{2}$ days, the length of the year being $365\frac{1}{4}$ days. As the sun moves in each season through the same angular distance, a right angle, and as the spring and summer make together more than half the year, and the spring is longer than the summer, it follows that the sun must, on the whole, be moving more slowly during the spring than in any other season, and that it must therefore pass through the apogee in the spring. If, therefore, in fig. 18, we draw two perpendicular lines Q E S, P E R to represent the directions of the sun at the solstices and equinoxes, P corresponding to the vernal equinox and R to the autumnal equinox, the apogee must lie at some point A between P and Q. So much can be seen without any mathematics: the actual calculation of the position of A and of the eccentricity is a matter of some complexity. The angle P E A was found to be about 65°, so that the sun would pass through its apogee about the beginning of June; and the eccentricity was estimated at $\frac{1}{24}$.

The motion being thus represented geometrically, it became merely a matter of not very difficult calculation to construct a table from which the position of the sun for any day in the year could be easily deduced. This was done by computing the so-called **equation of the centre,** the angle C S E of fig. 17, which is the excess of the actual longitude of the sun over the longitude which it would have had if moving uniformly.

Owing to the imperfection of the observations used (Hipparchus estimated that the times of the equinoxes and solstices could only be relied upon to within about half a day), the actual results obtained were not, according to modern ideas, very accurate, but the theory represented the sun's motion with an accuracy about as great as that of the observations. It is worth noticing that with the same theory, but with an improved value of the eccentricity, the motion of the sun can be represented so accurately that the error never exceeds about 1′, a quantity insensible to the naked eye.

The theory of Hipparchus represents the variations in the distance of the sun with much less accuracy, and whereas in fact the angular diameter of the sun varies by about $\frac{1}{30}$th part of itself, or by about 1′ in the course of the year, this variation according to Hipparchus should be about twice as great. But this error would also have been quite imperceptible with his instruments.

Hipparchus saw that the motion of the sun could equally well be represented by the other device suggested by Apollonius, the **epicycle.** The body the motion of which is to be represented is supposed to move uniformly round the circumference of one circle, called the epicycle, the centre of which in turn moves on another circle called the **deferent.** It is in fact evident that if a circle equal to the eccentric, but with its centre at E (fig. 19), be taken as the deferent, and if S′ be taken on this so that E S′ is parallel to C S, then S′ S is parallel and equal to E C; and that therefore the sun S, moving uniformly on the eccentric, may equally well be regarded as lying on a circle of radius S′ S, the centre S′ of which moves on the deferent. The two constructions lead in fact in this particular problem to exactly the same result, and Hipparchus chose the eccentric as being the simpler.

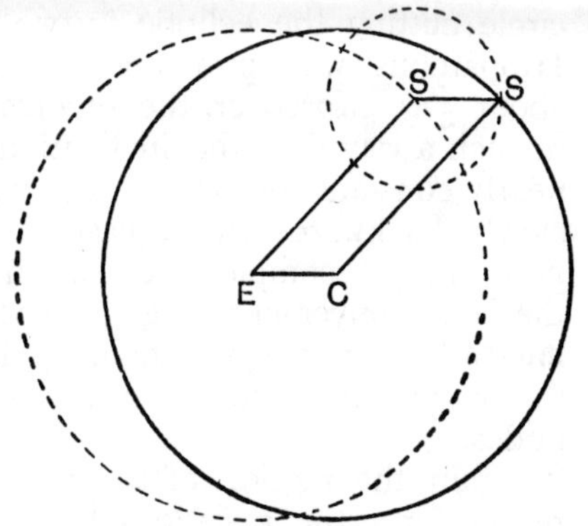

FIG. 19.—The epicycle and the deferent.

40. The motion of the moon being much more complicated than that of the sun has always presented difficulties to astronomers,* and Hipparchus required for it a more elaborate construction. Some further description of the

* At the present time there is still a small discrepancy between the observed and calculated places of the moon. See chapter XIII., § 290.

moon's motion is, however, necessary before discussing his theory.

We have already spoken (chapter I., § 16) of the lunar month as the period during which the moon returns to the same position with respect to the sun; more precisely this period (about $29\frac{1}{2}$ days) is spoken of as a **lunation** or **synodic month**: as, however, the sun moves eastward on the celestial sphere like the moon but more slowly, the moon returns to the same position with respect to the *stars* in a somewhat shorter time; this period (about 27 days 8 hours) is known as the **sidereal month.** Again, the moon's path on the celestial sphere is slightly inclined to the ecliptic, and may be regarded approximately as a great circle cutting the ecliptic in two **nodes,** at an angle which Hipparchus was probably the first to fix definitely at about 5°. Moreover, the moon's path is always changing in such a way that, the inclination to the ecliptic remaining nearly constant (but cf. chapter V., § 111), the nodes move slowly backwards (from east to west) along the ecliptic, performing a complete revolution in about 19 years. It is therefore convenient to give a special name, the **draconitic month,*** to the period (about 27 days 5 hours) during which the moon returns to the same position with respect to the nodes.

Again, the motion of the moon, like that of the sun, is not uniform, the variations being greater than in the case of the sun. Hipparchus appears to have been the first to discover that the part of the moon's path in which the motion is most rapid is not always in the same position on the celestial sphere, but moves continuously; or, in other words, that the line of apses (§ 39) of the moon's path moves. The motion is an advance, and a complete circuit is described in about nine years. Hence arises a fourth kind of month, the **anomalistic month,** which is the period in which the moon returns to apogee or perigee.

To Hipparchus is due the credit of fixing with greater

* The name is interesting as a remnant of a very early superstition. Eclipses, which always occur near the nodes, were at one time supposed to be caused by a dragon which devoured the sun or moon. The symbols ☊ ☋ still used to denote the two nodes are supposed to represent the head and tail of the dragon.

exactitude than before the lengths of each of these months. In order to determine them with accuracy he recognised the importance of comparing observations of the moon taken at as great a distance of time as possible, and saw that the most satisfactory results could be obtained by using Chaldaean and other eclipse observations, which, as eclipses only take place near the moon's nodes, were simultaneous records of the position of the moon, the nodes, and the sun.

To represent this complicated set of motions, Hipparchus used, as in the case of the sun, an eccentric, the centre of which described a circle round the earth in about nine years (corresponding to the motion of the apses), the plane of the eccentric being inclined to the ecliptic at an angle of 5°, and sliding back, so as to represent the motion of the nodes already described.

The result cannot, however, have been as satisfactory as in the case of the sun. The variation in the rate at which the moon moves is not only greater than in the case of the sun, but follows a less simple law, and cannot be adequately represented by means of a single eccentric; so that though Hipparchus' work would have represented the motion of the moon in certain parts of her orbit with fair accuracy, there must necessarily have been elsewhere discrepancies between the calculated and observed places. There is some indication that Hipparchus was aware of these, but was not able to reconstruct his theory so as to account for them.

41. In the case of the planets Hipparchus found so small a supply of satisfactory observations by his predecessors, that he made no attempt to construct a system of epicycles or eccentrics to represent their motion, but collected fresh observations for the use of his successors. He also made use of these observations to determine with more accuracy than before the average times of revolution of the several planets.

He also made a satisfactory estimate of the size and distance of the moon, by an eclipse method, the leading idea of which was due to Aristarchus (§ 32); by observing the angular diameter of the earth's shadow (Q R) at the distance of the moon at the time of an eclipse, and comparing

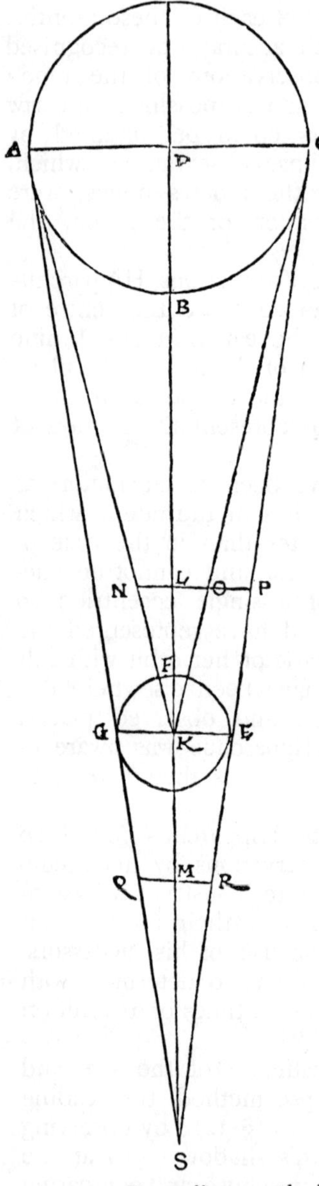

FIG. 20.—The eclipse method of connecting the distances of the sun and moon.

it with the known angular diameters of the sun and moon, he obtained, by a simple calculation,* a relation between the distances of the sun and moon, which gives either when

* In the figure, which is taken from the *De Revolutionibus* of Coppernicus (chapter IV., § 85), let D, K, M represent respectively the centres of the sun, earth, and moon, at the time of an eclipse of the moon, and let SQG, SRE denote the boundaries of the shadow-cone cast by the earth; then QR, drawn at right angles to the axis of the cone, is the breadth of the shadow at the distance of the moon. We have then at once from similar triangles

$$GK - QM : AD - GK :: MK : KD.$$

Hence if $KD = n \cdot MK$ and $\therefore$ also $AD = n$. (radius of moon), n being 19 according to Aristarchus,

$$GK - QM : n \cdot (\text{radius of moon}) - GK :: 1 : n$$

$$n \cdot (\text{radius of moon}) - GK = n\,GK - n\,QM$$

$\therefore$ radius of moon + radius of shadow

$$= \left(1 + \frac{1}{n}\right) (\text{radius of earth}).$$

By observation the angular radius of the shadow was found to be about 40′ and that of the moon to be 15′, so that

radius of shadow $= \frac{8}{3}$ radius of moon;

$\therefore$ radius of moon

$$= \frac{3}{11}\left(1 + \frac{1}{n}\right) (\text{radius of earth}).$$

But the angular radius of the moon being 15′, its distance is necessarily about 220 times its radius,

and $\therefore$ distance of the moon

$$= 60\left(1 + \frac{1}{n}\right) (\text{radius of the earth}),$$

which is roughly Hipparchus's result, if n be *any* fairly large number.

the other is known. Hipparchus knew that the sun was very much more distant than the moon, and appears to have tried more than one distance, that of Aristarchus among them, and the result obtained in each case shewed that the distance of the moon was nearly 59 times the radius of the earth. Combining the estimates of Hipparchus and Aristarchus, we find the distance of the sun to be about 1,200 times the radius of the earth—a number which remained substantially unchanged for many centuries (chapter VIII., § 161).

42. The appearance in 134 B.C. of a new star in the Scorpion is said to have suggested to Hipparchus the construction of a new catalogue of the stars. He included 1,080 stars, and not only gave the (celestial) latitude and longitude of each star, but divided them according to their brightness into six magnitudes. The constellations to which he refers are nearly identical with those of Eudoxus (§ 26), and the list has undergone few alterations up to the present day, except for the addition of a number of southern constellations, invisible in the civilised countries of the ancient world. Hipparchus recorded also a number of cases in which three or more stars appeared to be in line with one another, or, more exactly, lay on the same great circle, his object being to enable subsequent observers to detect more easily possible changes in the positions of the stars. The catalogue remained, with slight alterations, the standard one for nearly sixteen centuries (cf. chapter III., § 63).

The construction of this catalogue led to a notable discovery, the best known probably of all those which Hipparchus made. In comparing his observations of certain stars with those of Timocharis and Aristyllus (§ 33), made about a century and a half earlier, Hipparchus found that their distances from the equinoctial points had changed. Thus, in the case of the bright star Spica, the distance from the equinoctial points (measured eastwards) had increased by about 2° in 150 years, or at the rate of 48″ per annum. Further inquiry showed that, though the roughness of the observations produced considerable variations in the case of different stars, there was evidence of a general increase in the longitude of the stars (measured from west to east), unaccompanied by any change of latitude, the amount of the change being estimated by Hipparchus as at least 36″ annually, and possibly more. The agreement between the motions of different stars was enough to justify him in concluding that the change could be accounted for, not as a motion of individual stars, but rather as a change in the position of the equinoctial points, from which longitudes were measured. Now these points are the intersection of the equator and the ecliptic: consequently one or another of these two circles must have changed. But the fact that the latitudes of the stars had undergone no change shewed that the ecliptic must have retained its position and that the change had been caused

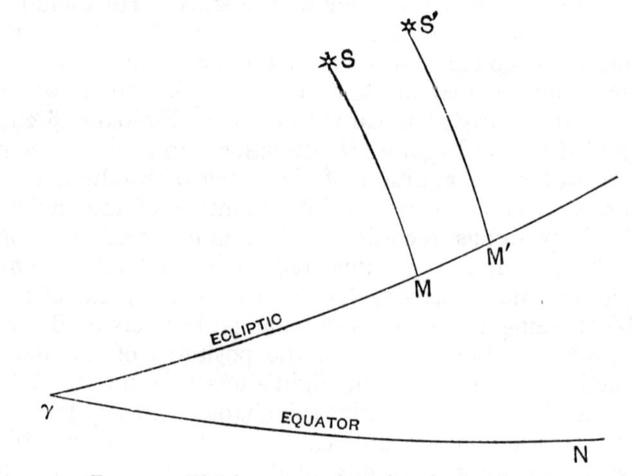

FIG. 21.—The increase of the longitude of a star.

by a motion of the equator. Again, Hipparchus measured the obliquity of the ecliptic as several of his predecessors had done, and the results indicated no appreciable change. Hipparchus accordingly inferred that the equator was, as it were, slowly sliding backwards (*i.e.* from east to west), keeping a constant inclination to the ecliptic.

The argument may be made clearer by figures. In fig. 21 let ♈M denote the ecliptic, ♈N the equator, S a star as seen by Timocharis, S M a great circle drawn perpendicular to the ecliptic. Then S M is the latitude, ♈M the longitude. Let S′ denote the star as seen by Hipparchus;

then he found that S′ M was equal to the former S M, but that ♈ M′ was greater than the former ♈ M, or that M′

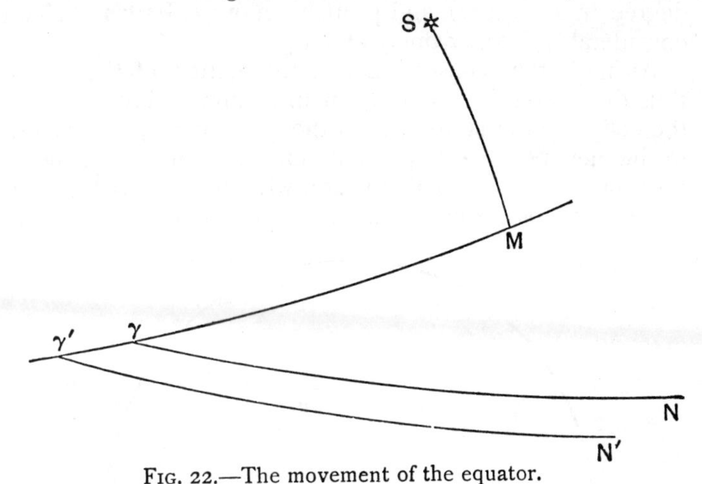

FIG. 22.—The movement of the equator.

was slightly to the east of M. This change M M′ being nearly the same for all stars, it was simpler to attribute it to an equal motion in the opposite direction of the point ♈, say from ♈ to ♈′ (fig. 22), *i.e.* by a motion of the equator from ♈ N to ♈′ N′, its inclination N′ ♈′ M remaining equal to its former amount N ♈ M. The general effect of this change is shewn in a different way in fig. 23, where ♈ ♈′ ♎ ♎′ being the ecliptic, A B C D represents the equator as it appeared in the time of Timocharis, A′ B′ C′ D′ (printed **bold**) the same in the time of Hipparchus, ♈, ♎ being the earlier positions of the two equinoctial points, and ♈′, ♎′ the later positions.

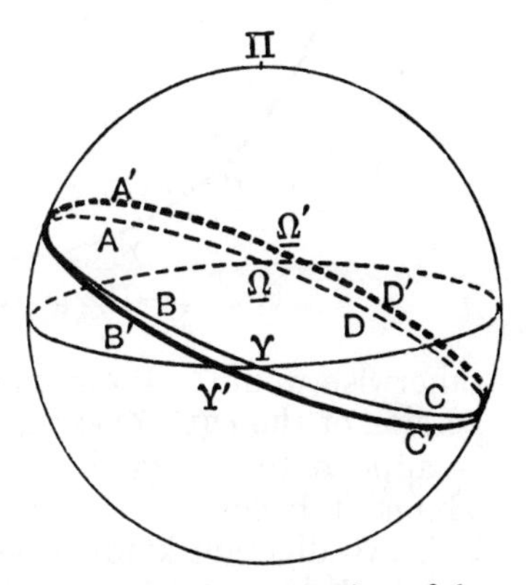

FIG. 23.—The precession of the equinoxes.

The annual motion ♈ ♈′ was, as has been stated, estimated by Hipparchus as being at least 36″ (equivalent to one degree in a century), and probably more. Its true value is considerably more, namely about 50″.

An important consequence of the motion of the equator thus discovered is that the sun in its annual journey round the ecliptic, after starting from the equinoctial point, returns to the new position of the equinoctial point a little before returning to its original position with respect to the stars, and the successive equinoxes occur slightly earlier than they

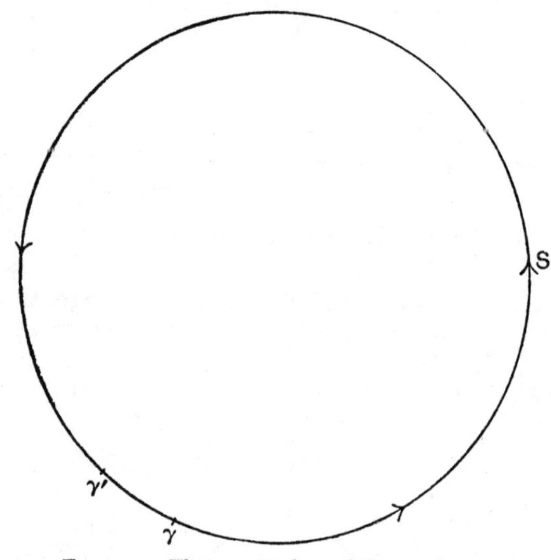

Fig. 24.—The precession of the equinoxes.

otherwise would. From this fact is derived the name **precession of the equinoxes**, or more shortly, **precession**, which is applied to the motion that we have been considering. Hence it becomes necessary to recognise, as Hipparchus did, two different kinds of year, the **tropical year** or period required by the sun to return to the same position with respect to the equinoctial points, and the **sidereal year** or period of return to the same position with respect to the stars. If ♈ ♈′ denote the motion of the equinoctial point during a tropical year, then the sun after starting from the equinoctial point at ♈ arrives—at the end of a tropical year—at the new equinoctial point at ♈′; but the sidereal year is only complete when the sun has further described the arc ♈′♈ and returned to its original starting-point ♈. Hence, taking the modern estimate 50″ of the arc ♈♈′, the sun, in the sidereal year, describes an arc of 360°, in the tropical year an arc less by 50″, or 359° 59′ 10″; the lengths of the two years are therefore in this proportion, and the amount by which the sidereal year exceeds the tropical year bears to either the same ratio as 50″ to 360° (or 1,296,000″), and is therefore $\frac{365\frac{1}{4} \times 50}{1296000}$ days or about 20 minutes.

Another way of expressing the amount of the precession is to say that the equinoctial point will describe the complete circuit of the ecliptic and return to the same position after about 26,000 years.

The length of each kind of year was also fixed by Hipparchus with considerable accuracy. That of the tropical year was obtained by comparing the times of solstices and equinoxes observed by earlier astronomers with those observed by himself. He found, for example, by comparison of the date of the summer solstice of 280 B.C., observed by Aristarchus of Samos, with that of the year 135 B.C., that the current estimate of $365\frac{1}{4}$ days for the length of the year had to be diminished by $\frac{1}{300}$th of a day or about five minutes, an estimate confirmed roughly by other cases. It is interesting to note as an illustration of his scientific method that he discusses with some care the possible error of the observations, and concludes that the time of a solstice may be erroneous to the extent of about $\frac{3}{4}$ day, while that of an equinox may be expected to be within $\frac{1}{4}$ day of the truth. In the illustration given, this would indicate a possible error of $1\frac{1}{2}$ days in a period of 145 years, or about 15 minutes in a year. Actually his estimate of the length of the year is about six minutes too great, and the error is thus much less than that which he indicated as possible. In the course of this work he considered also the possibility of a change in the length of the year, and arrived at the conclusion that, although his observations were not precise enough to show definitely the invariability of the year, there was no evidence to suppose that it had changed.

The length of the tropical year being thus evaluated at 365 days 5 hours 55 minutes, and the difference between the two kinds of year being given by the observations of precession, the sidereal year was ascertained to exceed $365\frac{1}{4}$ days by about 10 minutes, a result agreeing almost exactly with modern estimates. That the addition of two erroneous quantities, the length of the tropical year and the amount of the precession, gave such an accurate result was not, as at first sight appears, a mere accident. The chief source of error in each case being the erroneous times of the several equinoxes and solstices employed, the errors in them would tend to produce errors of opposite kinds in the tropical year and in precession, so that they would in part compensate one another. This estimate of the length of the sidereal year was probably also to some extent verified by Hipparchus by comparing eclipse observations made at different epochs.

43. The great improvements which Hipparchus effected in the theories of the sun and moon naturally enabled him to deal more successfully than any of his predecessors with a problem which in all ages has been of the greatest interest, the prediction of eclipses of the sun and moon.

That eclipses of the moon were caused by the passage of the moon through the shadow of the earth thrown by the sun, or, in other words, by the interposition of the earth between the sun and moon, and eclipses of the sun by the passage of the moon between the sun and the observer, was perfectly well known to Greek astronomers in the time of Aristotle (§ 29), and probably much earlier (chapter I., § 17), though the knowledge was probably confined to comparatively few people and superstitious terrors were long associated with eclipses.

The chief difficulty in dealing with eclipses depends on the fact that the moon's path does not coincide with the ecliptic. If the moon's path on the celestial sphere were identical with the ecliptic, then, once every month, at new moon, the moon (M) would pass exactly between the earth and the sun, and the latter would be eclipsed, and once every month also, at full moon, the moon (M′) would be in the opposite direction to the sun as seen from the earth, and would consequently be obscured by the shadow of the earth.

As, however, the moon's path is inclined to the ecliptic (§ 40), the latitudes of the sun and moon may differ by as much as 5°, either when they are in **conjunction**, *i.e.* when they have the same longitudes, or when they are

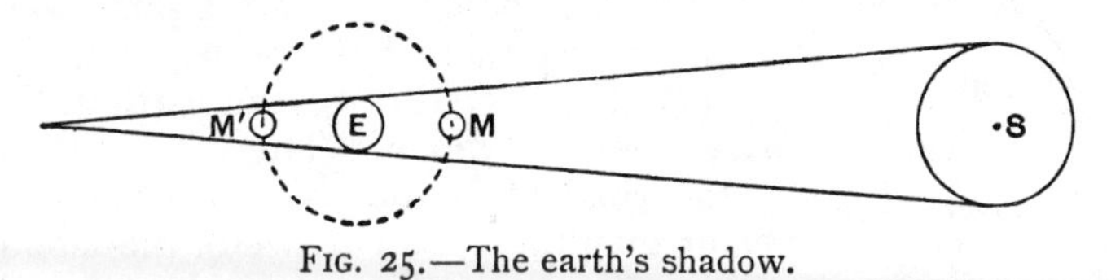

FIG. 25.—The earth's shadow.

in **opposition**, *i.e.* when their longitudes differ by 180°, and they will then in either case be too far apart for an eclipse to occur. Whether then at any full or new moon an eclipse will occur or not, will depend primarily on the latitude of the moon at the time, and hence upon her position with respect to the nodes of her orbit (§ 40). If conjunction takes place when the sun and moon happen

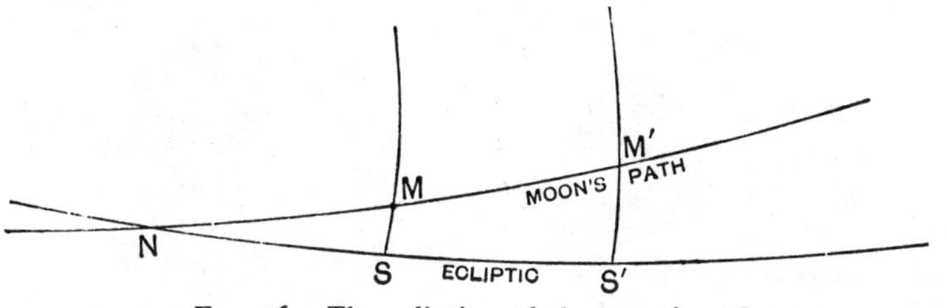

FIG. 26.—The ecliptic and the moon's path.

to be near one of the nodes (N), as at S M in fig. 26, the sun and moon will be so close together that an eclipse will occur; but if it occurs at a considerable distance from a node, as at S′ M′, their centres are so far apart that no eclipse takes place.

Now the apparent diameter of either sun or moon is, as we have seen (§ 32), about $\frac{1}{2}$°; consequently when their discs just touch, as in fig. 27, the distance between their centres is also about $\frac{1}{2}$°. If then at conjunction the distance between their centres is less than this amount, an eclipse of the sun will take place; if not, there will be no eclipse. It is an easy calculation to determine (in fig. 26) the length of the side N S or N M of the triangle N M S, when S M has this value, and hence to determine the greatest distance from the node at which conjunction can take place if an eclipse is to occur. An eclipse of the moon can be treated in the same way, except that we there have to deal with the moon and the shadow of the earth at the distance of the moon. The apparent size of the shadow is, however, considerably greater than the apparent size of the moon, and an eclipse of the moon takes place if the distance between the centre of the moon and the centre of the shadow is less than about 1°. As before, it is easy to compute the distance of the moon or of the centre of the shadow from the node when opposition occurs, if an eclipse just takes place. As, however, the apparent sizes of both sun and moon, and consequently also that of the earth's shadow, vary according to the distances of the sun and

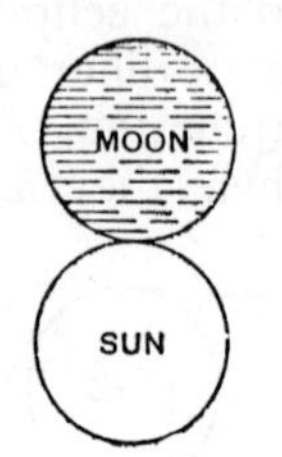

FIG. 27.—The sun and moon.

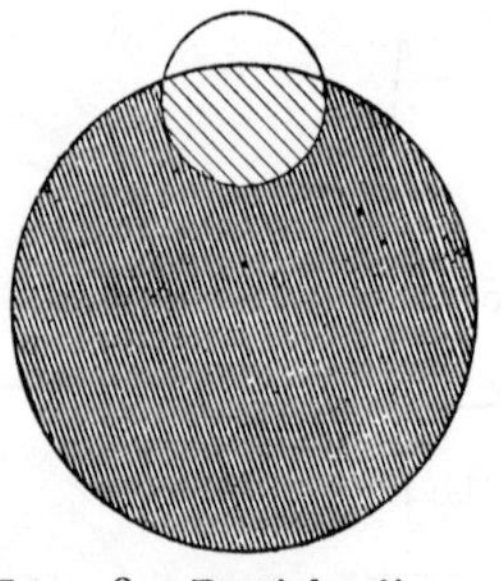
FIG. 28.—Partial eclipse of the moon.

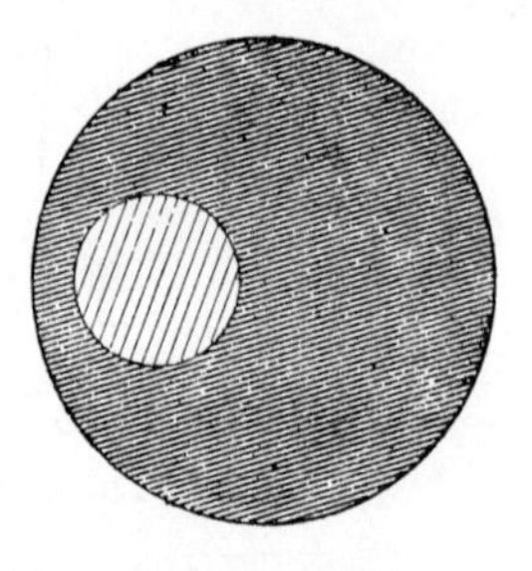
FIG. 29.—Total eclipse of the moon.

moon, a variation of which Hipparchus had no accurate knowledge, the calculation becomes really a good deal more complicated than at first sight appears, and was only dealt with imperfectly by him.

Eclipses of the moon are divided into **partial** or **total**, the former occurring when the moon and the earth's shadow only overlap partially (as in fig. 28), the latter when the moon's disc is completely immersed in the shadow (fig. 29). In the same way an eclipse of the sun may be partial or total; but as the sun's disc may be at times slightly larger than that of the moon, it sometimes happens also that the whole disc of the sun is hidden by the moon, except a narrow ring round the edge (as in fig. 30): such an eclipse is called **annular.** As the earth's shadow at the distance of the moon is always larger than the moon's disc, annular eclipses of the moon cannot occur.

FIG. 30.—Annular eclipse of the sun.

Thus eclipses take place if, and only if, the distance of the moon from a node at the time of conjunction or opposition lies within certain limits approximately known; and the problem of predicting eclipses could be roughly solved by such knowledge of the motion of the moon and of the nodes as Hipparchus possessed. Moreover, the length of the synodic and draconitic months (§ 40) being once ascertained, it became merely a matter of arithmetic to compute one or more periods after which eclipses would recur nearly in the same manner. For if any period of time contains an exact number of each kind of month, and if at any time an eclipse occurs, then after the lapse of the period, conjunction (or opposition) again takes place, and the moon is at the same distance as before from the node and the eclipse recurs very much as before. The saros, for example (chapter I., § 17), contained very nearly 223 synodic or 242 draconitic months, differing from either by less than an hour. Hipparchus saw that this period was not completely reliable as a means of predicting eclipses, and showed how to allow for the irregularities in the moon's and sun's motion (§§ 39, 40) which were ignored by it, but was unable to deal fully with the difficulties arising from the variations in the apparent diameters of the sun or moon.

An important complication, however, arises in the case of eclipses of the sun, which had been noticed by earlier writers, but which Hipparchus was the first to deal with. Since an eclipse of the moon is an actual darkening of the moon, it is visible to anybody, wherever situated, who can see the moon at all; for example, to possible inhabitants of other planets, just as we on the earth can see precisely similar eclipses of Jupiter's moons. An eclipse of the sun is, however, merely the screening off of the sun's light from a particular observer, and the sun may therefore be eclipsed to one observer while to another elsewhere it is visible as usual. Hence in computing an eclipse of the sun it is necessary to take into account the position of the observer on the earth. The simplest way of doing this is to make allowance for the difference of direction of the moon as seen by an observer at the place in question, and by an observer in some standard position on the earth, preferably

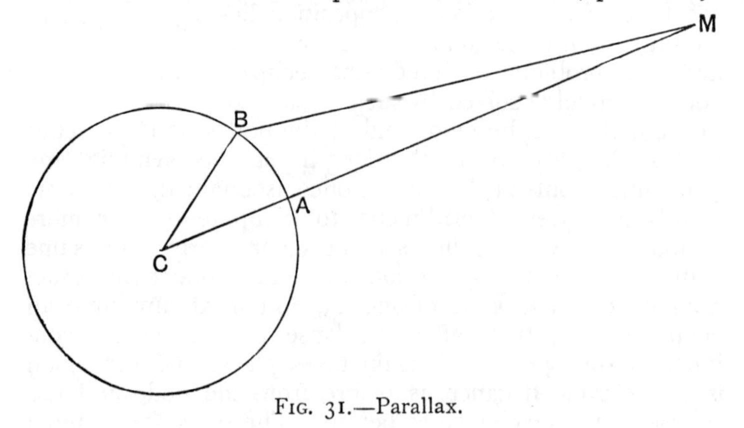

FIG. 31.—Parallax.

an ideal observer at the centre of the earth. If, in fig. 31, M denote the moon, C the centre of the earth, A a point on the earth between C and M (at which therefore the moon is overhead), and B any other point on the earth, then observers at C (or A) and B see the moon in slightly different directions, C M, B M, the difference between which is an angle known as the **parallax,** which is equal to the angle B M C and depends on the distance of the moon, the size of the earth, and the position of the observer at B. In the case of the sun, owing to its great distance, even as estimated by the Greeks, the parallax was in all cases too small to be taken into account, but in the case of the moon the parallax might be as much as 1° and could not be neglected.

If then the path of the moon, as seen from the centre of the earth, were known, then the path of the moon as seen from any particular station on the earth could be deduced by allowing for parallax, and the conditions of an eclipse of the sun visible there could be computed accordingly.

From the time of Hipparchus onwards lunar eclipses could easily be predicted to within an hour or two by any ordinary astronomer; solar eclipses probably with less accuracy; and in both cases the prediction of the extent of the eclipse, *i.e.* of what portion of the sun or moon would be obscured, probably left very much to be desired.

44. The great services rendered to astronomy by Hipparchus can hardly be better expressed than in the words of the great French historian of astronomy, Delambre, who is in general no lenient critic of the work of his predecessors:—

"When we consider all that Hipparchus invented or perfected, and reflect upon the number of his works and the mass of calculations which they imply, we must regard him as one of the most astonishing men of antiquity, and as the greatest of all in the sciences which are not purely speculative, and which require a combination of geometrical knowledge with a knowledge of phenomena, to be observed only by diligent attention and refined instruments." *

45. For nearly three centuries after the death of Hipparchus, the history of astronomy is almost a blank. Several textbooks written during this period are extant, shewing the gradual popularisation of his great discoveries. Among the few things of interest in these books may be noticed a statement that the stars are not necessarily on the surface of a sphere, but may be at different distances from us, which, however, there are no means of estimating; a conjecture that the sun and stars are so far off that the earth would be a mere point seen from the sun and invisible from the stars; and a re-statement of an old opinion traditionally attributed to the Egyptians (whether of the Alexandrine period or earlier is uncertain), that Venus and Mercury revolve round the sun. It seems also that in this period some attempts were made to explain the planetary

* *Histoire de l'Astronomie Ancienne*, Vol. I., p. 185.

motions by means of epicycles, but whether these attempts marked any advance on what had been done by Apollonius and Hipparchus is uncertain.

It is interesting also to find in *Pliny* (A.D. 23–79) the well-known modern argument for the spherical form of the earth, that when a ship sails away the masts, etc., remain visible after the hull has disappeared from view.

A new measurement of the circumference of the earth by *Posidonius* (born about the end of Hipparchus's life) may also be noticed; he adopted a method similar to that of Eratosthenes (§ 36), and arrived at two different results. The later estimate, to which he seems to have attached most weight, was 180,000 stadia, a result which was about as much below the truth as that of Eratosthenes was above it.

46. The last great name in Greek astronomy is that of Claudius Ptolemaeus, commonly known as *Ptolemy*, of whose life nothing is known except that he lived in Alexandria about the middle of the 2nd century A.D. His reputation rests chiefly on his great astronomical treatise, known as the *Almagest*,* which is the source from which by far the greater part of our knowledge of Greek astronomy is derived, and which may be fairly regarded as the astronomical Bible of the Middle Ages. Several other minor astronomical and astrological treatises are attributed to him, some of which are probably not genuine, and he was also the author of an important work on geography, and possibly of a treatise on *Optics*, which is, however, not certainly authentic and maybe of Arabian origin. The *Optics* discusses, among other topics, the **refraction** or bending of light, by the atmosphere on the earth: it is pointed out that the light of a star or other heavenly body S, on entering our atmosphere (at A) and on penetrating to the lower and denser portions of it, must be gradually bent or **refracted**, the result being that the

* The chief MS. bears the title *μεγάλη σύνταξις*, or great composition ,though the author refers to his book elsewhere as *μαθηματικὴ σύνταξις* (mathematical composition). The Arabian translators, either through admiration or carelessness, converted *μεγάλη*, great, into *μεγίστη*, greatest, and hence it became known by the Arabs as *Al Magisti*, whence the Latin *Almagestum* and our *Almagest*.

star appears to the observer at B nearer to the zenith Z than it actually is, *i.e.* the light appears to come from S′ instead of from S; it is shewn further that this effect must be greater for bodies near the horizon than for those near the zenith, the light from the former travelling through a greater extent of atmosphere; and these results are shewn to account for certain observed deviations in the daily paths of the stars, by which they appear unduly raised up when near the horizon. Refraction also explains the well-known flattened appearance of the sun or moon when rising or setting, the lower edge being raised by

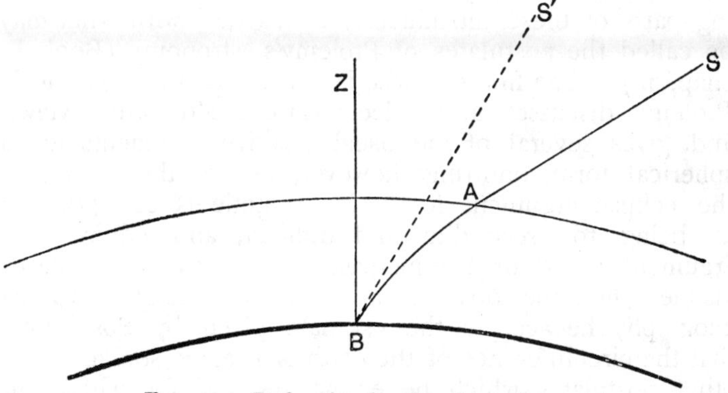

FIG. 32.—Refraction by the atmosphere.

refraction more than the upper, so that a contraction of the vertical diameter results, the horizontal contraction being much less.*

47. The *Almagest* is avowedly based largely on the work of earlier astronomers, and in particular on that of Hipparchus, for whom Ptolemy continually expresses the greatest admiration and respect. Many of its contents have therefore already been dealt with by anticipation, and need not be discussed again in detail. The book plays, however, such an important part in astronomical history, that it may be worth while to give a short outline of its contents,

* The better known apparent enlargement of the sun or moon when rising or setting has nothing to do with refraction. It is an optical illusion not very satisfactorily explained, but probably due to the lesser brilliancy of the sun at the time.

in addition to dealing more fully with the parts in which Ptolemy made important advances.

The *Almagest* consists altogether of 13 books. The first two deal with the simpler observed facts, such as the daily motion of the celestial sphere, and the general motions of the sun, moon, and planets, and also with a number of topics connected with the celestial sphere and its motion, such as the length of the day and the times of rising and setting of the stars in different zones of the earth ; there are also given the solutions of some important mathematical problems,* and a mathematical table † of considerable accuracy and extent. But the most interesting parts of these introductory books deal with what may be called the postulates of Ptolemy's astronomy (Book I., chap. ii.). The first of these is that the earth is spherical; Ptolemy discusses and rejects various alternative views, and gives several of the usual positive arguments for a spherical form, omitting, however, one of the strongest, the eclipse argument found in Aristotle (§ 29), possibly as being too recondite and difficult, and adding the argument based on the increase in the area of the earth visible when the observer ascends to a height. In his geography he accepts the estimate given by Posidonius that the circumference of the earth is 180,000 stadia. The other postulates which he enunciates and for which he argues are, that the heavens are spherical and revolve like a sphere; that the earth is in the centre of the heavens, and is merely a point in comparison with the distance of the fixed stars, and that it has no motion. The position of these postulates in the treatise and Ptolemy's general method of procedure suggest that he was treating them, not so much as important results to be established by the best possible evidence, but rather as assumptions, more probable than any others with which the author was acquainted, on which to base mathematical calculations which should explain observed phenomena.‡ His attitude is thus

* In spherical trigonometry.

† A table of chords (or double sines of half-angles) for every $\frac{1}{2}^{\circ}$ from 0° to 180°.

‡ His procedure may be compared with that of a political economist of the school of Ricardo, who, in order to establish some

essentially different from that either of the early Greeks, such as Pythagoras, or of the controversialists of the 16th and early 17th centuries, such as Galilei (chapter VI.), for whom the truth or falsity of postulates analogous to those of Ptolemy was of the very essence of astronomy and was among the final objects of inquiry. The arguments which Ptolemy produces in support of his postulates, arguments which were probably the commonplaces of the astronomical writing of his time, appear to us, except in the case of the shape of the earth, loose and of no great value. The other postulates were, in fact, scarcely capable of either proof or disproof with the evidence which Ptolemy had at command. His argument in favour of the immobility of the earth is interesting, as it shews his clear perception that the more obvious appearances can be explained equally well by a motion of the stars or by a motion of the earth; he concludes, however, that it is easier to attribute motion to bodies like the stars which seem to be of the nature of fire than to the solid earth, and points out also the difficulty of conceiving the earth to have a rapid motion of which we are entirely unconscious. He does not, however, discuss seriously the possibility that the earth or even Venus and Mercury may revolve round the sun.

The third book of the *Almagest* deals with the length of the year and theory of the sun, but adds nothing of importance to the work of Hipparchus.

48. The fourth book of the *Almagest*, which treats of the length of the month and of the theory of the moon, contains one of Ptolemy's most important discoveries. We have seen that, apart from the motion of the moon's orbit as a whole, and the revolution of the line of apses, the chief irregularity or inequality was the so-called equation of the centre (§§ 39, 40), represented fairly accurately by

rough explanation of economic phenomena, starts with certain simple assumptions as to human nature, which at any rate are more plausible than any other equally simple set, and deduces from them a number of abstract conclusions, the applicability of which to real life has to be considered in individual cases. But the perfunctory discussion which such a writer gives of the qualities of the "economic man" cannot of course be regarded as his deliberate and final estimate of human nature.

means of an eccentric, and depending only on the position of the moon with respect to its apogee. Ptolemy, however, discovered, what Hipparchus only suspected, that there was a further inequality in the moon's motion—to which the name **evection** was afterwards given—and that this depended partly on its position with respect to the sun. Ptolemy compared the observed positions of the moon with those calculated by Hipparchus in various positions relative to the sun and apogee, and found that, although there was a satisfactory agreement at new and full moon, there was a considerable error when the moon was half-full, provided it was also not very near perigee or apogee. Hipparchus based his theory of the moon chiefly on observations of eclipses, *i.e.* on observations taken necessarily at full or new moon (§ 43), and Ptolemy's discovery is due to the fact that he checked Hipparchus's theory by observations taken at other times. To represent this new inequality, it was found necessary to use an epicycle and a deferent, the latter being itself a moving eccentric circle, the centre of which revolved round the earth. To account, to some extent, for certain remaining discrepancies between theory and observation, which occurred neither at new and full moon, nor at the **quadratures** (half-moon), Ptolemy introduced further a certain small to-and-fro oscillation of the epicycle, an oscillation to which he gave the name of **prosneusis.***

* The equation of the centre and the evection may be expressed trigonometrically by two terms in the expression for the moon's longitude, $a \sin \theta + b \sin (2\phi - \theta)$, where a, b are two numerical quantities, in round numbers 6° and 1°, θ is the angular distance of the moon from perigee, and ϕ is the angular distance from the sun. At conjunction and opposition ϕ is 0° or 180°, and the two terms reduce to $(a-b) \sin \theta$. This would be the form in which the equation of the centre would have presented itself to Hipparchus. Ptolemy's correction is therefore equivalent to adding on

$$b\,[\sin \theta + \sin (2\phi - \theta)], \text{ or } 2\,b \sin \phi \cos (\phi - \theta),$$

which vanishes at conjunction or opposition, but reduces at the quadratures to $2\,b \sin \theta$, which again vanishes if the moon is at apogee or perigee ($\theta = 0°$ or $180°$), but has its greatest value half-way between, when $\theta = 90°$. Ptolemy's construction gave rise also to a still smaller term of the type,

$$c \sin 2\phi\,[\cos (2\phi + \theta) + 2 \cos (2\phi - \theta)],$$

which, it will be observed, vanishes at quadratures as well as at conjunction and opposition.

Ptolemy thus succeeded in fitting his theory on to his observations so well that the error seldom exceeded 10′, a small quantity in the astronomy of the time, and on the basis of this construction he calculated tables from which the position of the moon at any required time could be easily deduced.

One of the inherent weaknesses of the system of epicycles occurred in this theory in an aggravated form. It has already been noticed in connection with the theory of the sun (§ 39), that the eccentric or epicycle produced an erroneous variation in the distance of the sun, which was, however, imperceptible in Greek times. Ptolemy's system, however, represented the moon as being sometimes nearly twice as far off as at others, and consequently the apparent diameter ought at some times to have been not much more than half as great as at others—a conclusion obviously inconsistent with observation. It seems probable that Ptolemy noticed this difficulty, but was unable to deal with it; it is at any rate a significant fact that when he is dealing with eclipses, for which the apparent diameters of the sun and moon are of importance, he entirely rejects the estimates that might have been obtained from his lunar theory and appeals to direct observation (cf. also § 51, note).

49. The fifth book of the *Almagest* contains an account of the construction and use of Ptolemy's chief astronomical instrument, a combination of graduated circles known as the **astrolabe.***

Then follows a detailed discussion of the moon's parallax (§ 43), and of the distances of the sun and moon. Ptolemy obtains the distance of the moon by a parallax method which is substantially identical with that still in use. If we know the direction of the line C M (fig. 33) joining the centres of the earth and moon, or the direction of the moon as seen by an observer at A; and also the direction of the line B M, that is the direction of the moon as seen by an observer at B, then the angles of the triangle C B M are known, and the ratio of the sides C B, C M is known.

* Here, as elsewhere, I have given no detailed account of astronomical instruments, believing such descriptions to be in general neither interesting nor intelligible to those who have not the actual instruments before them, and to be of little use to those who have.

Ptolemy obtained the two directions required by means of observations of the moon, and hence found that C M was 59 times C B, or that the distance of the moon was equal to 59 times the radius of the earth. He then uses Hipparchus's eclipse method to deduce the distance of the sun from that of the moon thus ascertained, and finds the distance of the sun to be 1,210 times the radius of the earth. This number, which is substantially the same as that obtained by Hipparchus (§ 41), is, however, only

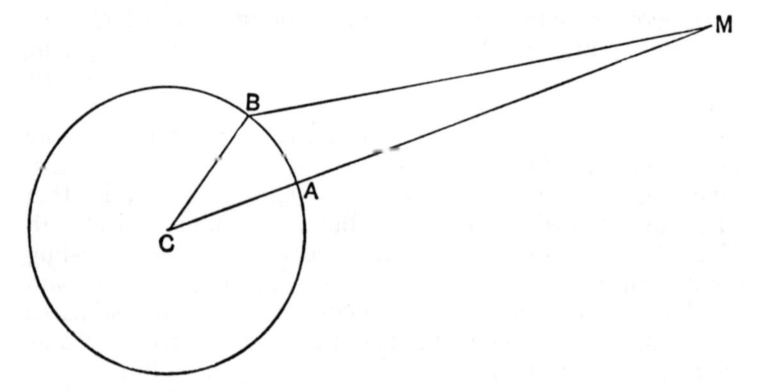

FIG. 33.—Parallax.

about $\frac{1}{20}$ of the true number, as indicated by modern work (chapter XIII., § 284).

The sixth book is devoted to eclipses, and contains no substantial additions to the work of Hipparchus.

50. The seventh and eighth books contain a catalogue of stars, and a discussion of precession (§ 42). The catalogue, which contains 1,028 stars (three of which are duplicates), appears to be nearly identical with that of Hipparchus. It contains none of the stars which were visible to Ptolemy at Alexandria, but not to Hipparchus at Rhodes. Moreover, Ptolemy professes to deduce from a comparison of his observations with those of Hipparchus and others the (erroneous) value 36″ for the precession, which Hipparchus had given as the least possible value, and which Ptolemy regards as his final estimate. But an examination of

the positions assigned to the stars in Ptolemy's catalogue agrees better with their actual positions in the time of Hipparchus, *corrected for precession at the supposed rate of 36″ annually*, than with their actual positions in Ptolemy's time. It is therefore probable that the catalogue as a whole does not represent genuine observations made by Ptolemy, but is substantially the catalogue of Hipparchus corrected for precession and only occasionally modified by new observations by Ptolemy or others.

51. The last five books deal with the theory of the planets, the most important of Ptolemy's original contributions to astronomy. The problem of giving a satisfactory explanation of the motions of the planets was, on account of their far greater irregularity, a much more difficult one than the corresponding problem for the sun or moon. The motions of the latter are so nearly uniform that their irregularities may usually be regarded as of the nature of small corrections, and for many purposes may be ignored. The planets, however, as we have seen (chapter I., § 14), do not even always move from west to east, but stop at intervals, move in the reverse direction for a time, stop again, and then move again in the original direction. It was probably recognised in early times, at latest by Eudoxus (§ 26), that in the case of three of the planets, Mars, Jupiter, and Saturn, these motions could be represented roughly by supposing each planet to oscillate to and fro on each side of a fictitious planet, moving uniformly round the celestial sphere in or near the ecliptic, and that Venus and Mercury could similarly be regarded as oscillating to and fro on each side of the sun. These rough motions could easily be interpreted by means of revolving spheres or of epicycles, as was done by Eudoxus and probably again with more precision by Apollonius. In the case of Jupiter, for example, we may regard the planet as moving on an epicycle, the centre of which, j, describes uniformly a deferent, the centre of which is the earth. The planet will then as seen from the earth appear alternately to the east (as at J_1) and to the west (as at J_2) of the fictitious planet j; and the extent of the oscillation on each side, and the interval between successive appearances in the extreme positions (J_1, J_2) on either side, can be made right by choosing appropriately the size

and rapidity of motion of the epicycle. It is moreover evident that with this arrangement the apparent motion of Jupiter will vary considerably, as the two motions—that on the epicycle and that of the centre of the epicycle on the deferent—are sometimes in the same direction, so as to increase one another's effect, and at other times in opposite directions. Thus, when Jupiter is most distant from the earth, that is at J_3, the motion is most rapid, at J_1 and J_2 the motion as seen from the earth is nearly the same as that of *j*; while at J_4 the two motions are in opposite directions, and the size and motion of the epicycle having been chosen in the way indicated above, it is found in fact that the motion of the planet in the epicycle is the greater of the two motions, and that therefore the planet when in this position appears to be moving from east to west (from left to right in the figure), as is actually the case. As then at J_1 and J_2 the planet appears to be moving from west to east, and at J_4 in the opposite direction, and sudden changes of motion do not occur in astronomy, there must be a position between J_1 and J_4, and another between J_4 and J_2, at which the planet is just reversing its direction of motion, and therefore appears for the instant at rest. We thus arrive at an explanation of the stationary points (chapter I., § 14). An exactly similar scheme explains roughly the motion of Mercury and Venus, except that the centre of the epicycle must always be in the direction of the sun.

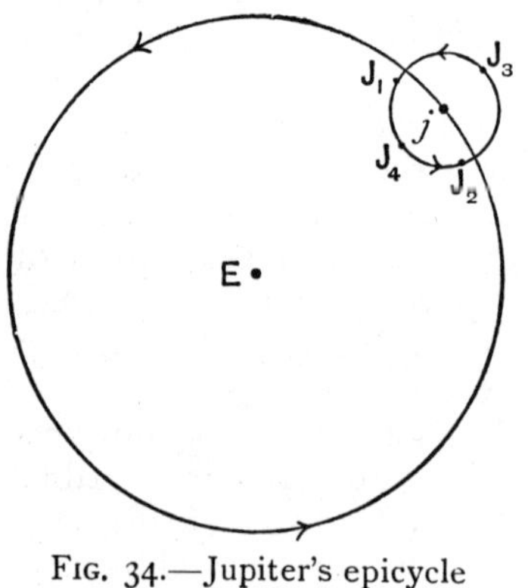

FIG. 34.—Jupiter's epicycle and deferent.

Hipparchus, as we have seen (§ 41), found the current representations of the planetary motions inaccurate, and collected a number of fresh observations. These, with fresh observations of his own, Ptolemy now employed in order to construct an improved planetary system.

As in the case of the moon, he used as deferent an eccentric circle (centre C), but instead of making the centre *j* of the epicycle move uniformly in the deferent, he introduced a new point called an **equant** (E′), situated at the same distance from the centre of the deferent as the earth but on the opposite side, and regulated the motion of *j* by the condition that the apparent motion *as seen from the equant* should be uniform; in other words, the angle A E′ *j* was made to increase uniformly. In the case of Mercury (the motions of which have been found troublesome by astronomers of all periods), the relation of the equant to the centre of the epicycle was different, and the latter was made to move in a small circle. The deviations of the planets from the ecliptic (chapter I., §§ 13, 14) were accounted for by tilting up the planes of the several deferents and epicycles so that they were inclined to the ecliptic at various small angles.

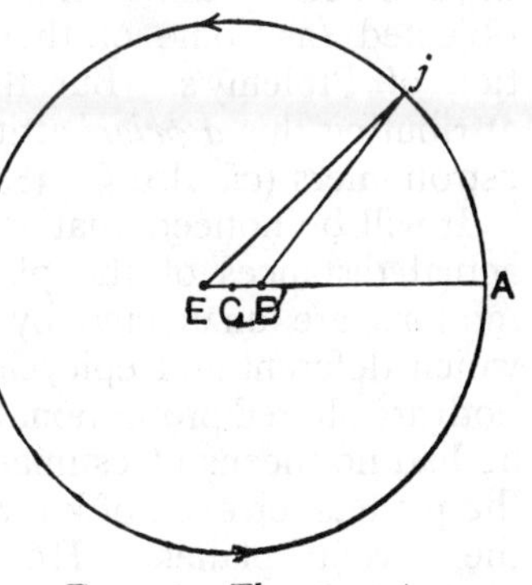

FIG. 35.—The equant.

By means of a system of this kind, worked out with great care, and evidently at the cost of enormous labour, Ptolemy was able to represent with very fair exactitude the motions of the planets, as given by the observations in his possession.

It has been pointed out by modern critics, as well as by some mediaeval writers, that the use of the equant (which played also a small part in Ptolemy's lunar theory) was a violation of the principle of employing only uniform circular motions, on which the systems of Hipparchus and Ptolemy were supposed to be based, and that Ptolemy himself appeared unconscious of his inconsistency. It may, however, fairly be doubted whether Hipparchus or Ptolemy ever had an abstract belief in the exclusive virtue of such motions, except as a convenient and easily intelligible way of representing certain more complicated motions, and it is difficult to conceive that Hipparchus would have scrupled any more than his great follower, in using an equant to represent an irregular motion, if he had found that the motion was thereby represented with accuracy. The criticism appears to me in fact to be an anachronism. The earlier Greeks, whose astronomy was speculative rather than scientific, and again many astronomers of the Middle Ages, felt that it was on *a priori* grounds necessary to represent the "perfection" of the heavenly motions by the most "perfect" or regular of geometrical schemes; so that it is highly probable that Pythagoras or Plato, or even Aristotle, would have objected, and certain that the astronomers of the 14th and 15th centuries ought to have objected (as some of them actually did), to this innovation of Ptolemy's. But there seems no good reason for attributing this *a priori* attitude to the later scientific Greek astronomers (cf. also §§ 38, 47).*

It will be noticed that nothing has been said as to the actual distances of the planets, and in fact the apparent motions are unaffected by any alteration in the scale on which deferent and epicycle are constructed, provided that both are altered proportionally. Ptolemy expressly states that he had no means of estimating numerically the distances of the planets, or even of knowing the order of the distance of the several planets. He followed tradition in accepting conjecturally rapidity of motion as a test of nearness, and placed Mars, Jupiter, Saturn (which perform the circuit of the celestial sphere in about 2, 12, and 29 years respectively) beyond the sun in that order. As Venus and

* The advantage derived from the use of the equant can be made clearer by a mathematical comparison with the elliptic motion introduced by Kepler. In elliptic motion the angular motion and distance are represented approximately by the formulæ $nt + 2e \sin nt$, $a(1 - e \cos nt)$ respectively; the corresponding formulæ given by the use of the simple eccentric are $nt + e' \sin nt$, $a(1 - e' \cos nt)$. To make the angular motions agree we must therefore take $e' = 2e$, but to make the distances agree we must take $e' = e$; the two conditions are therefore inconsistent. But by the introduction of an equant the formulae become $nt + 2e' \sin nt$, $a(1 - e' \cos nt)$, and *both* agree if we take $e' = e$. Ptolemy's lunar theory could have been nearly freed from the serious difficulty already noticed (§ 48) if he had used an equant to represent the chief inequality of the moon; and his planetary theory would have been made accurate to the first order of small quantities by the use of an equant both for the deferent and the epicycle.

Mercury accompany the sun, and may therefore be regarded as on the average performing their revolutions in a year, the test to some extent failed in their case, but Ptolemy again accepted the opinion of the "ancient mathematicians" (*i.e.* probably the Chaldaeans) that Mercury and Venus lie between the sun and moon, Mercury being the nearer to us. (Cf. chapter I., § 15.)

52. There has been much difference of opinion among astronomers as to the merits of Ptolemy. Throughout the Middle Ages his authority was regarded as almost final on astronomical matters, except where it was outweighed by the even greater authority assigned to Aristotle. Modern criticism has made clear, a fact which indeed he never conceals, that his work is to a large extent based on that of Hipparchus; and that his observations, if not actually fictitious, were at any rate in most cases poor. On the other hand his work shews clearly that he was an accomplished and original mathematician.* The most important of his positive contributions to astronomy were the discovery of evection and his planetary theory, but we ought probably to rank above these, important as they are, the services which he rendered by preserving and developing the great ideas of Hipparchus—ideas which the other astronomers of the time were probably incapable of appreciating, and which might easily have been lost to us if they had not been embodied in the *Almagest*.

53. The history of Greek astronomy practically ceases with Ptolemy. The practice of observation died out so completely that only eight observations are known to have been made during the eight and a half centuries which separate him from Albategnius (chapter III., § 59). The only Greek writers after Ptolemy's time are compilers and commentators, such as *Theon* (*fl.* A.D. 365), to none of whom original ideas of any importance can be attributed. The murder of his daughter Hypatia (A.D. 415), herself also a writer on astronomy, marks an epoch in the decay of the Alexandrine school; and the end came in A.D. 640, when Alexandria was captured by the Arabs.†

* De Morgan classes him as a geometer with Archimedes, Euclid, and Apollonius, the three great geometers of antiquity.

† The legend that the books in the library served for six months as

54. It remains to attempt to estimate briefly the value of the contributions to astronomy made by the Greeks and of their method of investigation. It is obviously unreasonable to expect to find a brief formula which will characterise the scientific attitude of a series of astronomers whose lives extend over a period of eight centuries; and it is futile to explain the inferiority of Greek astronomy to our own on some such ground as that they had not discovered the method of induction, that they were not careful enough to obtain facts, or even that their ideas were not clear. In habits of thought and scientific aims the contrast between Pythagoras and Hipparchus is probably greater than that between Hipparchus on the one hand and Coppernicus or even Newton on the other, while it is not unfair to say that the fanciful ideas which pervade the work of even so great a discoverer as Kepler (chapter VII., §§ 144, 151) place his scientific method in some respects behind that of his great Greek predecessor.

The Greeks inherited from their predecessors a number of observations, many of them executed with considerable accuracy, which were nearly sufficient for the requirements of practical life, but in the matter of astronomical theory and speculation, in which their best thinkers were very much more interested than in the detailed facts, they received virtually a blank sheet on which they had to write (at first with indifferent success) their speculative ideas. A considerable interval of time was obviously necessary to bridge over the gulf separating such data as the eclipse observations of the Chaldaeans from such ideas as the harmonical spheres of Pythagoras; and the necessary theoretical structure could not be erected without the use of mathematical methods which had gradually to be invented. That the Greeks, particularly in early times, paid little attention to making observations, is true enough, but it may fairly be doubted whether the collection of fresh material for observations would really have carried astronomy much beyond the point reached by the Chaldaean observers. When once speculative ideas, made

fuel for the furnaces of the public baths is rejected by Gibbon and others. One good reason for not accepting it is that by this time there were probably very few books left to burn.

definite by the aid of geometry, had been sufficiently developed to be capable of comparison with observation, rapid progress was made. The Greek astronomers of the scientific period, such as Aristarchus, Eratosthenes, and above all Hipparchus, appear moreover to have followed in their researches the method which has always been fruitful in physical science—namely, to frame provisional hypotheses, to deduce their mathematical consequences, and to compare these with the results of observation. There are few better illustrations of genuine scientific caution than the way in which Hipparchus, having tested the planetary theories handed down to him and having discovered their insufficiency, deliberately abstained from building up a new theory on data which he knew to be insufficient, and patiently collected fresh material, never to be used by himself, that some future astronomer might thereby be able to arrive at an improved theory.

Of positive additions to our astronomical knowledge made by the Greeks the most striking in some ways is the discovery of the approximately spherical form of the earth, a result which later work has only slightly modified. But their explanation of the chief motions of the solar system and their resolution of them into a comparatively small number of simpler motions was, in reality, a far more important contribution, though the Greek epicyclic scheme has been so remodelled, that at first sight it is difficult to recognise the relation between it and our modern views. The subsequent history will, however, show how completely each stage in the progress of astronomical science has depended on those that preceded.

When we study the great conflict in the time of Coppernicus between the ancient and modern ideas, our sympathies naturally go out towards those who supported the latter, which are now known to be more accurate, and we are apt to forget that those who then spoke in the name of the ancient astronomy and quoted Ptolemy were indeed believers in the doctrines which they had derived from the Greeks, but that their methods of thought, their frequent refusal to face facts, and their appeals to authority, were all entirely foreign to the spirit of the great men whose disciples they believed themselves to be.

CHAPTER III.

THE MIDDLE AGES.

> "The lamp burns low, and through the casement bars
> Grey morning glimmers feebly."
>
> BROWNING'S *Paracelsus.*

55. ABOUT fourteen centuries elapsed between the publication of the *Almagest* and the death of Coppernicus (1543), a date which is in astronomy a convenient landmark on the boundary between the Middle Ages and the modern world. In this period, nearly twice as long as that which separated Thales from Ptolemy, almost four times as long as that which has now elapsed since the death of Coppernicus, no astronomical discovery of first-rate importance was made. There were some important advances in mathematics, and the art of observation was improved; but theoretical astronomy made scarcely any progress, and in some respects even went backward, the current doctrines, if in some points slightly more correct than those of Ptolemy, being less intelligently held.

In the Western World we have already seen that there was little to record for nearly five centuries after Ptolemy. After that time ensued an almost total blank, and several more centuries elapsed before there was any appreciable revival of the interest once felt in astronomy.

56. Meanwhile a remarkable development of science had taken place in the East during the 7th century. The descendants of the wild Arabs who had carried the banner of Mahomet over so large a part of the Roman empire, as well as over lands lying farther east, soon began to feel the influence of the civilisation of the peoples whom they had subjugated, and Bagdad, which in the 8th century became

the capital of the Caliphs, rapidly developed into a centre of literary and scientific activity. Al Mansur, who reigned from A.D. 754 to 775, was noted as a patron of science, and collected round him learned men both from India and the West. In particular we are told of the arrival at his court in 772 of a scholar from India bearing with him an Indian treatise on astronomy,* which was translated into Arabic by order of the Caliph, and remained the standard treatise for nearly half a century. From Al Mansur's time onwards a body of scholars, in the first instance chiefly Syrian Christians, were at work at the court of the Caliphs translating Greek writings, often through the medium of Syriac, into Arabic. The first translations made were of the medical treatises of Hippocrates and Galen; the Aristotelian ideas contained in the latter appear to have stimulated interest in the writings of Aristotle himself, and thus to have enlarged the range of subjects regarded as worthy of study. Astronomy soon followed medicine, and became the favourite science of the Arabians, partly no doubt out of genuine scientific interest, but probably still more for the sake of its practical applications. Certain Mahometan ceremonial observances required a knowledge of the direction of Mecca, and though many worshippers, living anywhere between the Indus and the Straits of Gibraltar, must have satisfied themselves with rough-and-ready solutions of this problem, the assistance which astronomy could give in fixing the true direction was welcome in larger centres of population. The Mahometan calendar, a lunar one, also required some attention in order that fasts and feasts should be kept at the proper times. Moreover the belief in the possibility of predicting the future by means of the stars, which had flourished among the Chaldaeans (chapter I., § 18), but which remained to a great extent in abeyance among the Greeks, now revived rapidly on a congenial oriental soil, and the Caliphs were probably quite as much interested in seeing that the learned men of

* The data as to Indian astronomy are so uncertain, and the evidence of any important original contributions is so slight, that I have not thought it worth while to enter into the subject in any detail. The chief Indian treatises, including the one referred to in the text, bear strong marks of having been based on Greek writings.

their courts were proficient in astrology as in astronomy proper.

The first translation of the *Almagest* was made by order of Al Mansur's successor Harun al Rasid (A.D. 765 or 766 –A.D. 809), the hero of the *Arabian Nights*. It seems, however, to have been found difficult to translate; fresh attempts were made by *Honein ben Ishak* (?–873) and by his son *Ishak ben Honein* (?–910 or 911), and a final version by *Tabit ben Korra* (836–901) appeared towards the end of the 9th century. Ishak ben Honein translated also a number of other astronomical and mathematical books, so that by the end of the 9th century, after which translations almost ceased, most of the more important Greek books on these subjects, as well as many minor treatises, had been translated. To this activity we owe our knowledge of several books of which the Greek originals have perished.

57. During the period in which the Caliphs lived at Damascus an observatory was erected there, and another on a more magnificent scale was built at Bagdad in 829 by the Caliph Al Mamun. The instruments used were superior both in size and in workmanship to those of the Greeks, though substantially of the same type. The Arab astronomers introduced moreover the excellent practice of making regular and as far as possible nearly continuous observations of the chief heavenly bodies, as well as the custom of noting the positions of known stars at the beginning and end of an eclipse, so as to have afterwards an exact record of the times of their occurrence. So much importance was attached to correct observations that we are told that those of special interest were recorded in formal documents signed on oath by a mixed body of astronomers and lawyers.

Al Mamun ordered Ptolemy's estimate of the size of the earth to be verified by his astronomers. Two separate measurements of a portion of a meridian were made, which, however, agreed so closely with one another and with the erroneous estimate of Ptolemy that they can hardly have been independent and careful measurements, but rather rough verifications of Ptolemy's figures.

58. The careful observations of the Arabs soon shewed

the defects in the Greek astronomical tables, and new tables were from time to time issued, based on much the same principles as those in the *Almagest*, but with changes in such numerical data as the relative sizes of the various circles, the positions of the apogees, and the inclinations of the planes, etc.

To Tabit ben Korra, mentioned above as the translator of the *Almagest*, belongs the doubtful honour of the discovery of a supposed variation in the amount of the precession (chapter II., §§ 42, 50). To account for this he devised a complicated mechanism which produced a certain alteration in the position of the ecliptic, thus introducing a purely imaginary complication, known as the **trepidation**, which confused and obscured most of the astronomical tables issued during the next five or six centuries.

59. A far greater astronomer than any of those mentioned in the preceding articles was the Arab prince called from his birthplace Al Battani, and better known by the Latinised name *Albategnius*, who carried on observations from 878 to 918 and died in 929. He tested many of Ptolemy's results by fresh observations, and obtained more accurate values of the obliquity of the ecliptic (chapter I., § 11) and of precession. He wrote also a treatise on astronomy which contained improved tables of the sun and moon, and included his most notable discovery—namely, that the direction of the point in the sun's orbit at which it is farthest from the earth (the apogee), or, in other words, the direction of the centre of the eccentric representing the sun's motion (chapter II., § 39), was not the same as that given in the *Almagest*; from which change, too great to be attributed to mere errors of observation or calculation, it might fairly be inferred that the apogee was slowly moving, a result which, however, he did not explicitly state. Albategnius was also a good mathematician, and the author of some notable improvements in methods of calculation.*

60. The last of the Bagdad astronomers was *Abul Wafa*

* He introduced into trigonometry the use of *sines*, and made also some little use of *tangents*, without apparently realising their importance: he also used some new formulæ for the solution of spherical triangles.

(939 or 940–998), the author of a voluminous treatise on astronomy also known as the *Almagest*, which contained some new ideas and was written on a different plan from Ptolemy's book, of which it has sometimes been supposed to be a translation. In discussing the theory of the moon Abul Wafa found that, after allowing for the equation of the centre and for the evection, there remained a further irregularity in the moon's motion which was imperceptible at conjunction, opposition, and quadrature, but appreciable at the intermediate points. It is possible that Abul Wafa here detected an inequality rediscovered by Tycho Brahe (chapter v., § 111) and known as the **variation**, but it is equally likely that he was merely restating Ptolemy's prosneusis (chapter II., § 48).* In either case Abul Wafa's discovery appears to have been entirely ignored by his successors and to have borne no fruit. He also carried further some of the mathematical improvements of his predecessors.

Another nearly contemporary astronomer, commonly known as *Ibn Yunos* (?–1008), worked at Cairo under the patronage of the Mahometan rulers of Egypt. He published a set of astronomical and mathematical tables, the *Hakemite Tables*, which remained the standard ones for about two centuries, and he embodied in the same book a number of his own observations as well as an extensive series by earlier Arabian astronomers.

61. About this time astronomy, in common with other branches of knowledge, had made some progress in the Mahometan dominions in Spain and the opposite coast of Africa. A great library and an academy were founded at Cordova about 970, and centres of education and learning were established in rapid succession at Cordova, Toledo, Seville, and Morocco.

The most important work produced by the astronomers of these places was the volume of astronomical tables published under the direction of *Arzachel* in 1080, and known as the *Toletan Tables*, because calculated for an observer at Toledo, where Arzachel probably lived. To

* A prolonged but indecisive controversy has been carried on, chiefly by French scholars, with regard to the relations of Ptolemy, Abul Wafa, and Tycho in this matter.

the same school are due some improvements in instruments and in methods of calculation, and several writings were published in criticism of Ptolemy, without, however, suggesting any improvements on his ideas.

Gradually, however, the Spanish Christians began to drive back their Mahometan neighbours. Cordova and Seville were captured in 1236 and 1248 respectively, and with their fall Arab astronomy disappeared from history.

62. Before we pass on to consider the progress of astronomy in Europe, two more astronomical schools of the East deserve mention, both of which illustrate an extraordinarily rapid growth of scientific interests among barbarous peoples. Hulagu Khan, a grandson of the Mongol conqueror Genghis Khan, captured Bagdad in 1258 and ended the rule of the Caliphs there. Some years before this he had received into favour, partly as a political adviser, the astronomer *Nassir Eddin* (born in 1201 at Tus in Khorassan), and subsequently provided funds for the establishment of a magnificent observatory at Meraga, near the north-west frontier of modern Persia. Here a number of astronomers worked under the general superintendence of Nassir Eddin. The instruments they used were remarkable for their size and careful construction, and were probably better than any used in Europe in the time of Coppernicus, being surpassed first by those of Tycho Brahe (chapter v.).

Nassir Eddin and his assistants translated or commented on nearly all the more important available Greek writings on astronomy and allied subjects, including Euclid's *Elements*, several books by Archimedes, and the *Almagest*. Nassir Eddin also wrote an abstract of astronomy, marked by some little originality, and a treatise on geometry. He does not appear to have accepted the authority of Ptolemy without question, and objected in particular to the use of the equant (chapter II., § 51), which he replaced by a new combination of spheres. Many of these treatises had for a long time a great reputation in the East, and became in their turn the subject-matter of commentary.

But the great work of the Meraga astronomers, which occupied them 12 years, was the issue of a revised set of astronomical tables, based on the Hakemite Tables of Ibn

Yunos (§ 60), and called in honour of their patron the *Ilkhanic Tables*. They contained not only the usual tables for computing the motions of the planets, etc., but also a star catalogue, based to some extent on new observations.

An important result of the observations of fixed stars made at Meraga was that the precession (chapter II., § 42) was fixed at 51″, or within about 1″ of its true value. Nassir Eddin also discussed the supposed trepidation (§ 58), but seems to have been a little doubtful of its reality. He died in 1273, soon after his patron, and with him the Meraga School came to an end as rapidly as it was formed.

63. Nearly two centuries later *Ulugh Begh* (born in 1394), a grandson of the savage Tartar Tamerlane, developed a great personal interest in astronomy, and built about 1420 an observatory at Samarcand (in the present Russian Turkestan), where he worked with assistants. He published fresh tables of the planets, etc., but his most important work was a star catalogue, embracing nearly the same stars as that of Ptolemy, but observed afresh. This was probably the first substantially independent catalogue made since Hipparchus. The places of the stars were given with unusual precision, the minutes as well as the degrees of celestial longitude and latitude being recorded; and although a comparison with modern observation shews that there were usually errors of several minutes, it is probable that the instruments used were extremely good. Ulugh Begh was murdered by his son in 1449, and with him Tartar astronomy ceased.

64. No great original idea can be attributed to any of the Arab and other astronomers whose work we have sketched. They had, however, a remarkable aptitude for absorbing foreign ideas, and carrying them slightly further. They were patient and accurate observers, and skilful calculators. We owe to them a long series of observations, and the invention or introduction of several important improvements in mathematical methods.* Among the most important of their services to mathematics, and hence to astronomy, must be counted the introduction, from India,

* For example, the practice of treating the trigonometrical functions as *algebraic* quantities to be manipulated by formulæ, not merely as geometrical lines.

of our present system of writing numbers, by which the value of a numeral is altered by its position, and fresh symbols are not wanted, as in the clumsy Greek and Roman systems, for higher numbers. An immense simplification was thereby introduced into arithmetical work.* More important than the actual original contributions of the Arabs to astronomy was the service that they performed in keeping alive interest in the science and preserving the discoveries of their Greek predecessors.

Some curious relics of the time when the Arabs were the great masters in astronomy have been preserved in astronomical language. Thus we have derived from them, usually in very corrupt forms, the current names of many individual stars, *e.g.* Aldebaran, Altair, Betelgeux, Rigel, Vega (the constellations being mostly known by Latin translations of the Greek names), and some common astronomical terms such as zenith and **nadir** (the invisible point on the celestial sphere opposite the zenith); while at least one such word, almanack, has passed into common language.

65. In Europe the period of confusion following the break-up of the Roman empire and preceding the definite formation of feudal Europe is almost a blank as regards astronomy, or indeed any other natural science. The best intellects that were not absorbed in practical life were occupied with theology. A few men, such as the Venerable Bede (672–735), living for the most part in secluded monasteries, were noted for their learning, which included in general some portions of mathematics and astronomy; none were noted for their additions to scientific knowledge. Some advance was made by Charlemagne (742–814), who, in addition to introducing something like order into his extensive dominions, made energetic attempts to develop education and learning. In 782 he summoned to his court our learned countryman *Alcuin* (735–804) to give instruction in astronomy, arithmetic, and rhetoric, as well as in other subjects, and invited other scholars to join him, forming thus a kind of Academy of which Alcuin was the head.

* Any one who has not realised this may do so by performing with Roman numerals the simple operation of multiplying by itself a number such as MDCCCXCVIII.

Charlemagne not only founded a higher school at his own court, but was also successful in urging the ecclesiastical authorities in all parts of his dominions to do the same. In these schools were taught the seven liberal arts, divided into the so-called trivium (grammar, rhetoric, and dialectic) and quadrivium, which included astronomy in addition to arithmetic, geometry, and music.

66. In the 10th century the fame of the Arab learning began slowly to spread through Spain into other parts of Europe, and the immense learning of *Gerbert*, the most famous scholar of the century, who occupied the papal chair as Sylvester II. from 999 to 1003, was attributed in large part to the time which he spent in Spain, either in or near the Moorish dominions. He was an ardent student, indefatigable in collecting and reading rare books, and was especially interested in mathematics and astronomy. His skill in making astrolabes (chapter II., § 49) and other instruments was such that he was popularly supposed to have acquired his powers by selling his soul to the Evil One. Other scholars shewed a similar interest in Arabic learning, but it was not till the lapse of another century that the Mahometan influence became important.

At the beginning of the 12th century began a series of translations from Arabic into Latin of scientific and philosophic treatises, partly original works of the Arabs, partly Arabic translations of the Greek books. One of the most active of the translators was *Plato of Tivoli*, who studied Arabic in Spain about 1116, and translated Albategnius's *Astronomy* (§ 59), as well as other astronomical books. At about the same time Euclid's *Elements*, among other books, was translated by *Athelard of Bath*. *Gherardo of Cremona* (1114–1187) was even more industrious, and is said to have made translations of about 70 scientific treatises, including the *Almagest*, and the *Toletan Tables* of Arzachel (§ 61). The beginning of the 13th century was marked by the foundation of several Universities, and at that of Naples (founded in 1224) the Emperor Frederick II., who had come into contact with the Mahometan learning in Sicily, gathered together a number of scholars whom he directed to make a fresh series of translations from the Arabic.

Aristotle's writings on logic had been preserved in Latin translations from classical times, and were already much esteemed by the scholars of the 11th and 12th centuries. His other writings were first met with in Arabic versions, and were translated into Latin during the end of the 12th and during the 13th centuries; in one or two cases translations were also made from the original Greek. The influence of Aristotle over mediæval thought, already considerable, soon became almost supreme, and his works were by many scholars regarded with a reverence equal to or greater than that felt for the Christian Fathers.

Western knowledge of Arab astronomy was very much increased by the activity of *Alfonso X.* of Leon and Castile (1223–1284), who collected at Toledo, a recent conquest from the Arabs, a body of scholars, Jews and Christians, who calculated under his general superintendence a set of new astronomical tables to supersede the *Toletan Tables.* These *Alfonsine Tables* were published in 1252, on the day of Alfonso's accession, and spread rapidly through Europe. They embodied no new ideas, but several numerical data, notably the length of the year, were given with greater accuracy than before. To Alfonso is due also the publication of the *Libros del Saber*, a voluminous encyclopædia of the astronomical knowledge of the time, which, though compiled largely from Arab sources, was not, as has sometimes been thought, a mere collection of translations. One of the curiosities in this book is a diagram representing Mercury's orbit as an ellipse, the earth being in the *centre* (cf. chapter VII., § 140), this being probably the first trace of the idea of representing the celestial motions by means of curves other than circles.

67. To the 13th century belong also several of the great scholars, such as *Albertus Magnus*, *Roger Bacon*, and *Cecco d'Ascoli* (from whom Dante learnt), who took all knowledge for their province. Roger Bacon, who was born in Somersetshire about 1214 and died about 1294, wrote three principal books, called respectively the *Opus Majus*, *Opus Minus*, and *Opus Tertium*, which contained not only treatises on most existing branches of knowledge, but also some extremely interesting discussions of their relative importance and of the right method for the advancement

of learning. He inveighs warmly against excessive adherence to authority, especially to that of Aristotle, whose books he wishes burnt, and speaks strongly of the importance of experiment and of mathematical reasoning in scientific inquiries. He evidently had a good knowledge of optics and has been supposed to have been acquainted with the telescope, a supposition which we can hardly regard as confirmed by his story that the invention was known to Caesar, who when about to invade Britain surveyed the new country from the opposite shores of Gaul with a telescope!

Another famous book of this period was written by the Yorkshireman John Halifax or Holywood, better known by his Latinised name *Sacrobosco*, who was for some time a well-known teacher of mathematics at Paris, where he died about 1256. His *Sphaera Mundi* was an elementary treatise on the easier parts of current astronomy, dealing in fact with little but the more obvious results of the daily motion of the celestial sphere. It enjoyed immense popularity for three or four centuries, and was frequently re-edited, translated, and commented on: it was one of the very first astronomical books ever printed; 25 editions appeared between 1472 and the end of the century, and 40 more by the middle of the 17th century.

68. The European writers of the Middle Ages whom we have hitherto mentioned, with the exception of Alfonso and his assistants, had contented themselves with collecting and rearranging such portions of the astronomical knowledge of the Greeks and Arabs as they could master; there were no serious attempts at making progress, and no observations of importance were made. A new school, however, grew up in Germany during the 15th century which succeeded in making some additions to knowledge, not in themselves of first-rate importance, but significant of the greater independence that was beginning to inspire scientific work. *George Purbach*, born in 1423, became in 1450 professor of astronomy and mathematics at the University of Vienna, which had soon after its foundation (1365) become a centre for these subjects. He there began an *Epitome of Astronomy* based on the *Almagest*, and also a Latin version of Ptolemy's planetary theory, intended partly

as a supplement to Sacrobosco's textbook, from which this part of the subject had been omitted, but in part also as a treatise of a higher order; but he was hindered in both undertakings by the badness of the only available versions of the *Almagest*—Latin translations which had been made not directly from the Greek, but through the medium at any rate of Arabic and very possibly of Syriac as well (cf. § 56), and which consequently swarmed with mistakes. He was assisted in this work by his more famous pupil John Müller of Königsberg (in Franconia), hence known as *Regiomontanus*, who was attracted to Vienna at the age of 16 (1452) by Purbach's reputation. The two astronomers made some observations, and were strengthened in their conviction of the necessity of astronomical reforms by the serious inaccuracies which they discovered in the *Alfonsine Tables*, now two centuries old; an eclipse of the moon, for example, occurring an hour late and Mars being seen 2° from its calculated place. Purbach and Regiomontanus were invited to Rome by one of the Cardinals, largely with a view to studying a copy of the *Almagest* contained among the Greek manuscripts which since the fall of Constantinople (1453) had come into Italy in considerable numbers, and they were on the point of starting when the elder man suddenly died (1461).

Regiomontanus, who decided on going notwithstanding Purbach's death, was altogether seven years in Italy; he there acquired a good knowledge of Greek, which he had already begun to study in Vienna, and was thus able to read the *Almagest* and other treatises in the original; he completed Purbach's *Epitome of Astronomy*, made some observations, lectured, wrote a mathematical treatise * of considerable merit, and finally returned to Vienna in 1468 with originals or copies of several important Greek manuscripts. He was for a short time professor there, but then accepted an invitation from the King of Hungary to arrange a valuable collection of Greek manuscripts. The king, however, soon

* On trigonometry. He reintroduced the *sine*, which had been forgotten; and made some use of the *tangent*, but like Albategnius (§ 59 *n.*) did not realise its importance, and thus remained behind Ibn Yunos and Abul Wafa. An important contribution to mathematics was a table of sines calculated for every minute from 0° to 90°.

turned his attention from Greek to fighting, and Regiomontanus moved once more, settling this time in Nürnberg, then one of the most flourishing cities in Germany, a special attraction of which was that one of the early printing presses was established there. The Nürnberg citizens received Regiomontanus with great honour, and one rich man in particular, *Bernard Walther* (1430–1504), not only supplied him with funds, but, though an older man, became his pupil and worked with him. The skilled artisans of Nürnberg were employed in constructing astronomical instruments of an accuracy hitherto unknown in Europe, though probably still inferior to those of Nassir Eddin and Ulugh Begh (§§ 62, 63). A number of observations were made, among the most interesting being those of the comet of 1472, the first comet which appears to have been regarded as a subject for scientific study rather than for superstitious terror. Regiomontanus recognised at once the importance for his work of the new invention of printing, and, finding probably that the existing presses were unable to meet the special requirements of astronomy, started a printing press of his own. Here he brought out in 1472 or 1473 an edition of Purbach's book on planetary theory, which soon became popular and was frequently reprinted. This book indicates clearly the discrepancy already being felt between the views of Aristotle and those of Ptolemy. Aristotle's original view was that sun, moon, the five planets, and the fixed stars were attached respectively to eight spheres, one inside the other; and that the outer one, which contained the fixed stars, by its revolution was the primary cause of the apparent daily motion of all the celestial bodies. The discovery of precession required on the part of those who carried on the Aristotelian tradition the addition of another sphere. According to this scheme, which was probably due to some of the translators or commentators at Bagdad (§ 56), the fixed stars were on a sphere, often called the **firmament**, and outside this was a ninth sphere, known as the **primum mobile**, which moved all the others; another sphere was added by Tabit ben Korra to account for trepidation (§ 58), and accepted by Alfonso and his school; an eleventh sphere was added towards the end of the Middle Ages to account for the supposed changes in the obliquity of the ecliptic. A few writers invented a larger number. Outside these spheres mediaeval thought usually placed the Empyrean or Heaven. The accompanying diagram illustrates the whole arrangement.

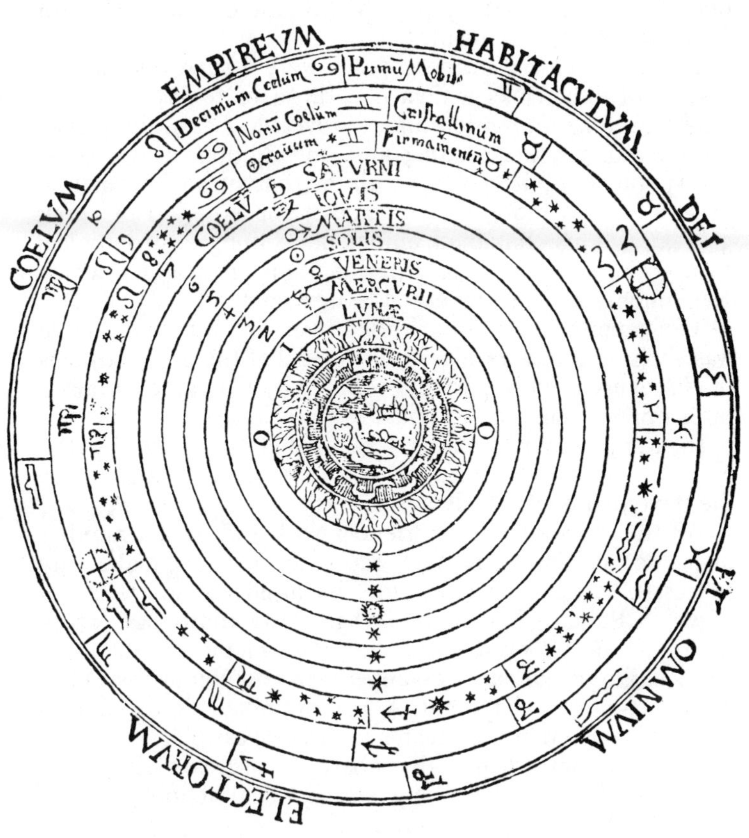

Fig. 36.—The celestial spheres. From Apian's *Cosmographia.*

These spheres, which were almost entirely fanciful and in no serious way even professed to account for the details of the celestial motions, are of course quite different from the circles known as deferents and epicycles, which Hipparchus and Ptolemy used. These were mere geometrical abstractions, which enabled the planetary motions to be represented with tolerable accuracy. Each planet moved freely in space, its motion being represented or described (not *controlled*) by a particular geometrical arrangement of circles. Purbach suggested a compromise by hollowing out Aristotle's crystal spheres till there was room for Ptolemy's epicycles inside!

From the new Nürnberg press were issued also a succession of almanacks which, like those of to-day, gave the public useful information about moveable feasts, the phases of the moon, eclipses, etc.; and, in addition, a volume of less popular *Ephemerides*, with astronomical information of a fuller and more exact character for a period of about 30 years. This contained, among other things, astronomical data for finding latitude and longitude at sea, for which Regiomontanus had invented a new method.*

The superiority of these tables over any others available was such that they were used on several of the great voyages of discovery of this period, probably by Columbus himself on his first voyage to America.

In 1475 Regiomontanus was invited to Rome by the Pope to assist in a reform of the calendar, but died there the next year at the early age of forty.

Walther carried on his friend's work and took a number of good observations; he was the first to make any successful attempt to allow for the atmospheric refraction of which Ptolemy had probably had some knowledge (chapter II., § 46); to him is due also the practice of obtaining the position of the sun by comparison with Venus instead of with the moon (chapter II., § 39), the much slower motion of the planet rendering greater accuracy possible.

After Walther's death other observers of less merit carried on the work, and a Nürnberg astronomical school of some kind lasted into the 17th century.

69. A few minor discoveries in astronomy belong to this or to a slightly later period and may conveniently be dealt with here.

Lionardo da Vinci (1452–1519), who was not only a great painter and sculptor, but also an anatomist, engineer, mechanician, physicist, and mathematician, was the first

* That of "lunar distances."

to explain correctly the dim illumination seen over the rest of the surface of the moon when the bright part is only a thin crescent. He pointed out that when the moon was nearly new the half of the earth which was then illuminated by the sun was turned nearly directly towards the moon, and that the moon was in consequence illuminated slightly by this **earthshine**, just as we are by moonshine. The explanation is interesting in itself, and was also of some value as shewing an analogy between the earth and moon which tended to break down the supposed barrier between terrestrial and celestial bodies (chapter VI., § 119).

Jerome Fracastor (1483–1543) and *Peter Apian* (1495–1552), two voluminous writers on astronomy, made observations of comets of some interest, both noticing that a comet's tail continually points away from the sun, as the comet changes its position, a fact which has been used in modern times to throw some light on the structure of comets (chapter XIII., § 304).

Peter Nonius (1492–1577) deserves mention on account of the knowledge of twilight which he possessed; several problems as to the duration of twilight, its variation in different latitudes, etc., were correctly solved by him; but otherwise his numerous books are of no great interest.*

A new determination of the size of the earth, the first since the time of the Caliph Al Mamun (§ 57), was made about 1528 by the French doctor *John Fernel* (1497–1558), who arrived at a result the error in which (less than 1 per cent.) was far less than could reasonably have been expected from the rough methods employed.

The life of Regiomontanus overlapped that of Coppernicus by three years; the four writers last named were nearly his contemporaries; and we may therefore be said to have come to the end of the comparatively stationary period dealt with in this chapter.

* He did not invent the measuring instrument called the *vernier*, often attributed to him, but something quite different and of very inferior value.

CHAPTER IV.

COPPERNICUS.

"But in this our age, one rare witte (seeing the continuall errors that from time to time more and more continually have been discovered, besides the infinite absurdities in their Theoricks, which they have been forced to admit that would not confesse any Mobilitie in the ball of the Earth) hath by long studye, payntull practise, and rare invention delivered a new Theorick or Model of the world, shewing that the Earth resteth not in the Center of the whole world or globe of elements, which encircled and enclosed in the Moone's orbit, and together with the whole globe of mortality is carried yearly round about the Sunne, which like a king in the middest of all, rayneth and giveth laws of motion to all the rest, sphaerically dispersing his glorious beames of light through all this sacred coelestiall Temple."

THOMAS DIGGES, 1590.

70. THE growing interest in astronomy shewn by the work of such men as Regiomontanus was one of the early results in the region of science of the great movement of thought to different aspects of which are given the names of Revival of Learning, Renaissance, and Reformation. The movement may be regarded primarily as a general quickening of intelligence and of interest in matters of thought and knowledge. The invention of printing early in the 15th century, the stimulus to the study of the Greek authors, due in part to the scholars who were driven westwards after the capture of Constantinople by the Turks (1453), and the discovery of America by Columbus in 1492, all helped on a movement the beginning of which has to be looked for much earlier.

Every stimulus to the intelligence naturally brings with it a tendency towards inquiry into opinions received through tradition and based on some great authority. The effective discovery and the study of Greek philosophers other than Aristotle naturally did much to shake the supreme authority of that great philosopher, just as the Reformers shook the authority of the Church by pointing out what they considered to be inconsistencies between its doctrines and those of the Bible. At first there was little avowed opposition to the principle that truth was to be derived from some authority, rather than to be sought independently by the light of reason; the new scholars replaced the authority of Aristotle by that of Plato or of Greek and Roman antiquity in general, and the religious Reformers replaced the Church by the Bible. Naturally, however, the conflict between authorities produced in some minds scepticism as to the principle of authority itself; when freedom of judgment had to be exercised to the extent of deciding between authorities, it was but a step further —a step, it is true, that comparatively few took—to use the individual judgment on the matter at issue itself.

In astronomy the conflict between authorities had already arisen, partly in connection with certain divergencies between Ptolemy and Aristotle, partly in connection with the various astronomical tables which, though on substantially the same lines, differed in minor points. The time was therefore ripe for some fundamental criticism of the traditional astronomy, and for its reconstruction on a new basis.

Such a fundamental change was planned and worked out by the great astronomer whose work has next to be considered.

71. *Nicholas* Coppernic or *Coppernicus* * was born on February 19th, 1473, in a house still pointed out in the little trading town of Thorn on the Vistula. Thorn now lies just within the eastern frontier of the present kingdom of Prussia; in the time of Coppernicus it lay in a region over which the King of Poland had some sort of suzerainty, the

* The name is spelled in a large number of different ways both by Coppernicus and by his contemporaries. He himself usually wrote his name Coppernic, and in learned productions commonly used the Latin form Coppernicus. The spelling Copernicus is so much less commonly used by him that I have thought it better to discard it, even at the risk of appearing pedantic.

precise nature of which was a continual subject of quarrel between him, the citizens, and the order of Teutonic knights, who claimed a good deal of the neighbouring country. The astronomer's father (whose name was most commonly written Koppernigk) was a merchant who came to Thorn from Cracow, then the capital of Poland, in 1462. Whether Coppernicus should be counted as a Pole or as a German is an intricate question, over which his biographers have fought at great length and with some acrimony, but which is not worth further discussion here.

Nicholas, after the death of his father in 1483, was under the care of his uncle, Lucas Watzelrode, afterwards bishop of the neighbouring diocese of Ermland, and was destined by him from a very early date for an ecclesiastical career. He attended the school at Thorn, and at the age of 17 entered the University of Cracow. Here he seems to have first acquired (or shewn) a decided taste for astronomy and mathematics, subjects in which he probably received help from Albert Brudzewski, who had a great reputation as a learned and stimulating teacher; the lecture lists of the University show that the comparatively modern treatises of Purbach and Regiomontanus (chapter III., § 68) were the standard textbooks used. Coppernicus had no intention of graduating at Cracow, and probably left after three years (1494). During the next year or two he lived partly at home, partly at his uncle's palace at Heilsberg, and spent some of the time in an unsuccessful candidature for a canonry at Frauenburg, the cathedral city of his uncle's diocese.

The next nine or ten years of his life (from 1496 to 1505 or 1506) were devoted to studying in Italy, his stay there being broken only by a short visit to Frauenburg in 1501. He worked chiefly at Bologna and Padua, but graduated at Ferrara, and also spent some time at Rome, where his astronomical knowledge evidently made a favourable impression. Although he was supposed to be in Italy primarily with a view to studying law and medicine, it is evident that much of his best work was being put into mathematics and astronomy, while he also paid a good deal of attention to Greek.

During his absence he was appointed (about 1497) to

COPPERNICUS. [To face p. 94.

a canonry at Frauenburg, and at some uncertain date he also received a sinecure ecclesiastical appointment at Breslau.

72. On returning to Frauenburg from Italy Coppernicus almost immediately obtained fresh leave of absence, and joined his uncle at Heilsberg, ostensibly as his medical adviser and really as his companion.

It was probably during the quiet years spent at Heilsberg that he first put into shape his new ideas about astronomy, and wrote the first draft of his book. He kept the manuscript by him, revising and rewriting from time to time, partly from a desire to make his work as perfect as possible, partly from complete indifference to reputation, coupled with dislike of the controversy to which the publication of his book would almost certainly give rise. In 1509 he published at Cracow his first book, a Latin translation of a set of Greek letters by Theophylactus, interesting as being probably the first translation from the Greek ever published in Poland or the adjacent districts. In 1512, on the death of his uncle, he finally settled in Frauenburg, in a set of rooms which he occupied, with short intervals, for the next 31 years. Once fairly in residence, he took his share in conducting the business of the Chapter: he acted, for example, more than once as their representative in various quarrels with the King of Poland and the Teutonic knights; in 1523 he was general administrator of the diocese for a few months after the death of the bishop; and for two periods, amounting altogether to six years (1516–1519 and 1520–1521), he lived at the castle of Allenstein, administering some of the outlying property of the Chapter. In 1521 he was commissioned to draw up a statement of the grievances of the Chapter against the Teutonic knights for presentation to the Prussian Estates, and in the following year wrote a memorandum on the debased and confused state of the coinage in the district, a paper which was also laid before the Estates, and was afterwards rewritten in Latin at the special request of the bishop. He also gave a certain amount of medical advice to his friends as well as to the poor of Frauenburg, though he never practised regularly as a physician; but notwithstanding these various occupations it is probable that a very large part of his time during the last 30 years of his life was devoted to astronomy.

73. We are so accustomed to associate the revival of astronomy, as of other branches of natural science, with increased care in the collection of observed facts, and to think of Coppernicus as the chief agent in the revival, that it is worth while here to emphasise the fact that he was in no sense a great observer. His instruments, which were mostly of his own construction, were far inferior to those of Nassir Eddin and of Ulugh Begh (chapter III., §§ 62, 63), and not even as good as those which he could have procured if he had wished from the workshops of Nürnberg; his observations were not at all numerous (only 27, which occur in his book, and a dozen or two besides being known), and he appears to have made no serious attempt to secure great accuracy. His determination of the position of one star, which was extensively used by him as a standard of reference and was therefore of special importance, was in error to the extent of nearly 40′ (more than the apparent breadth of the sun or moon), an error which Hipparchus would have considered very serious. His pupil Rheticus (§ 74) reports an interesting discussion between his master and himself, in which the pupil urged the importance of making observations with all imaginable accuracy; Coppernicus answered that minute accuracy was not to be looked for at that time, and that a rough agreement between theory and observation was all that he could hope to attain. Coppernicus moreover points out in more than one place that the high latitude of Frauenburg and the thickness of the air were so detrimental to good observation that, for example, though he had occasionally been able to see the planet Mercury, he had never been able to observe it properly.

Although he published nothing of importance till towards the end of his life, his reputation as an astronomer and mathematician appears to have been established among experts from the date of his leaving Italy, and to have steadily increased as time went on.

In 1515 he was consulted by a committee appointed by the Lateran Council to consider the reform of the calendar, which had now fallen into some confusion (chapter II.,

§ 22), but he declined to give any advice on the ground that the motions of the sun and moon were as yet too imperfectly known for a satisfactory reform to be possible. A few years later (1524) he wrote an open letter, intended for publication, to one of his Cracow friends, in reply to a tract on precession, in which, after the manner of the time, he used strong language about the errors of his opponent.*

It was meanwhile gradually becoming known that he held the novel doctrine that the earth was in motion and the sun and stars at rest, a doctrine which was sufficiently startling to attract notice outside astronomical circles. About 1531 he had the distinction of being ridiculed on the stage at some popular performance in the neighbourhood; and it is interesting to note (especially in view of the famous persecution of Galilei at Rome a century later) that Luther in his *Table Talk* frankly described Coppernicus as a fool for holding such opinions, which were obviously contrary to the Bible, and that Melanchthon, perhaps the most learned of the Reformers, added to a somewhat similar criticism a broad hint that such opinions should not be tolerated. Coppernicus appears to have taken no notice of these or similar attacks, and still continued to publish nothing. No observation made later than 1529 occurs in his great book, which seems to have been nearly in its final form by that date; and to about this time belongs an extremely interesting paper, known as the *Commentariolus*, which contains a short account of his system of the world, with some of the evidence for it, but without any calculations. It was apparently written to be shewn or lent to friends, and was not published; the manuscript disappeared after the death of the author and was only rediscovered in 1878. The *Commentariolus* was probably the basis of a lecture on the ideas of Coppernicus given in 1533 by one of the Roman astronomers at the request of Pope Clement VII. Three years later Cardinal Schomberg wrote to ask Coppernicus for further information as to his views, the letter showing that the chief features were already pretty accurately known.

* *Nullo demum loco ineptior est quam . . . ubi nimis pueriliter hallucinatur:* Nowhere is he more foolish than . . . where he suffers from delusions of too childish a character.

74. Similar requests must have been made by others, but his final decision to publish his ideas seems to have been due to the arrival at Frauenburg in 1539 of the enthusiastic young astronomer generally known as *Rheticus.** Born in 1514, he studied astronomy under Schoner at Nürnberg, and was appointed in 1536 to one of the chairs of mathematics created by the influence of Melanchthon at Wittenberg, at that time the chief Protestant University.

Having heard, probably through the *Commentariolus*, of Coppernicus and his doctrines, he was so much interested in them that he decided to visit the great astronomer at Frauenburg. Coppernicus received him with extreme kindness, and the visit, which was originally intended to last a few days or weeks, extended over nearly two years. Rheticus set to work to study Coppernicus's manuscript, and wrote within a few weeks of his arrival an extremely interesting and valuable account of it, known as the First Narrative (*Prima Narratio*), in the form of an open letter to his old master Schoner, a letter which was printed in the following spring and was the first easily accessible account of the new doctrines.†

When Rheticus returned to Wittenberg, towards the end of 1541, he took with him a copy of a purely mathematical section of the great book, and had it printed as a textbook of the subject (Trigonometry); it had probably been already settled that he was to superintend the printing of the complete book itself. Coppernicus, who was now an old man and would naturally feel that his end was approaching, sent the manuscript to his friend Giese, Bishop of Kulm, to do what he pleased with. Giese sent it at once to Rheticus, who made arrangements for having it printed at Nürnberg. Unfortunately Rheticus was not able to see it all through the press, and the work had to be entrusted to Osiander, a Lutheran preacher interested in astronomy. Osiander

* His real name was Georg Joachim, that by which he is known having been made up by himself from the Latin name of the district where he was born (Rhætia).

† The *Commentariolus* and the *Prima Narratio* give most readers a better idea of what Coppernicus did than his larger book, in which it is comparatively difficult to disentangle his leading ideas from the mass of calculations based on them.

appears to have been much alarmed at the thought of the disturbance which the heretical ideas of Coppernicus would cause, and added a prefatory note of his own (which he omitted to sign), praising the book in a vulgar way, and declaring (what was quite contrary to the views of the author) that the fundamental principles laid down in it were merely abstract hypotheses convenient for purposes of calculation; he also gave the book the title *De Revolutionibus Orbium Celestium* (On the Revolutions of the Celestial Spheres), the last two words of which were probably his own addition. The printing was finished in the winter 1542–3, and the author received a copy of his book on the day of his death (May 24th, 1543), when his memory and mental vigour had already gone.

75. The central idea with which the name of Coppernicus is associated, and which makes the *De Revolutionibus* one of the most important books in all astronomical literature, by the side of which perhaps only the *Almagest* and Newton's *Principia* (chapter IX., §§ 177 *seqq.*) can be placed, is that the apparent motions of the celestial bodies are to a great extent not real motions, but are due to the motion of the earth carrying the observer with it. Coppernicus tells us that he had long been struck by the unsatisfactory nature of the current explanations of astronomical observations, and that, while searching in philosophical writings for some better explanation, he had found a reference of Cicero to the opinion of Hicetas that the earth turned round on its axis daily. He found similar views held by other Pythagoreans, while Philolaus and Aristarchus of Samos had also held that the earth not only rotates, but moves bodily round the sun or some other centre (cf. chapter II., § 24). The opinion that the earth is not the sole centre of motion, but that Venus and Mercury revolve round the sun, he found to be an old Egyptian belief, supported also by *Martianus Capella*, who wrote a compendium of science and philosophy in the 5th or 6th century A.D. A more modern authority, *Nicholas of Cusa* (1401–1464), a mystic writer who refers to a possible motion of the earth, was ignored or not noticed by Coppernicus. None of the writers here named, with the possible exception of Aristarchus of Samos, to whom Coppernicus apparently

paid little attention, presented the opinions quoted as more than vague speculations; none of them gave any substantial reasons for, much less a proof of, their views; and Coppernicus, though he may have been glad, after the fashion of the age, to have the support of recognised authorities, had practically to make a fresh start and elaborate his own evidence for his opinions.

It has sometimes been said that Coppernicus *proved* what earlier writers had guessed at or suggested; it would perhaps be truer to say that he took up certain floating ideas, which were extremely vague and had never been worked out scientifically, based on them certain definite fundamental principles, and from these principles developed mathematically an astronomical system which he shewed to be at least as capable of explaining the observed celestial motions as any existing variety of the traditional Ptolemaic system. The Coppernican system, as it left the hands of the author, was in fact decidedly superior to its rivals as an explanation of ordinary observations, an advantage which it owed quite as much to the mathematical skill with which it was developed as to its first principles; it was in many respects very much simpler; and it avoided certain fundamental difficulties of the older system. It was however liable to certain serious objections, which were only overcome by fresh evidence which was subsequently brought to light. For the predecessors of Coppernicus there was, apart from variations of minor importance, but one scientific system which made any serious attempt to account for known facts; for his immediate successors there were two, the newer of which would to an impartial mind appear on the whole the more satisfactory, and the further study of the two systems, with a view to the discovery of fresh arguments or fresh observations tending to support the one or the other, was immediately suggested as an inquiry of first-rate importance.

76. The plan of the *De Revolutionibus* bears a general resemblance to that of the *Almagest*. In form at least the book is not primarily an argument in favour of the motion of the earth, and it is possible to read much of it without ever noticing the presence of this doctrine.

Coppernicus, like Ptolemy, begins with certain first principles or postulates, but on account of their novelty takes a little more trouble than his predecessor (cf. chapter II., § 47) to make them at once appear probable. With these postulates as a basis he proceeds to develop, by means of elaborate and rather tedious mathematical reasoning, aided here and there by references to observations, detailed schemes of the various celestial motions; and it is by the agreement of these calculations with observations, far more than by the general reasoning given at the beginning, that the various postulates are in effect justified.

His first postulate, that the universe is spherical, is supported by vague and inconclusive reasons similar to those given by Ptolemy and others; for the spherical form of the earth he gives several of the usual valid arguments, one of his proofs for its curvature from east to west being the fact that eclipses visible at one place are not visible at another. A third postulate, that the motions of the celestial bodies are uniform circular motions or are compounded of such motions, is, as might be expected, supported only by reasons of the most unsatisfactory character. He argues, for example, that any want of uniformity in motion

"must arise either from irregularity in the moving power, whether this be within the body or foreign to it, or from some inequality of the body in revolution. . . . Both of which things the intellect shrinks from with horror, it being unworthy to hold such a view about bodies which are constituted in the most perfect order."

77. The discussion of the possibility that the earth may move, and may even have more than one motion, then follows, and is more satisfactory though by no means conclusive. Coppernicus has a firm grasp of the principle, which Aristotle had also enunciated, sometimes known as that of relative motion, which he states somewhat as follows:—

"For all change in position which is seen is due to a motion either of the observer or of the thing looked at, or to changes in the position of both, provided that these are different. For when things are moved equally relatively to the same things, no motion is perceived, as between the object seen and the observer." *

Coppernicus gives no proof of this principle, regarding it probably as sufficiently obvious, when once stated, to the mathematicians and astronomers for whom he was writing. It is, however, so fundamental that it may be worth while to discuss it a little more fully.

Let, for example, the observer be at A and an object at B, then whether the object move from B to B′, the observer remaining at rest, or the observer move an *equal* distance in the *opposite* direction, from A to A′, the object remaining at rest, the effect is to the eye exactly the same, since in

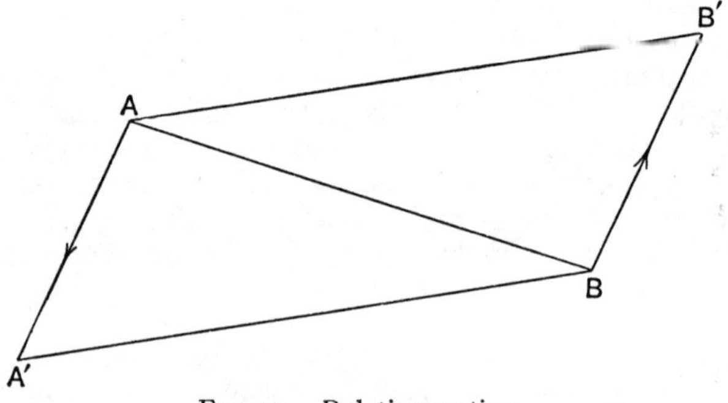

FIG. 37.—Relative motion.

either case the distance between the observer and object and the direction in which the object is seen, represented in the first case by A B′ and in the second by A′ B, are the same.

Thus if in the course of a year *either* the sun passes successively through the positions A, B, C, D (fig. 38), the earth remaining at rest at E, *or* if the sun is at rest and the earth passes successively through the positions *a*, *b*, *c*, *d*,

* *Omnis enim quæ videtur secundum locum mutatio, aut est propter locum mutatio, aut est propter spectatæ rei motum, aut videntis, aut certe disparem utriusque mutationem. Nam inter mota æqualiter ad eadem non percipitur motus, inter rem visam dico, et videntem* (De Rev., I. v.).

I have tried to remove some of the crabbedness of the original passage by translating freely.

at the corresponding times, the sun remaining at rest at S, exactly the same effect is produced on the eye, provided that the lines *a* S, *b* S, *c* S, *d* S are, as in the figure, equal in length and parallel in direction to E A, E B, E C, E D respectively. The same being true of intermediate points, exactly the same apparent effect is produced whether the sun describe the circle A B C D, or the earth describe at the same rate the equal circle *a b c d*. It will be noticed further that, although the corresponding motions in the two cases are at the same times in *opposite* directions (as at A and *a*), yet each circle as a whole is described,

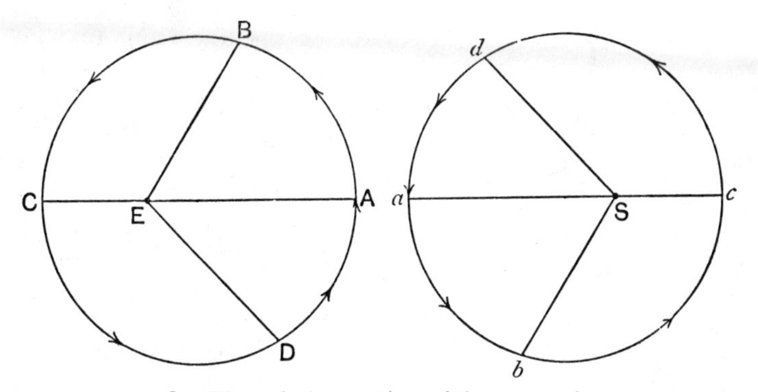

FIG. 38.—The relative motion of the sun and moon.

as indicated by the arrow-heads, in the *same* direction (contrary to that of the motion of the hands of a clock, in the figures given). It follows in the same sort of way that an apparent motion (as of a planet) may be explained as due partially to the motion of the object, partially to that of the observer.

Coppernicus gives the familiar illustration of the passenger in a boat who sees the land apparently moving away from him, by quoting and explaining Virgil's line :—

"Provehimur portu, terræque urbesque recedunt."

78. The application of the same ideas to an apparent rotation round the observer, as in the case of the apparent daily motion of the celestial sphere, is a little more difficult. It must be remembered that the eye has no means of

judging the direction of an object taken by itself; it can only judge the difference between the direction of the object and some other direction, whether that of another object or a direction fixed in some way by the body of the observer. Thus when after looking at a star twice at an interval of time we decide that it has moved, this means that its direction has changed relatively to, say, some tree or house which we had noticed nearly in its direction, or that its direction has changed relatively to the direction in which we are directing our eyes or holding our bodies. Such a change can evidently be interpreted as a change of direction, either of the star or of the line from the eye to the tree which we used as a line of reference. To apply this to the case of the celestial sphere, let us suppose that S represents a star on the celestial sphere, which (for simplicity) is overhead to an observer on the earth at A, this being determined by comparison with a line A B drawn upright on the earth. Next, earth and celestial sphere being supposed to have a common centre at O, let us suppose *firstly* that the celestial sphere turns round (in the direction of the hands of a clock) till S comes to S′, and that the observer now sees the star on his horizon or in a direction at right angles to the original direction A B, the angle turned through by the celestial sphere being S O S′; and *secondly* that, the celestial sphere being unchanged, the earth turns round in the opposite direction, till A B comes to A′ B′, and the star is again seen by the observer on his horizon. Whichever of these motions has taken place, the observer sees exactly the same apparent motion in the sky; and the figure shews at once that the angle S O S′ through which the celestial sphere was supposed to turn in the first case is equal to the angle A O A′ through which

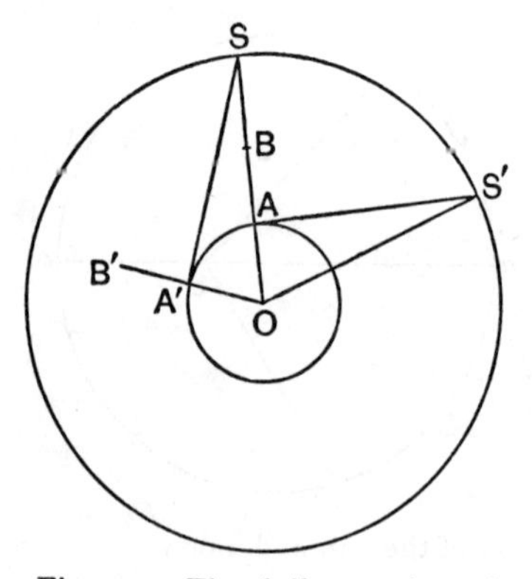

Fig. 39.—The daily rotation of the earth.

the earth turns in the second case, but that the two rotations are in opposite directions. A similar explanation evidently applies to more complicated cases.

Hence the apparent daily rotation of the celestial sphere about an axis through the poles would be produced equally well, either by an actual rotation of this character, or by a rotation of the earth about an axis also passing through the poles, and at the same rate, but in the opposite direction, *i.e.* from west to east. This is the first motion which Coppernicus assigns to the earth.

79. The apparent annual motion of the sun, in accordance with which it appears to revolve round the earth in a path which is nearly a circle, can be equally well explained by supposing the sun to be at rest, and the earth to describe an exactly equal path round the sun, the direction of the revolution being the same. This is virtually the second motion which Coppernicus gives to the earth, though, on account of a peculiarity in his geometrical method, he resolves this motion into two others, and combines with one of these a further small motion which is required for precession.*

80. Coppernicus's conception then is that the earth revolves round the sun in the plane of the ecliptic, while rotating daily on an axis which continually points to the poles of the celestial sphere, and therefore retains (save for precession) a fixed direction in space.

It should be noticed that the two motions thus assigned to the earth are perfectly distinct; each requires its own proof, and explains a different set of appearances. It was quite possible, with perfect consistency, to believe in one motion without believing in the other, as in fact a very few of the 16th-century astronomers did (chapter v., § 105).

In giving his reasons for believing in the motion of the

* To Coppernicus, as to many of his contemporaries, as well as to the Greeks, the simplest form of a revolution of one body round another was a motion in which the revolving body moved as if rigidly attached to the central body. Thus in the case of the earth the second motion was such that the axis of the earth remained inclined at a constant angle to the line joining earth and sun, and therefore changed its direction in space. In order then to make the axis retain a (nearly) fixed direction in space, it was necessary to add a *third* motion.

earth Coppernicus discusses the chief objections which had been urged by Ptolemy. To the objection that if the earth had a rapid motion of rotation about its axis, the earth would be in danger of flying to pieces, and the air, as well as loose objects on the surface, would be left behind, he replies that if such a motion were dangerous to the solid earth, it must be much more so to the celestial sphere, which, on account of its vastly greater size, would have to move enormously faster than the earth to complete its daily rotation; he enters also into an obscure discussion of difference between a "natural" and an "artificial" motion, of which the former might be expected not to disturb anything on the earth.

Coppernicus shews that the earth is very small compared to the sphere of the stars, because wherever the observer is on the earth the horizon appears to divide the celestial sphere into two equal parts and the observer appears always to be at the centre of the sphere, so that any distance through which the observer moves on the earth is imperceptible as compared with the distance of the stars.

81. He goes on to argue that the chief irregularity in the motion of the planets, in virtue of which they move backwards at intervals (chapter I., § 14, and chapter II., § 51), can readily be explained in general by the motion of the earth and by a motion of each planet round the sun, in its own time and at its own distance. From the fact that Venus and Mercury were never seen very far from the sun, it could be inferred that their paths were nearer to the sun than that of the earth, Mercury being the nearer to the sun of the two, because never seen so far from it in the sky as Venus. The other three planets, being seen at times in a direction opposite to that of the sun, must necessarily evolve round the sun in orbits larger than that of the earth, a view confirmed by the fact that they were brightest when opposite the sun (in which positions they would be nearest to us). The order of their respective distances from the sun could be at once inferred from the disturbing effects produced on their apparent motions by the motion of the earth; Saturn being least affected must on the whole be farthest from the earth, Jupiter next, and Mars next. The earth thus became one of six planets revolving round

the sun, the order of distance—Mercury, Venus, Earth, Mars, Jupiter, Saturn—being also in accordance with the rates of motion round the sun, Mercury performing its revolution most rapidly (in about 88 days *), Saturn most slowly (in about 30 years). On the Coppernican system

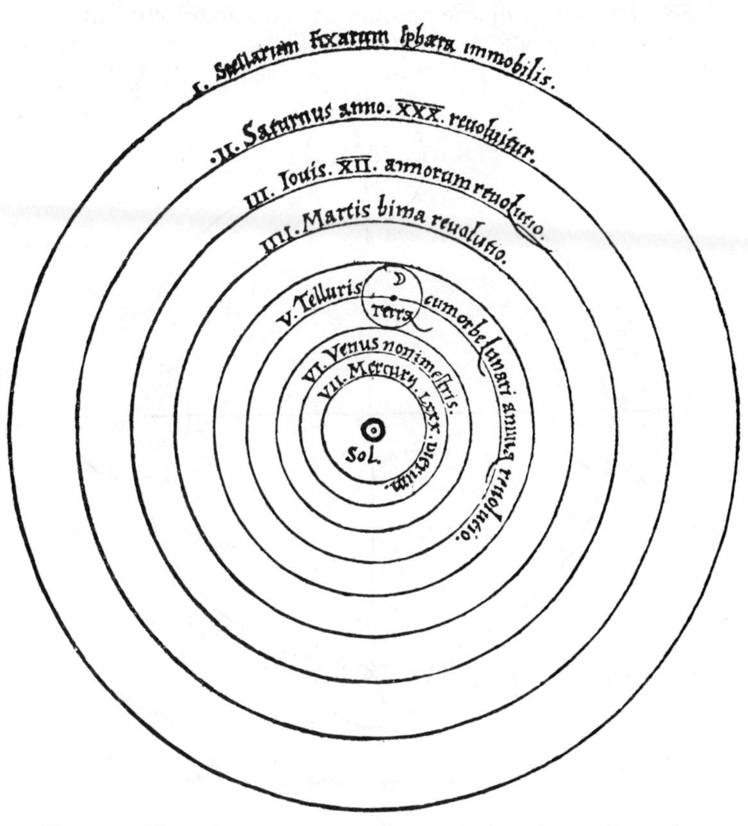

Fig. 40.—The solar system according to Coppernicus. From the *De Revolutionibus.*

the moon alone still revolved round the earth, being the only celestial body the status of which was substantially

* In this preliminary discussion, as in fig. 40, Coppernicus gives 80 days; but in the more detailed treatment given in Book V. he corrects this to 88 days.

unchanged; and thus Coppernicus was able to give the accompanying diagram of the solar system (fig. 40), representing his view of its general arrangement (though not of the right proportions of the different parts) and of the various motions.

82. The effect of the motion of the earth round the sun on the length of the day and other seasonal effects is

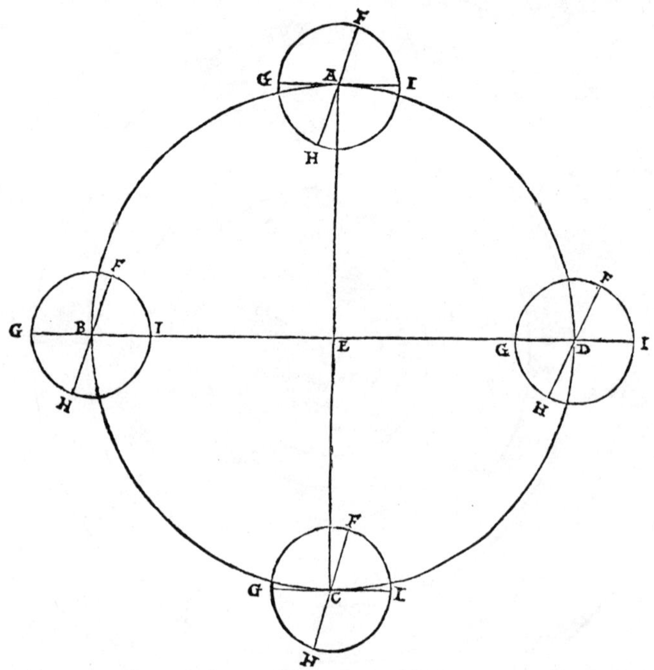

FIG. 41.—Coppernican explanation of the seasons. From the *De Revolutionibus*.

discussed in some detail, and illustrated by diagrams which are here reproduced.*

In fig. 41 A, B, C, D represent the centre of the earth in four positions, occupied by it about December 23rd, March 21st, June 22nd, and September 22nd respectively (*i.e.* at the

* Fig. 42 has been slightly altered, so as to make it agree with fig. 41.

beginnings of the four seasons, according to astronomical reckoning); the circle F G H I in each of its positions represents the equator of the earth, *i.e.* a great circle on the earth the plane of which is perpendicular to the axis of the earth and is consequently always parallel to the celestial equator. This circle is not in the plane of the ecliptic, but tilted up at an angle of $23\frac{1}{2}°$, so that F must always be supposed *below* and H *above* the plane of the paper (which represents the ecliptic); the equator cuts the ecliptic along G I. The diagram (in accordance with the common custom in astronomical diagrams) represents the various circles as seen from the north side of the equator and ecliptic. When the earth is at A, the north pole (as is shewn more clearly in fig. 42, in which P, P′ denote the north pole and south pole respectively) is turned away

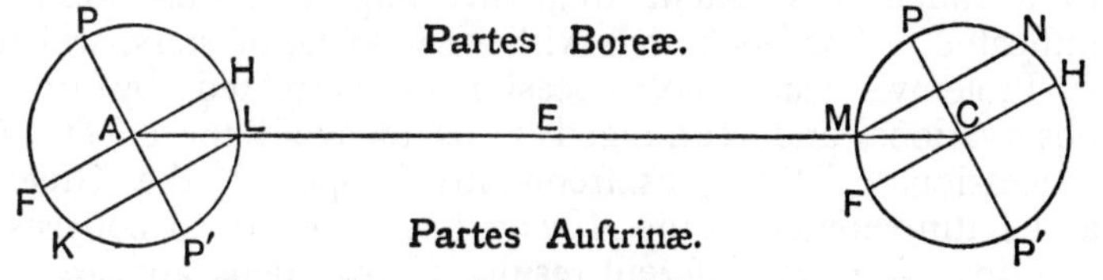

FIG. 42.—Coppernican explanation of the seasons. From the *De Revolutionibus.*

from the sun, E, which is on the lower or south side of the plane of the equator, and consequently inhabitants of the northern hemisphere see the sun for less than half the day, while those on the southern hemisphere see the sun for more than half the day, and those beyond the line K L (in fig. 42) see the sun during the whole day. Three months later, when the earth's centre is at B (fig. 41), the sun lies in the plane of the equator, the poles of the earth are turned neither towards nor away from the sun, but aside, and all over the earth daylight lasts for 12 hours and night for an equal time. Three months later still, when the earth's centre is at C, the sun is above the plane of the equator, and the inhabitants of the northern hemisphere see the sun for more than half the day, those on the southern hemisphere for less than half, while those in parts of the earth farther north than the line M N (in fig. 42) see the sun for the whole 24 hours. Finally, when, at the autumn equinox, the earth has reached D (fig. 41), the sun is again in the plane of the equator, and the day is everywhere equal to the night.

83. Coppernicus devotes the first eleven chapters of the first book to this preliminary sketch of his system; the remainder of this book he fills with some mathematical propositions and tables, which, as previously mentioned (§ 74), had already been separately printed by Rheticus. The second book contains chiefly a number of the usual results relating to the celestial sphere and its apparent daily motion, treated much as by earlier writers, but with greater mathematical skill. Incidentally Coppernicus gives his measurement of the obliquity of the ecliptic, and infers from a comparison with earlier observations that the obliquity had decreased, which was in fact the case, though to a much less extent than his imperfect observations indicated. The book ends with a catalogue of stars, which is Ptolemy's catalogue, occasionally corrected by fresh observations, and rearranged so as to avoid the effects of precession.* When, as frequently happened, the Greek and Latin versions of the *Almagest* gave, owing to copyists' or printers' errors, different results, Coppernicus appears to have followed sometimes the Latin and sometimes the Greek version, without in general attempting to ascertain by fresh observations which was right.

84. The third book begins with an elaborate discussion of the precession of the equinoxes (chapter II., § 42). From a comparison of results obtained by Timocharis, by later Greek astronomers, and by Albategnius, Coppernicus infers that the amount of precession has varied, but that its average value is 50″·2 annually (almost exactly the true value), and accepts accordingly Tabit ben Korra's unhappy suggestion of the trepidation (chapter III., § 58). An examination of the data used by Coppernicus shews that the erroneous or fraudulent observations of Ptolemy (chapter II., § 50) are chiefly responsible for the perpetuation of this mistake.

* Coppernicus, instead of giving longitudes as measured from the first point of Aries (or vernal equinoctial point, chapter I., §§ 11, 13), which moves on account of precession, measured the longitudes from a standard fixed star (α *Arietis*) not far from this point.

Of much more interest than the detailed discussion of trepidation and of geometrical schemes for representing it is the interpretation of precession as the result of a motion of the earth's axis. Precession was originally recognised by Hipparchus as a motion of the celestial equator, in which its inclination to the ecliptic was sensibly unchanged. Now the ideas of Coppernicus make the celestial equator dependent on the equator of the earth, and hence on its axis; it is in fact a great circle of the celestial sphere which is always perpendicular to the axis about which the earth rotates daily. Hence precession, on the theory of Coppernicus, arises from a slow motion of the axis of the earth, which moves so as always to remain inclined at the same angle to the ecliptic, and to return to its original position after a period of about 26,000 years (since a motion of $50''\cdot 2$ annually is equivalent to 360° or a complete circuit in that period); in other words, the earth's axis has a slow conical motion, the central line (or axis) of the cone being at right angles to the plane of the ecliptic.

85. Precession being dealt with, the greater part of the remainder of the third book is devoted to a discussion in detail of the apparent annual motion of the sun round the earth, corresponding to the real annual motion of the earth round the sun. The geometrical theory of the *Almagest* was capable of being immediately applied to the new system, and Coppernicus, like Ptolemy, uses an eccentric. He makes the calculations afresh, arrives at a smaller and more accurate value of the eccentricity (about $\frac{1}{31}$ instead of $\frac{1}{24}$), fixes the position of the apogee and perigee (chapter II., § 39), or rather of the equivalent **aphelion** and **perihelion** (*i.e.* the points in the earth's orbit where it is respectively farthest from and nearest to the sun), and thus verifies Albategnius's discovery (chapter III., § 59) of the motion of the line of apses. The theory of the earth's motion is worked out in some detail, and tables are given whereby the apparent place of the sun at any time can be easily computed.

The fourth book deals with the theory of the moon. As has been already noticed, the moon was the only celestial body the position of which in the universe was substantially unchanged by Coppernicus, and it might hence have been expected that little alteration would have been required in the traditional theory. Actually, however, there is scarcely any part of the subject in which Coppernicus did more to diminish the discrepancies between theory and observation. He rejects Ptolemy's equant (chapter II., § 51), partly on the ground that it produces an irregular motion unsuitable for the heavenly bodies, partly on the more substantial ground that, as already pointed out (chapter II., § 48), Ptolemy's theory makes the apparent size of the moon at times twice as great as at others. By an arrangement of epicycles Coppernicus succeeded in representing the chief irregularities in the moon's motion, including evection, but without Ptolemy's prosneusis (chapter II., § 48) or Abul Wafa's inequality (chapter III., § 60), while he made the changes in the moon's distance, and consequently in its apparent size, not very much greater than those which actually take place, the difference being imperceptible by the rough methods of observation which he used.*

In discussing the distances and sizes of the sun and moon Coppernicus follows Ptolemy closely (chapter II., § 49; cf. also fig. 20); he arrives at substantially the same estimate of the distance of the moon, but makes the sun's distance 1,500 times the earth's radius, thus improving to some extent on the traditional estimate, which was based on Ptolemy's. He also develops in some detail the effect of parallax on the apparent place of the moon, and the variations in the apparent size, owing to the variations in distance; and the book ends with a discussion of eclipses.

86. The last two books (V. and VI.) deal at length with the motion of the planets.

In the cases of Mercury and Venus, Ptolemy's explanation of the motion could with little difficulty be rearranged so as to fit the ideas of Coppernicus. We have seen (chapter II., § 51) that, minor irregularities being ignored, the motion of either of these planets could be represented by means of an epicycle moving on a deferent, the centre of

* According to the theory of Coppernicus, the diameter of the moon when greatest was about $\frac{1}{8}$ greater than its average amount; modern observations make this fraction about $\frac{1}{13}$. Or, to put it otherwise, the diameter of the moon when greatest ought to exceed its value when least by about 8′ according to Coppernicus, and by about 5′ according to modern observations.

the epicycle being always in the direction of the sun, the ratio of the sizes of the epicycle and deferent being fixed, but the actual dimensions being practically arbitrary. Ptolemy preferred on the whole to regard the epicycles of both these planets as lying between the earth and the sun. The idea of making the sun a centre of motion having once been accepted, it was an obvious simplification to make the centre of the epicycle not merely lie in the direction of the sun, but actually be the sun. In fact, if the planet

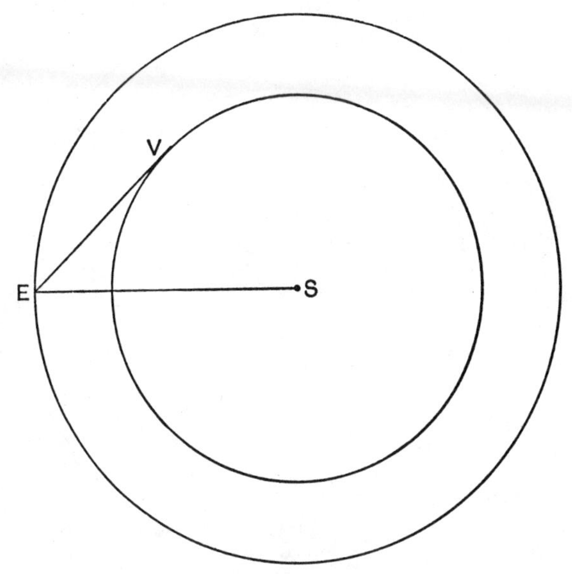

FIG. 43.—The orbits of Venus and of the earth.

in question revolved round the sun at the proper distance and at the proper rate, the same appearances would be produced as by Ptolemy's epicycle and deferent, the path of the planet round the sun replacing the epicycle, and the apparent path of the sun round the earth (or the path of the earth round the sun) replacing the deferent.

In discussing the time of revolution of a planet a distinction has to be made, as in the case of the moon (chapter II., § 40), between the synodic and sidereal periods of revolution. Venus, for example, is seen as an evening star at its greatest angular distance from the sun (as at V in fig. 43) at intervals of about 584 days. This is therefore the time which Venus takes to return to the same position relatively to the sun, as seen from the earth, or relatively to the earth, as seen from the sun; this time is called the **synodic period**. But as during this time the line E S has changed its direction, Venus is no longer in the same position relatively to the stars, as seen either from the sun or from the earth. If at first Venus and the

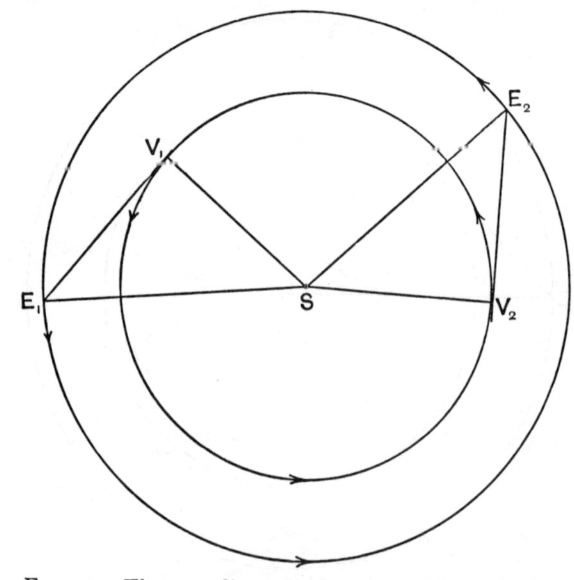

FIG. 44.—The synodic and sidereal periods of Venus.

earth are at V_1, E_1 respectively, after 584 days (or about a year and seven months) the earth will have performed rather more than a revolution and a half round the sun and will be at E_2; Venus being again at the greatest distance from the sun will therefore be at V_2, but will evidently be seen in quite a different part of the sky, and will not have performed an exact revolution round the sun. It is important to know how long the line S V_1 takes to return to the same position, *i.e.* how long Venus takes to return to the same position with respect to the stars, as seen from the sun, an interval of time known as the **sidereal period.** This can evidently be calculated by a simple rule-of-three sum from the data given. For Venus has in 584 days gained a complete revolution on the earth, or has gone as far as the earth would have gone in 584 + 365 or 949 days (fractions of days being omitted for simplicity); hence Venus goes in $584 \times \frac{365}{949}$ days as far as the earth in 365 days, *i.e.* Venus completes a revolution in $584 \times \frac{365}{949}$ or 225 days. This is therefore the sidereal period of Venus. The process used by Coppernicus was different, as he saw the advantage of using a long period of time, so as to diminish the error due to minor irregularities, and he therefore obtained two observations of Venus at a considerable interval of time, in which Venus occupied very nearly the same position both with respect to the sun and to the stars, so that the interval of time contained very nearly an exact number of sidereal periods as well as of synodic periods. By dividing therefore the observed interval of time by the number of sidereal periods (which being a whole number could readily be estimated), the sidereal period was easily obtained. A similar process shewed that the synodic period of Mercury was about 116 days, and the sidereal period about 88 days.

The comparative sizes of the orbits of Venus and Mercury as compared with that of the earth could easily be ascertained from observations of the position of either planet when most distant from the sun. Venus, for example, appears at its greatest distance from the sun when at a point V_1 (fig. 44) such that $V_1 E_1$ touches the circle in which Venus moves, and the angle $E_1 V_1 S$ is then (by a known property of a circle) a right angle. The angle $S E_1 V_1$ being observed, the shape of the triangle $S E_1 V_1$ is known, and the ratio of its sides can be readily calculated. Thus Coppernicus found that the average distance of Venus from the sun was about 72 and that of Mercury about 36, the distance of the earth from the sun being taken to be 100; the corresponding modern figures are 72·3 and 38·7.

87. In the case of the superior planets, Mars, Jupiter, and Saturn, it was much more difficult to recognise that their motions could be explained by supposing them to

revolve round the sun, since the centre of the epicycle did not always lie in the direction of the sun, but might be anywhere in the ecliptic. One peculiarity, however, in the motion of any of the superior planets might easily have suggested their motion round the sun, and was either completely overlooked by Ptolemy or not recognised by him as important. It is possible that it was one of the clues which led Coppernicus to his system. This peculiarity is that the radius of the epicycle of the planet, *j* J, is always parallel to the line E S joining the earth and sun, and consequently performs a complete revolution in a year. This connection between the motion of the planet and that of the sun received no explanation from Ptolemy's theory. Now if we draw E J′ parallel to *j* J and equal to it in length, it is easily seen * that the line J′ J is equal and parallel to E *j*, that consequently J describes a circle round J′ just as *j* round E. Hence the motion of the planet can equally well be represented by supposing it to move in an epicycle (represented by the large dotted circle in the figure) of which J′ is the centre and J′ J the radius, while the centre of the epicycle, remaining always in the direction of the sun, describes a deferent (represented by the small circle round E) of which the earth is the centre. By this method of representation the motion of the superior planet is exactly like that of an inferior planet, except that its epicycle is larger than its deferent; the same reasoning as before shows that the motion can be represented simply by supposing the centre J′ of the epicycle to be actually the sun. Ptolemy's epicycle and deferent are therefore capable of being replaced, without affecting the position of the planet in the sky, by a

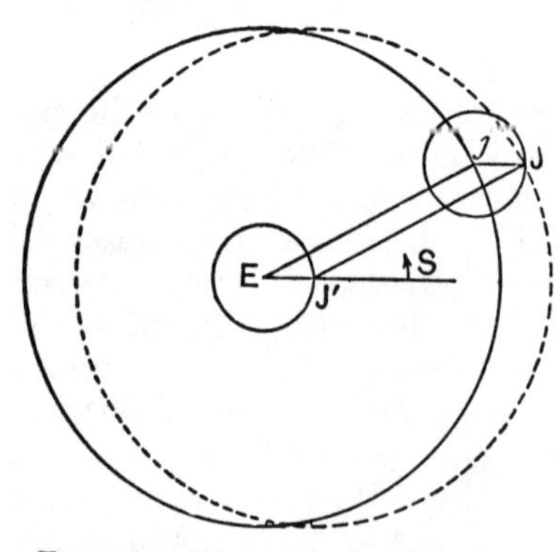

FIG. 45.—The epicycle of Jupiter.

* Euclid, I. 33.

motion of the planet in a circle round the sun, while the sun moves round the earth, or, more simply, the earth round the sun.

The synodic period of a superior planet could best be determined by observing when the planet was in opposition, *i.e.* when it was (nearly) opposite the sun, or, more accurately (since a planet does not move exactly in the ecliptic), when the longitudes of the planet and sun differed by 180° (or two right angles, chapter II., § 43). The

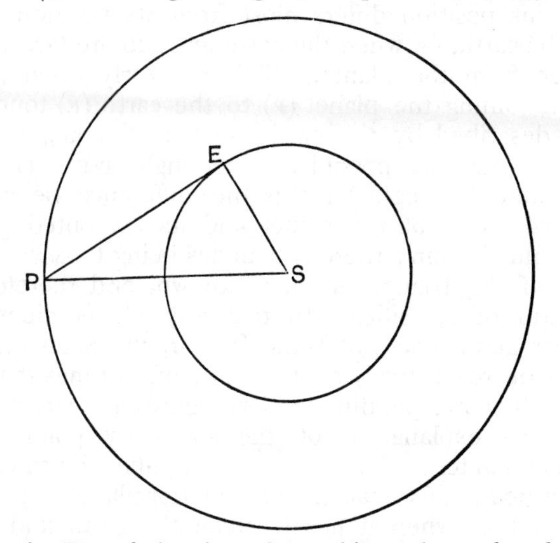

Fig. 46.—The relative sizes of the orbits ot the earth and of a superior planet.

sidereal period could then be deduced nearly as in the case of an inferior planet, with this difference, that the superior planet moves more slowly than the earth, and therefore *loses* one complete revolution in each synodic period; or the sidereal period might be found as before by observing when oppositions occurred nearly in the same part of the sky.* Coppernicus thus obtained very fairly accurate

* If P be the synodic period of a planet (in years), and S the sidereal period, then we evidently have $\frac{1}{P} + 1 = \frac{1}{S}$ for an inferior planet, and $1 - \frac{1}{P} = \frac{1}{S}$ for a superior planet.

values for the synodic and sidereal periods, *viz.* 780 days and 687 days respectively for Mars, 399 days and about 12 years for Jupiter, 378 days and 30 years for Saturn (cf. fig. 40).

The calculation of the distance of a superior planet from the sun is a good deal more complicated than that of Venus or Mercury. If we ignore various details, the process followed by Coppernicus is to compute the position of the planet as seen from the sun, and then to notice when this position differs most from its position as seen from the earth, *i.e.* when the earth and sun are farthest apart as seen from the planet. This is clearly when (fig. 46) the line joining the planet (P) to the earth (E) touches the circle described by the earth, so that the angle S P E is then as great as possible. The angle P E S is a right angle, and the angle S P E is the difference between the observed place of the planet and its computed place as seen from the sun; these two angles being thus known, the shape of the triangle S P E is known, and therefore also the ratio of its sides. In this way Coppernicus found the average distances of Mars, Jupiter, and Saturn from the sun to be respectively about $1\frac{1}{2}$, 5, and 9 times that of the earth; the corresponding modern figures are 1·5, 5·2, 9·5.

88. The explanation of the stationary points of the planets (chapter I., § 14) is much simplified by the ideas of Coppernicus. If we take first an inferior planet, say Mercury (fig. 47), then when it lies between the earth and sun, as at M (or as on Sept. 5 in fig. 7), both the earth and Mercury are moving in the same direction, but a comparison of the sizes of the paths of Mercury and the earth, and of their respective times of performing complete circuits, shews that Mercury is moving faster than the earth. Consequently to the observer at E, Mercury appears to be moving from left to right (in the figure), or from east to west; but this is contrary to the general direction of motion of the planets, *i.e.* Mercury appears to be retrograding. On the other hand, when Mercury appears at the greatest distance from the sun, as at M_1 and M_2, its own motion is directly towards or away from the earth, and is therefore imperceptible; but the earth is moving towards the observer's right, and therefore Mercury appears to be moving towards the left,

or from west to east. Hence between M_1 and M its motion has changed from direct to retrograde, and therefore at some intermediate point, say m_1 (about Aug. 23 in fig. 7), Mercury appears for the moment to be stationary, and similarly it appears to be stationary again when at some point m_2 between M and M_2 (about Sept. 13 in fig. 7).

In the case of a superior planet, say Jupiter, the argument

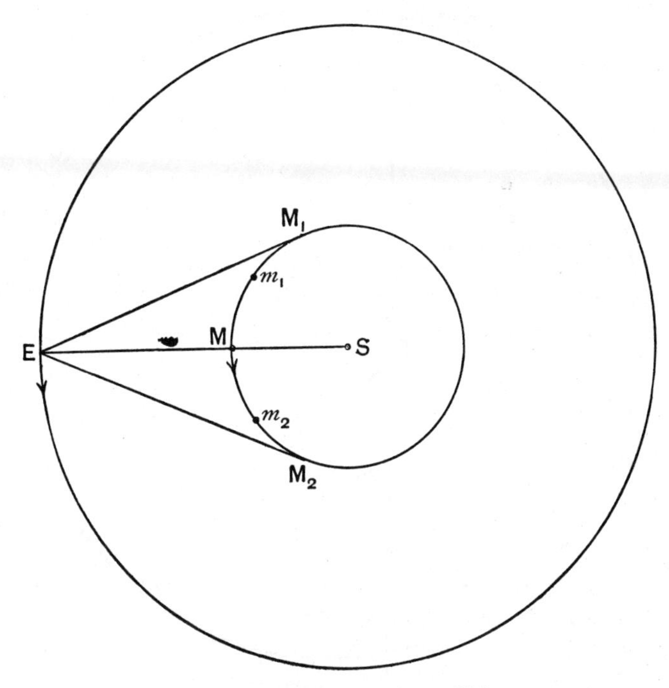

Fig. 47.—The stationary points of Mercury.

is nearly the same. When in opposition at J (as on Mar. 26 in fig. 6), Jupiter moves more slowly than the earth, and in the same direction, and therefore appears to be moving in the opposite direction to the earth, *i.e.* as seen from E (fig. 48), from left to right, or from east to west, that is in the retrograde direction. But when Jupiter is in either of the positions J_1 or J (in which the earth appears to the observer on Jupiter to be at its greatest distance from the sun), the motion of the earth itself being directly to or from Jupiter produces no effect on the apparent motion of Jupiter (since any displacement directly to or from the observer makes no difference in the object's place on the celestial sphere); but Jupiter itself is actually moving towards the left, and therefore the motion of

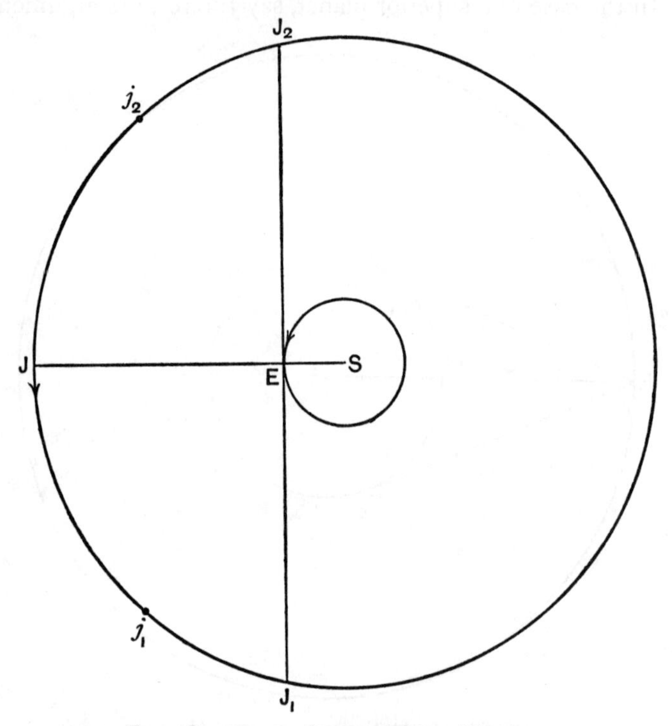

FIG. 48.—The stationary points of Jupiter.

Jupiter appears to be also from right to left, or from west to east. Hence, as before, between J_1 and J and between J and J_2 there must be points j_1, j_2 (Jan. 24 and May 27, in fig. 6) at which Jupiter appears for the moment to be stationary.

The actual discussion of the stationary points given by Coppernicus is a good deal more elaborate and more technical than the outline given here, as he not only shews that the stationary points must exist, but shews how to calculate their exact positions.

89. So far the theory of the planets has only been sketched very roughly, in order to bring into prominence the essential differences between the Coppernican and the Ptolemaic explanations of their motions, and no account has been taken of the minor irregularities for which Ptolemy devised his system of equants, eccentrics, etc., nor of the motion in latitude, *i.e.* to and from the ecliptic. Coppernicus, as already mentioned, rejected the equant, as being productive of an irregularity "unworthy" of the celestial bodies, and constructed for each planet a fairly complicated system of epicycles. For the motion in latitude discussed in Book VI. he supposed the orbit of each planet round the sun to be inclined to the ecliptic at a small angle, different for each planet, but found it necessary, in order that his theory should agree with observation, to introduce the wholly imaginary complication of a regular increase and decrease in the inclinations of the orbits of the planets to the ecliptic.

The actual details of the epicycles employed are of no great interest now, but it may be worth while to notice that for the motions of the moon, earth, and five other planets Coppernicus required altogether 34 circles, *viz.* four for the moon, three for the earth, seven for Mercury (the motion of which is peculiarly irregular), and five for each of the other planets; this number being a good deal less than that required in most versions of Ptolemy's system: Fracastor (chapter III., § 69), for example, writing in 1538, required 79 spheres, of which six were required for the fixed stars.

90. The planetary theory of Coppernicus necessarily suffered from one of the essential defects of the system of epicycles. It is, in fact, always possible to choose a system of epicycles in such a way as to make *either* the direction of any body *or* its distance vary in any required manner, but not to satisfy both requirements at once. In the case of the motion of the moon round the earth, or of the earth round the sun, cases in which variations in distance could not readily be observed, epicycles might therefore be expected to give a satisfactory result, at any rate until methods of

observation were sufficiently improved to measure with some accuracy the apparent sizes of the sun and moon, and so check the variations in their distances. But any variation in the distance of the earth from the sun would affect not merely the distance, but also the direction in which a planet would be seen; in the figure, for example, when the planet is at P and the sun at S, the apparent position of the planet, as seen from the earth, will be different according as the earth is at E or E′. Hence the epicycles and eccentrics of Coppernicus, which had to be adjusted in such a way that

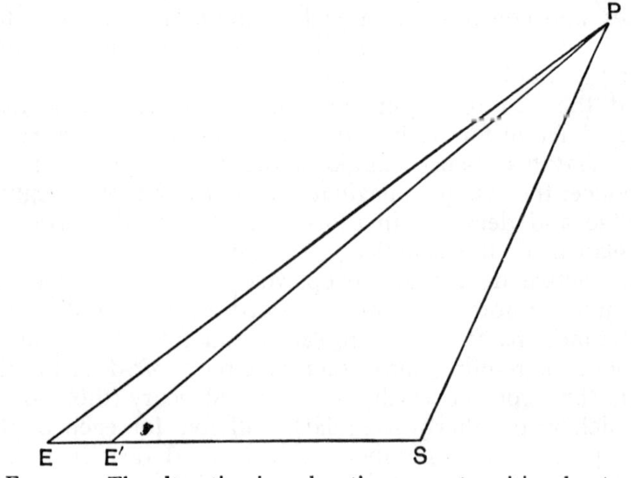

FIG. 49.—The alteration in a planet's apparent position due to an alteration in the earth's distance from the sun.

they necessarily involved incorrect values of the distances between the sun and earth, gave rise to corresponding errors in the observed places of the planets. The observations which Coppernicus used were hardly extensive or accurate enough to show this discrepancy clearly; but a crucial test was thus virtually suggested by means of which, when further observations of the planets had been made, a decision could be taken between an epicyclic representation of the motion of the planets and some other geometrical scheme.

91. The merits of Coppernicus are so great, and the part

which he played in the overthrow of the Ptolemaic system is so conspicuous, that we are sometimes liable to forget that, so far from rejecting the epicycles and eccentrics of the Greeks, he used no other geometrical devices, and was even a more orthodox "epicyclist" than Ptolemy himself, as he rejected the equants of the latter.* Milton's famous description (*Par. Lost*, VIII. 82-5) of

> "The Sphere
> With Centric and Eccentric scribbled o'er,
> Cycle and Epicycle, Orb in Orb,"

applies therefore just as well to the astronomy of Coppernicus as to that of his predecessors; and it was Kepler (chapter VII.), writing more than half a century later, not Coppernicus, to whom the rejection of the epicycle and eccentric is due.

92. One point which was of importance in later controversies deserves special mention here. The basis of the Coppernican system was that a motion of the earth carrying the observer with it produced an apparent motion of other bodies. The apparent motions of the sun and planets were thus shewn to be in great part explicable as the result of the motion of the earth round the sun. Similar reasoning ought apparently to lead to the conclusion that the fixed stars would also appear to have an annual motion. There would, in fact, be a displacement of the apparent position of a star due to the alteration of the earth's position in its orbit, closely resembling the alteration in the apparent position of the moon due to the alteration of the observer's position on the earth which had long been studied under the name of parallax (chapter II., § 43). As such a displacement had never been observed, Coppernicus explained the apparent contradiction by supposing the fixed stars so

* Recent biographers have called attention to a cancelled passage in the manuscript of the *De Revolutionibus* in which Coppernicus shews that an ellipse can be generated by a combination of circular motions. The proposition is, however, only a piece of pure mathematics, and has no relation to the motions of the planets round the sun. It cannot, therefore, fairly be regarded as in any way an anticipation of the ideas of Kepler (chapter VII.).

far off that any motion due to this cause was too small to be noticed. If, for example, the earth moves in six months from E to E′, the change in direction of a star at S′ is the angle E′ S′ E, which is less than that of a nearer star at S; and by supposing the star S′ sufficiently remote, the angle E′ S′ E can be made as small as may be required. For instance, if the distance of the star were 300 times the distance E E′, *i.e.* 600 times as far from the earth as

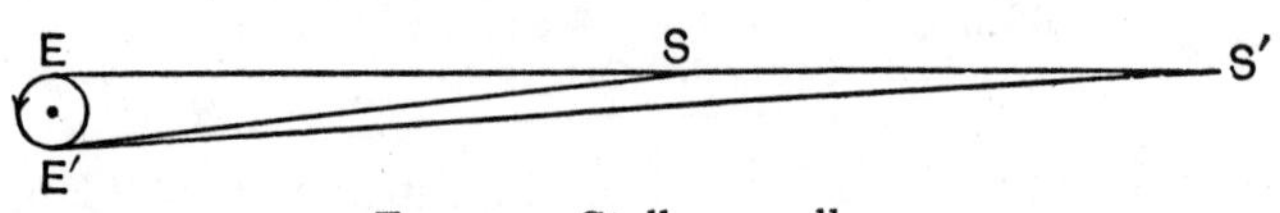

FIG. 50.—Stellar parallax.

the sun is, the angle E S′ E′ would be less than 12′, a quantity which the instruments of the time were barely capable of detecting.* But more accurate observations of the fixed stars might be expected to throw further light on this problem.

* It may be noticed that the differential method of parallax (chapter VI., § 129), by which such a quantity as 12′ could have been noticed, was put out of court by the general supposition, shared by Coppernicus, that the stars were all at the same distance from us.

CHAPTER V.

THE RECEPTION OF THE COPPERNICAN THEORY AND THE PROGRESS OF OBSERVATION.

> "Preposterous wits that cannot row at ease
> On the smooth channel of our common seas;
> And such are those, in my conceit at least,
> Those clerks that think—think how absurd a jest!—
> That neither heavens nor stars do turn at all,
> Nor dance about this great round Earthly Ball,
> But the Earth itself, this massy globe of ours,
> Turns round about once every twice twelve hours!"
>
> Du Bartas (Sylvester's translation).

93. The publication of the *De Revolutionibus* appears to have been received much more quietly than might have been expected from the startling nature of its contents. The book, in fact, was so written as to be unintelligible except to mathematicians of considerable knowledge and ability, and could not have been read at all generally. Moreover the preface, inserted by Osiander but generally supposed to be by the author himself, must have done a good deal to disarm the hostile criticism due to prejudice and custom, by representing the fundamental principles of Coppernicus as mere geometrical abstractions, convenient for calculating the celestial motions. Although, as we have seen (chapter IV., § 73), the contradiction between the opinions of Coppernicus and the common interpretation of various passages in the Bible was promptly noticed by Luther, Melanchthon, and others, no objection was raised either by the Pope to whom the book was dedicated, or by his immediate successors.

The enthusiastic advocacy of the Coppernican views by Rheticus has already been referred to. The only other astronomer of note who at once accepted the new views was his friend and colleague *Erasmus Reinhold* (born at Saalfeld in 1511), who occupied the chief chair of mathematics and astronomy at Wittenberg from 1536 to 1553, and it thus happened, curiously enough, that the doctrines so emphatically condemned by two of the great Protestant leaders were championed principally in what was generally regarded as the very centre of Protestant thought.

94. Rheticus, after the publication of the *Narratio Prima* and of an Ephemeris or Almanack based on Coppernican principles (1550), occupied himself principally with the calculation of a very extensive set of mathematical tables, which he only succeeded in finishing just before his death in 1576.

Reinhold rendered to astronomy the extremely important service of calculating, on the basis of the *De Revolutionibus*, tables of the motions of the celestial bodies, which were published in 1551 at the expense of Duke Albert of Prussia and hence called *Tabulæ Prutenicæ*, or *Prussian Tables*. Reinhold revised most of the calculations made by Coppernicus, whose arithmetical work was occasionally at fault; but the chief object of the tables was the development in great detail of the work in the *De Revolutionibus*, in such a form that the places of the chief celestial bodies at any required time could be ascertained with ease. The author claimed for his tables that from them the places of all the heavenly bodies could be computed for the past 3,000 years, and would agree with all observations recorded during that period. The tables were indeed found to be on the whole decidedly superior to their predecessors the *Alfonsine Tables* (chapter III., § 66), and gradually came more and more into favour, until superseded three-quarters of a century later by the *Rudolphine Tables* of Kepler (chapter VII., § 148). This superiority of the new tables was only indirectly connected with the difference in the principles on which the two sets of tables were based, and was largely due to the facts that Reinhold was a much better computer than the assistants of Alfonso, and that Coppernicus, if not a better mathematician than Ptolemy, at any rate had better mathematical tools at command. Nevertheless the tables naturally had great weight in inducing the astronomical world gradually to recognise the merits of the Coppernican system, at any rate as a basis for calculating the places of the celestial bodies.

Reinhold was unfortunately cut off by the plague in 1553, and with him disappeared a commentary on the *De Revolutionibus* which he had prepared but not published.

95. Very soon afterwards we find the first signs that the Coppernican system had spread into England. In 1556 *John Field* published an almanack for the following year avowedly based on Coppernicus and Reinhold, and a passage in the *Whetstone of Witte* (1557) by *Robert Recorde* (1510–1558), our first writer on algebra, shews that the author regarded the doctrines of Coppernicus with favour, even if he did not believe in them entirely. A few years later *Thomas Digges* (?–1595), in his *Alae sive Scalae Mathematicae* (1573), an astronomical treatise of no great importance, gave warm praise to Coppernicus and his ideas.

96. For nearly half a century after the death of Reinhold no important contributions were made to the Coppernican controversy. Reinhold's tables were doubtless slowly doing their work in familiarising men's minds with the new ideas, but certain definite additions to knowledge had to be made before the evidence for them could be regarded as really conclusive.

The serious mechanical difficulties connected with the assumption of a rapid motion of the earth which is quite imperceptible to its inhabitants could only be met by further progress in mechanics, and specially in knowledge of the laws according to which the motion of bodies is produced, kept up, changed, or destroyed; in this direction no considerable progress was made before the time of Galilei, whose work falls chiefly into the early 17th century (cf. chapter VI., §§ 116, 130, 133).

The objection to the Coppernican scheme that the stars shewed no such apparent annual motions as the motion of the earth should produce (chapter IV., § 92) would also be either answered or strengthened according as improved methods of observation did or did not reveal the required motion.

Moreover the *Prussian Tables*, although more accurate

than the *Alfonsine*, hardly claimed, and certainly did not possess, minute accuracy. Coppernicus had once told Rheticus that he would be extravagantly pleased if he could make his theory agree with observation to within 10′; but as a matter of fact discrepancies of a much more serious character were noticed from time to time. The comparatively small number of observations available and their roughness made it extremely difficult, either to find the most satisfactory numerical data necessary for the detailed development of any theory, or to test the theory properly by comparison of calculated with observed places of the celestial bodies. Accordingly it became evident to more than one astronomer that one of the most pressing needs of the science was that observations should be taken on as large a scale as possible and with the utmost attainable accuracy. To meet this need two schools of observational astronomy, of very unequal excellence, developed during the latter half of the 16th century, and provided a mass of material for the use of the astronomers of the next generation. Fortunately too the same period was marked by rapid progress in algebra and allied branches of mathematics. Of the three great inventions which have so enormously diminished the labour of numerical calculations, one, the so-called Arabic notation (chapter III., § 64), was already familiar, the other two (decimal fractions and logarithms) were suggested in the 16th century and were in working order early in the 17th century.

97. The first important set of observations taken after the death of Regiomontanus and Walther (chapter III., § 68) were due to the energy of the Landgrave *William IV.* of Hesse (1532–1592). He was remarkable as a boy for his love of study, and is reported to have had his interest in astronomy created or stimulated when he was little more than 20 by a copy of Apian's beautiful *Astronomicum Caesareum*, the cardboard models in which he caused to be imitated and developed in metal-work. He went on with the subject seriously, and in 1561 had an observatory built at Cassel, which was remarkable as being the first which had a revolving roof, a device now almost universal. In this he made extensive observations (chiefly of fixed stars) during the next six years. The death of his father then compelled

him to devote most of his energy to the duties of govern ment, and his astronomical ardour abated. A few years later, however (1575), as the result of a short visit from the talented and enthusiastic young Danish astronomer Tycho Brahe (§ 99), he renewed his astronomical work, and secured shortly afterwards the services of two extremely able assistants, *Christian Rothmann* (in 1577) and *Joost Bürgi* (in 1579). Rothmann, of whose life extremely little is known, appears to have been a mathematician and theoretical astronomer of considerable ability, and was the author of several improvements in methods of dealing with various astronomical problems. He was at first a Coppernican, but shewed his independence by calling attention to the needless complication introduced by Coppernicus in resolving the motion of the earth into *three* motions when two sufficed (chapter IV., § 79). His faith in the system was, however, subsequently shaken by the errors which observation revealed in the *Prussian Tables*. Bürgi (1552–1632) was originally engaged by the Landgrave as a clockmaker, but his remarkable mechanical talents were soon turned to astronomical account, and it then appeared that he also possessed unusual ability as a mathematician.*

98. The chief work of the Cassel Observatory was the formation of a star catalogue. The positions of stars were compared with that of the sun, Venus or Jupiter being used as connecting links, and their positions relatively to the equator and the first point of Aries (♈) deduced; allowance was regularly made for the errors due to the refraction of light by the atmosphere, as well as for the parallax of the sun, but the most notable new departure was the use of a clock to record the time of observations and to measure the motion of the celestial sphere. The construction of clocks of sufficient accuracy for the purpose was rendered possible by the mechanical genius of Bürgi, and in particular by his discovery that a clock could be regulated by a pendulum, a discovery which he

* There is little doubt that he invented what were substantially legarithms independently of Napier, but, with characteristic inability or unwillingness to proclaim his discoveries, allowed the invention to die with him.

appears to have taken no steps to publish, and which had in consequence to be made again independently before it received general recognition.* By 1586 121 stars had been carefully observed, but a more extensive catalogue which was to have contained more than a thousand stars was never finished, owing to the unexpected disappearance of Rothmann in 1590 † and the death of the Landgrave two years later.

99. The work of the Cassel Observatory was, however, overshadowed by that carried out nearly at the same time by *Tycho* (*Tyge*) *Brahe*. He was born in 1546 at Knudstrup in the Danish province of Scania (now the southern extremity of Sweden), being the eldest child of a nobleman who was afterwards governor of Helsingborg Castle. He was adopted as an infant by an uncle, and brought up at his country estate. When only 13 he went to the University of Copenhagen, where he began to study rhetoric and philosophy, with a view to a political career. He was, however, very much interested by a small eclipse of the sun which he saw in 1560, and this stimulus, added to some taste for the astrological art of casting horoscopes, led him to devote the greater part of the remaining two years spent at Copenhagen to mathematics and astronomy. In 1562 he went on to the University of Leipzig, accompanied, according to the custom of the time, by a tutor, who appears to have made persevering but unsuccessful attempts to induce his pupil to devote himself to law. Tycho, however, was now as always a difficult person to divert from his purpose, and went on steadily with his astronomy. In 1563 he made his first recorded observation, of a close approach of Jupiter and Saturn, the time of which he noticed to be predicted a whole month wrong by the *Alfonsine Tables* (chapter III., § 66), while the *Prussian Tables* (§ 94) were several days in error. While at Leipzig he bought also a few rough instruments, and anticipated one of the great improvements afterwards carried out systematically,

* A similar discovery was in fact made twice again, by Galilei (chapter VI., § 114) and by Huygens (chapter VIII., § 157).

† He obtained leave of absence to pay a visit to Tycho Brahe and never returned to Cassel. He must have died between 1599 and 1608.

by trying to estimate and to allow for the errors of his instruments.

In 1565 Tycho returned to Copenhagen, probably on account of the war with Sweden which had just broken out, and stayed about a year, during the course of which he lost his uncle. He then set out again (1566) on his travels, and visited Wittenberg, Rostock, Basle, Ingolstadt, Augsburg, and other centres of learning, thus making acquaintance with several of the most notable astronomers of Germany. At Augsburg he met the brothers Hainzel, rich citizens with a taste for science, for one of whom he designed and had constructed an enormous **quadrant** (quarter-circle) with a radius of about 19 feet, the rim of which was graduated to single minutes; and he began also here the construction of his great celestial globe, five feet in diameter, on which he marked one by one the positions of the stars as he afterwards observed them.

In 1570 Tycho returned to his father at Helsingborg, and soon after the death of the latter (1571) went for a long visit to Steen Bille, an uncle with scientific tastes. During this visit he seems to have devoted most of his time to chemistry (or perhaps rather to alchemy), and his astronomical studies fell into abeyance for a time.

100. His interest in astronomy was fortunately revived by the sudden appearance, in November 1572, of a brilliant new star in the constellation Cassiopeia. Of this Tycho took a number of extremely careful observations; he noted the gradual changes in its brilliancy from its first appearance, when it rivalled Venus at her brightest, down to its final disappearance 16 months later. He repeatedly measured its angular distance from the chief stars in Cassiopeia, and applied a variety of methods to ascertain whether it had any perceptible parallax (chapter II., §§ 43, 49). No parallax could be definitely detected, and he deduced accordingly that the star must certainly be farther off than the moon; as moreover it had no share in the planetary motions, he inferred that it must belong to the region of the fixed stars. To us of to-day this result may appear fairly commonplace, but most astronomers of the time held so firmly to Aristotle's doctrine that the heavens generally, and the region of the fixed stars in particular, were incorruptible and unchange-

able, that new stars were, like comets, almost universally ascribed to the higher regions of our own atmosphere. Tycho wrote an account of the new star, which he was ultimately induced by his friends to publish (1573), together with some portions of a calendar for that year which he had prepared. His reluctance to publish appears to have been due in great part to a belief that it was unworthy of the dignity of a Danish nobleman to write books! The book in question (*De Nova . . . Stella*) compares very favourably with the numerous other writings which the star called forth, though it shews that Tycho held the common beliefs that comets were in our atmosphere, and that the planets were carried round by solid crystalline spheres, two delusions which his subsequent work did much to destroy. He also dealt at some length with the astrological importance of the star, and the great events which it foreshadowed, utterances on which Kepler subsequently made the very sensible criticism that "if that star did nothing else, at least it announced and produced a great astronomer."

In 1574 Tycho was requested to give some astronomical lectures at the University of Copenhagen, the first of which, dealing largely with astrology, was printed in 1610, after his death. When these were finished, he set off again on his travels (1575). After a short visit to Cassel (§ 97), during which he laid the foundation of a lifelong friendship with the Landgrave, he went on to Frankfort to buy books, thence to Basle (where he had serious thoughts of settling) and on to Venice, then back to Augsburg and to Regensburg, where he obtained a copy of the *Commentariolus* of Coppernicus (chapter IV., § 73), and finally came home by way of Saalfeld and Wittenberg.

101. The next year (1576) was the beginning of a new epoch in Tycho's career. The King of Denmark, Frederick II., who was a zealous patron of science and literature, determined to provide Tycho with endowments sufficient to enable him to carry out his astronomical work in the most effective way. He accordingly gave him for occupation the little island of Hveen in the Sound (now belonging to Sweden), promised money for building a house and observatory, and supplemented the income

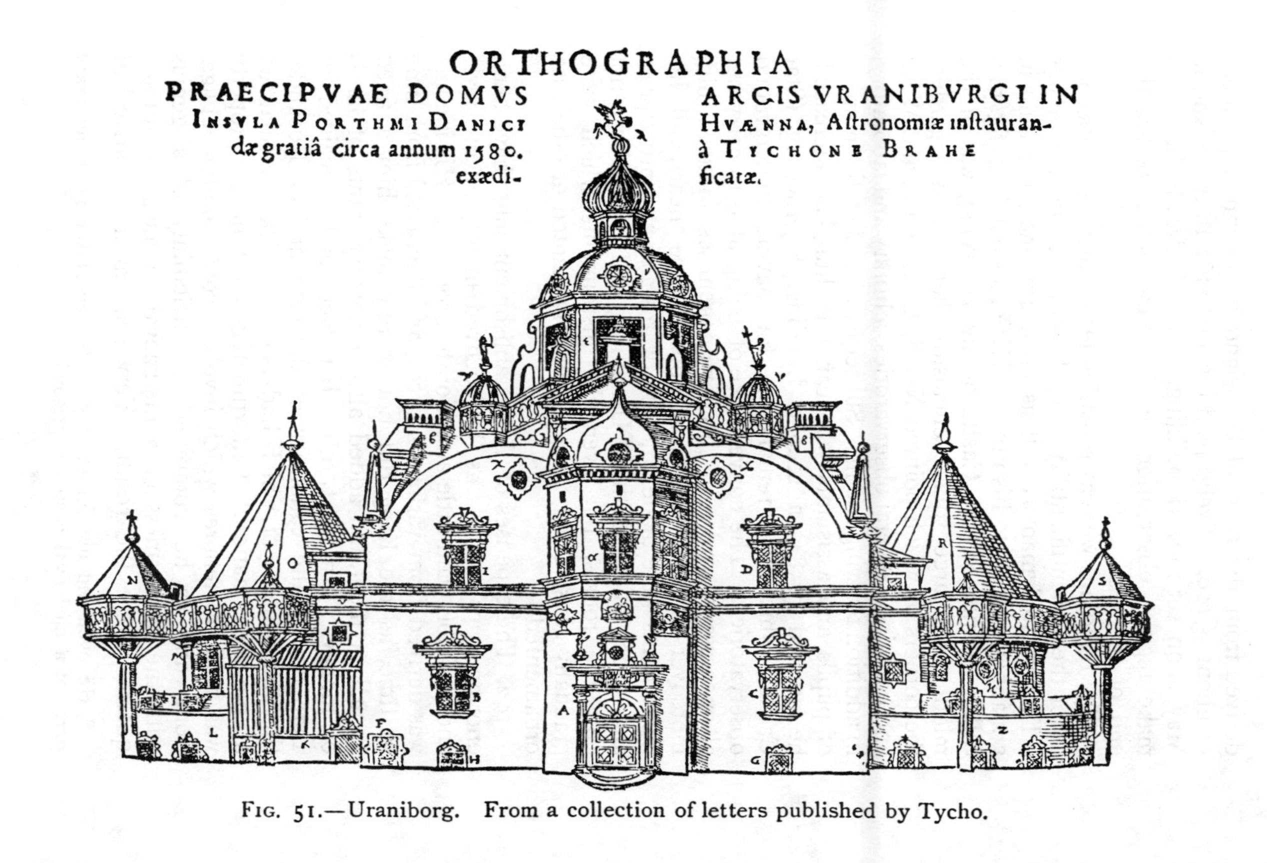

ORTHOGRAPHIA
PRAECIPVAE DOMVS
INSVLA PORTHMI DANICI
dægratiâ circa annum 1580.
exædi-
ARCIS VRANIBVRGI IN
HVÆNNA, Astronomiæ instauran-
à TYCHONE BRAHE
ficatæ.

FIG. 51.—Uraniborg. From a collection of letters published by Tycho.

derived from the rents of the island by an annual payment of about £100. Tycho paid his first visit to the island in May, soon set to work building, and had already begun to make regular observations in his new house before the end of the year.

The buildings were as remarkable for their magnificence as for their scientific utility. Tycho never forgot that he was a Danish nobleman as well as an astronomer, and built in a manner suitable to his rank.* His chief building (fig. 51), called Uraniborg (the Castle of the Heavens), was in the middle of a large square enclosure, laid out as a garden, the corners of which pointed North, East, South, and West, and contained several observatories, a library and laboratory, in addition to living rooms. Subsequently, when the number of pupils and assistants who came to him had increased, he erected (1584) a second building, Stjerneborg (Star Castle), which was remarkable for having underground observatories. The convenience of being able to carry out all necessary work on his own premises induced him moreover to establish workshops, where nearly all his instruments were made, and afterwards also a printing press and paper mill. Both at Uraniborg and Stjerneborg not only the rooms, but even the instruments which were gradually constructed, were elaborately painted or otherwise ornamented.

102. The expenses of the establishment must have been enormous, particularly as Tycho lived in magnificent style and probably paid little attention to economy. His income was derived from various sources, and fluctuated from time to time, as the King did not merely make him a fixed annual payment, but added also temporary grants of lands or money. Amongst other benefactions he received in 1579 one of the canonries of the cathedral of Roskilde, the endowments of which had been practically secularised at the Reformation. Unfortunately most of his property was held on tenures which involved corresponding obligations, and as he combined the irritability of a genius with the haughtiness of a mediaeval nobleman, continual quarrels were the result. Very soon after his arrival at

* He even did not forget to provide one of the most necessary parts of a mediæval castle, a prison!

Hveen his tenants complained of work which he illegally forced from them; chapel services which his canonry required him to keep up were neglected, and he entirely refused to make certain recognised payments to the widow of the previous canon. Further difficulties arose out of a lighthouse, the maintenance of which was a duty attached to one of his estates, but was regularly neglected. Nothing shews the King's good feeling towards Tycho more than the trouble which he took to settle these quarrels, often ending by paying the sum of money under dispute. Tycho was moreover extremely jealous of his scientific reputation, and on more than one occasion broke out into violent abuse of some assistant or visitor whom he accused of stealing his ideas and publishing them elsewhere.

In addition to the time thus spent in quarrelling, a good deal must have been occupied in entertaining the numerous visitors whom his fame attracted, and who included, in addition to astronomers, persons of rank such as several of the Danish royal family and James VI. of Scotland (afterwards James I. of England).

Notwithstanding these distractions, astronomical work made steady progress, and during the 21 years that Tycho spent at Hveen he accumulated, with the help of pupils and assistants, a magnificent series of observations, far transcending in accuracy and extent anything that had been accomplished by his predecessors. A good deal of attention was also given to alchemy, and some to medicine. He seems to have been much impressed with the idea of the unity of Nature, and to have been continually looking out for analogies or actual connection between the different subjects which he studied.

103. In 1577 appeared a brilliant comet, which Tycho observed with his customary care; and, although he had not at the time his full complement of instruments, his observations were exact enough to satisfy him that the comet was at least three times as far off as the moon, and thus to refute the popular belief, which he had himself held a few years before (§ 100), that comets were generated in our atmosphere. His observations led him also to the belief that the comet was revolving round the sun, at a distance from it greater than that of Venus, a conclusion

which interfered seriously with the common doctrine of the solid crystalline spheres. He had further opportunities of observing comets in 1580, 1582, 1585, 1590, and 1596, and one of his pupils also took observations of a comet seen in 1593. None of these comets attracted as much general attention as that of 1577, but Tycho's observations, as was natural, gradually improved in accuracy.

104. The valuable results obtained by means of the new star of 1572, and by the comets, suggested the propriety of undertaking a complete treatise on astronomy embodying these and other discoveries. According to the original plan, there were to be three preliminary volumes devoted respectively to the new star, to the comet of 1577, and to the later comets, while the main treatise was to consist of several more volumes dealing with the theories of the sun, moon, and planets. Of this magnificent plan comparatively little was ever executed. The first volume, called the *Astronomiae Instauratae Progymnasmata*, or Introduction to the New Astronomy, was hardly begun till 1588, and, although mostly printed by 1592, was never quite finished during Tycho's lifetime, and was actually published by Kepler in 1602. One question, in fact, led to another in such a way that Tycho felt himself unable to give a satisfactory account of the star of 1572 without dealing with a number of preliminary topics, such as the positions of the fixed stars, precession, and the annual motion of the sun, each of which necessitated an elaborate investigation. The second volume, dealing with the comet of 1577, called *De Mundi aetherei recentioribus Phaenomenis Liber secundus* (Second book about recent appearances in the Celestial World), was finished long before the first, and copies were sent to friends and correspondents in 1588, though it was not regularly published and on sale till 1603. The third volume was never written, though some material was collected for it, and the main treatise does not appear to have been touched.

105. The book on the comet of 1577 is of special interest, as containing an account of Tycho's system of the world, which was a compromise between those of Ptolemy and of Coppernicus. Tycho was too good an astronomer not to realise many of the simplifications which the

Coppernican system introduced, but was unable to answer two of the serious objections ; he regarded any motion of

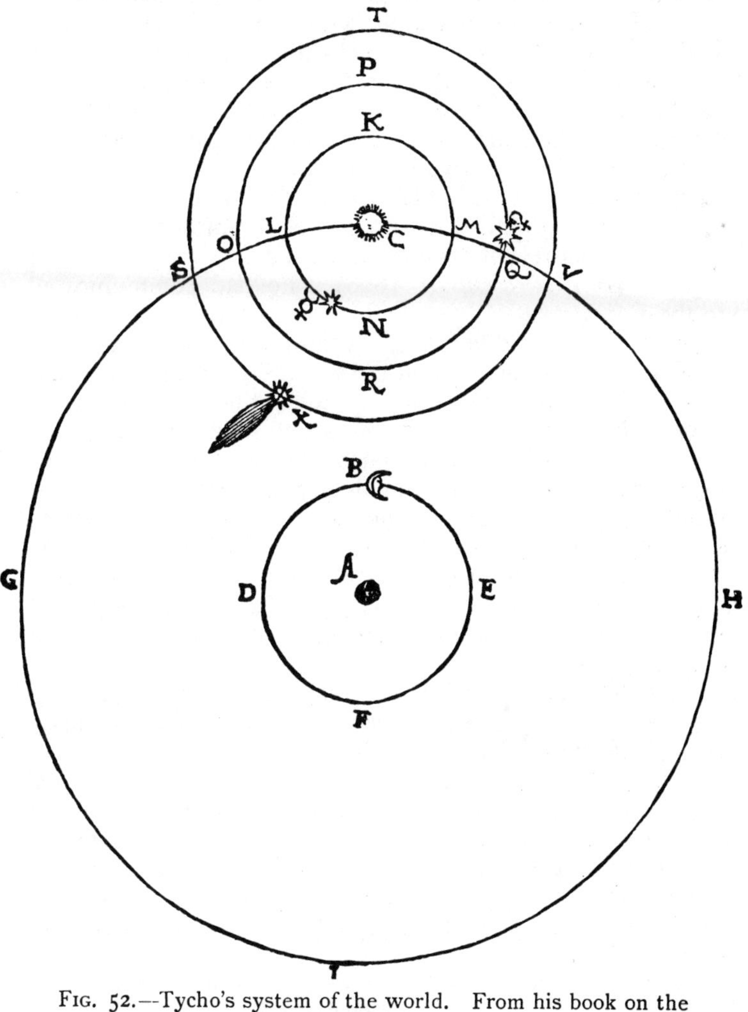

FIG. 52.—Tycho's system of the world. From his book on the comet of 1577.

"the sluggish and heavy earth" as contrary to "physical principles," and he objected to the great distance of the stars which the Coppernican system required, because a vast empty space would be left between them and the planets, a space which he regarded as wasteful.* Biblical difficulties † also had some weight with him. He accordingly devised (1583) a new system according to which the five planets revolved round the sun (C, in fig. 52), while the sun revolved annually round the earth (A), and the whole celestial sphere performed also a daily revolution round the earth. The system was never worked out in detail, and, like many compromises, met with little support; Tycho nevertheless was extremely proud of it, and one of the most violent and prolonged quarrels of his life (lasting a dozen years) was with *Reymers Bär* or *Ursus* (?–1600), who had communicated to the Landgrave in 1586 and published two years later a system of the world very like Tycho's. Reymers had been at Hveen for a short time in 1584, and Tycho had no hesitation in accusing him of having stolen the idea from some manuscript seen there. Reymers naturally retaliated with a counter-charge of theft against Tycho. There is, however, no good reason why the idea should not have occurred independently to each astronomer; and Reymers made in some respects a great improvement on Tycho's scheme by accepting the daily rotation of the earth, and so doing away with the daily rotation of the celestial sphere, which was certainly one of the weakest parts of the Ptolemaic scheme.

106. The same year (1588) which saw the publication of Tycho's book on the comet was also marked by the death of his patron, Frederick II. The new King Christian was a boy of 11, and for some years the country was managed by four leading statesmen. The new government seems to have been at first quite friendly to Tycho; a large sum was paid to him for expenses incurred at Hveen, and additional endowments were promised, but as time went on Tycho's usual quarrels with his tenants and others began to produce

* It would be interesting to know what use he assigned to the (presumably) still vaster space *beyond* the stars.

† Tycho makes in this connection the delightful remark that Moses must have been a skilled astronomer, because he refers to the moon as "the lesser light," notwithstanding the fact that the apparent diameters of sun and moon are very nearly equal!

TYCHO BRAHE. [*To face p.* 139.

their effect. In 1594 he lost one of his chief supporters at court, the Chancellor Kaas, and his successor, as well as two or three other important officials at court, were not very friendly, although the stories commonly told of violent personal animosities appear to have little foundation. As early as 1591 Tycho had hinted to a correspondent that he might not remain permanently in Denmark, and in 1594 he began a correspondence with representatives of the Emperor Rudolph II., who was a patron of science. But his scientific activity during these years was as great as ever; and in 1596 he completed the printing of an extremely interesting volume of scientific correspondence between the Landgrave, Rothmann, and himself. The accession of the young King to power in 1596 was at once followed by the withdrawal of one of Tycho's estates, and in the following year the annual payment which had been made since 1576 was stopped. It is difficult to blame the King for these economies; he was evidently not as much interested in astronomy as his father, and consequently regarded the heavy expenditure at Hveen as an extravagance, and it is also probable that he was seriously annoyed at Tycho's maltreatment of his tenants, and at other pieces of unruly conduct on his part. Tycho, however, regarded the forfeiture of his annual pension as the last straw, and left Hveen early in 1597, taking his more portable property with him. After a few months spent in Copenhagen, he took the decisive step of leaving Denmark for Germany, in return for which action the King deprived him of his canonry. Tycho thereupon wrote a remonstrance in which he pointed out the impossibility of carrying on his work without proper endowments, and offered to return if his services were properly appreciated. The King, however, was by this time seriously annoyed, and his reply was an enumeration of the various causes of complaint against Tycho which had arisen of late years. Although Tycho made some more attempts through various friends to regain royal favour, the breach remained final.

107. Tycho spent the winter 1597–8 with a friend near Hamburg, and, while there, issued, under the title of *Astronomiae Instauratae Mechanica*, a description of his instruments, together with a short autobiography and an

interesting account of his chief discoveries. About the same time he circulated manuscript copies of a catalogue of 1,000 fixed stars, of which only 777 had been properly observed, the rest having been added hurriedly to make up the traditional number. The catalogue and the *Mechanica* were both intended largely as evidence of his astronomical eminence, and were sent to various influential persons. Negotiations went on both with the Emperor and with the Prince of Orange, and after another year spent in various parts of Germany, Tycho definitely accepted an invitation of the Emperor and arrived at Prague in June 1599.

108. It was soon agreed that he should inhabit the castle of Benatek, some twenty miles from Prague, where he accordingly settled with his family and smaller instruments towards the end of 1599. He at once started observing, sent one of his sons to Hveen for his larger instruments, and began looking about for assistants. He secured one of the most able of his old assistants, and by good fortune was also able to attract a far greater man, *John Kepler*, to whose skilful use of the materials collected by Tycho the latter owes no inconsiderable part of his great reputation. Kepler, whose life and work will be dealt with at length in chapter VII., had recently published his first important work, the *Mysterium Cosmographicum* (§ 136), which had attracted the attention of Tycho among others, and was beginning to find his position at Gratz in Styria uncomfortable on account of impending religious disputes. After some hesitation he joined Tycho at Benatek early in 1600. He was soon set to work at the study of Mars for the planetary tables which Tycho was then preparing, and thus acquired special familiarity with the observations of this planet which Tycho had accumulated. The relations of the two astronomers were not altogether happy, Kepler being then as always anxious about money matters, and the disturbed state of the country rendering it difficult for Tycho to get payment from the Emperor. Consequently Kepler very soon left Benatek and returned to Prague, where he definitely settled after a short visit to Gratz; Tycho also moved there towards the end of 1600, and they then worked together harmoniously for

the short remainder of Tycho's life. Though he was by no means an old man, there were some indications that his health was failing, and towards the end of 1601 he was suddenly seized with an illness which terminated fatally after a few days (November 24th). It is characteristic of his devotion to the great work of his life that in the delirium which preceded his death he cried out again and again his hope that his life might not prove to have been fruitless (*Ne frustra vixisse videar*).

109. Partly owing to difficulties between Kepler and one of Tycho's family, partly owing to growing political disturbances, scarcely any use was made of Tycho's instruments after his death, and most of them perished during the Civil Wars in Bohemia. Kepler obtained possession of his observations; but they have never been published except in an imperfect form.

110. Anything like a satisfactory account of Tycho's services to astronomy would necessarily deal largely with technical details of methods of observing, which would be out of place here. It may, however, be worth while to attempt to give some general account of his characteristics as an observer before referring to special discoveries.

Tycho realised more fully than any of his predecessors the importance of obtaining observations which should not only be as accurate as possible, but should be taken so often as to preserve an almost continuous record of the positions and motions of the celestial bodies dealt with; whereas the prevailing custom (as illustrated for example by Coppernicus) was only to take observations now and then, either when an astronomical event of special interest such as an eclipse or a conjunction was occurring, or to supply some particular datum required for a point of theory. While Coppernicus, as has been already noticed (chapter IV., § 73), only used altogether a few dozen observations in his book, Tycho—to take one instance—observed the sun *daily* for many years, and must therefore have taken some thousands of observations of this one body, in addition to the many thousands which he took of other celestial bodies. It is true that the Arabs had some idea of observing continuously (cf. chapter III., § 57), but they had too little speculative power or originality to be able to make much use of their observations, few of which passed into the hands of European astronomers. Regiomontanus (chapter III., § 68), if he had lived, might probably have to a considerable extent anticipated Tycho, but his short life was too fully occupied with the study and interpretation of Greek astronomy for him to accomplish very much in other departments of the subject. The Landgrave and his staff, who were in constant communication with Tycho, were working in the same direction, though on the whole less effectively. Unlike the Arabs, Tycho was, however, fully impressed with the idea that observations were only a means to an end, and that mere observations without a hypothesis or theory to connect and interpret them were of little use.

The actual accuracy obtained by Tycho in his observations naturally varied considerably according to the nature of the observation, the care taken, and the period of his career at which it was made. The places which he assigned to nine stars which were fundamental in his star catalogue differ from their positions as deduced from the best modern observations by angles which are in most cases less than 1′, and in only one case as great as 2′ (this error being chiefly due to refraction (chapter II., § 46), Tycho's knowledge of which was necessarily imperfect). Other star places were presumably less accurate, but it will not be far from the truth if we assume that in most cases the errors in Tycho's observations did not exceed 1′ or 2′. Kepler in a famous passage speaks of an error of 8′ in a planetary observation by Tycho as impossible. This great increase in accuracy can only be assigned in part to the size and careful construction of the instruments used, the characteristics on which the Arabs and other observers had laid such stress. Tycho certainly used good instruments, but added very much to their efficiency, partly by minor mechanical devices, such as the use of specially constructed "sights" and of a particular method of graduation,* and partly by using instruments capable only of restricted motions, and therefore of much greater steadiness than instruments which were able to point to any part of the sky. Another extremely important idea

* By transversals.

was that of systematically allowing as far as possible for the inevitable mechanical imperfections of even the best constructed instruments, as well as for other permanent causes of error. It had been long known, for example, that the refraction of light through the atmosphere had the effect of slightly raising the apparent places of stars in the sky. Tycho took a series of observations to ascertain the amount of this displacement for different parts of the sky, hence constructed a table of refractions (a very imperfect one, it is true), and in future observations regularly allowed for the effect of refraction. Again, it was known that observations of the sun and planets were liable to be disturbed by the effect of parallax (chapter II., §§ 43, 49), though the amount of this correction was uncertain. In cases where special accuracy was required, Tycho accordingly observed the body in question at least twice, choosing positions in which parallax was known to produce nearly opposite effects, and thus by combining the observations obtained a result nearly free from this particular source of error. He was also one of the first to realise fully the importance of repeating the same observation many times under different conditions, in order that the various accidental sources of error in the separate observations should as far as possible neutralise one another.

111. Almost every astronomical quantity of importance was re-determined and generally corrected by him. The annual motion of the sun's apogee relative to ♈, for example, which Coppernicus had estimated at $24''$, Tycho fixed at $45''$, the modern value being $61''$; the length of the year he determined with an error of less than a second; and he constructed tables of the motion of the sun which gave its place to within $1'$, previous tables being occasionally $15'$ or $20'$ wrong. By an unfortunate omission he made no inquiry into the distance of the sun, but accepted the extremely inaccurate value which had been handed down, without substantial alteration, from astronomer to astronomer since the time of Hipparchus (chapter II., § 41).

In the theory of the moon Tycho made several important discoveries. He found that the irregularities in its movement were not fully represented by the equation of the centre and the evection (chapter II., §§ 39, 48), but that there was a further irregularity which vanished at opposition and conjunction as well as at quadratures, but in intermediate positions of the moon might be as great as 40′. This irregularity, known as the **variation**, was, as has been already mentioned (chapter III., § 60), very possibly discovered by Abul Wafa, though it had been entirely lost subsequently. At a later stage in his career, at latest during his visit to Wittenberg in 1598–9, Tycho found that it was necessary to introduce a further small inequality known as the **annual equation**, which depended on the position of the earth in its path round the sun; this, however, he never completely investigated. He also ascertained that the inclination of the moon's orbit to the ecliptic was not, as had been thought, fixed, but oscillated regularly, and that the motion of the moon's nodes (chapter II., § 40) was also variable.

112. Reference has already been made to the star catalogue. Its construction led to a study of precession, the amount of which was determined with considerable accuracy; the same investigation led Tycho to reject the supposed irregularity in precession which, under the name of trepidation (chapter III., § 58), had confused astronomy for several centuries, but from this time forward rapidly lost its popularity.

The planets were always a favourite subject of study with Tycho, but although he made a magnificent series of observations, of immense value to his successors, he died before he could construct any satisfactory theory of the planetary motions. He easily discovered, however, that their motions deviated considerably from those assigned by any of the planetary tables, and got as far as detecting some regularity in these deviations.

CHAPTER VI.

GALILEI.

"Dans la Science nous sommes tous disciples de Galilée."—TROUESSART.

"Bacon pointed out at a distance the road to true philosophy: Galileo both pointed it out to others, and made himself considerable advances in it."—DAVID HUME.

113. To the generation which succeeded Tycho belonged two of the best known of all astronomers, Galilei and Kepler. Although they were nearly contemporaries, Galilei having been born seven years earlier than Kepler, and surviving him by twelve years, their methods of work and their contributions to astronomy were so different in character, and their influence on one another so slight, that it is convenient to make some departure from strict chronological order, and to devote this chapter exclusively to Galilei, leaving Kepler to the next.

Galileo Galilei was born in 1564, at Pisa, at that time in the Grand Duchy of Tuscany, on the day of Michel Angelo's death and in the year of Shakespeare's birth. His father, Vincenzo, was an impoverished member of a good Florentine family, and was distinguished by his skill in music and mathematics. Galileo's talents shewed themselves early, and although it was originally intended that he should earn his living by trade, Vincenzo was wise enough to see that his son's ability and tastes rendered him much more fit for a professional career, and accordingly he sent him in 1581 to study medicine at the University of Pisa. Here his unusual gifts soon made him conspicuous, and he became noted in particular for his unwillingness to accept without question the dogmatic statements of his teachers, which were based not on direct evidence, but on the authority of the great writers of the past. This valuable characteristic, which marked him throughout his life, coupled with his skill in argument, earned for him the dislike of some of his professors, and from his fellow-students the nickname of The Wrangler.

114. In 1582 his keen observation led to his first scientific discovery. Happening one day in the Cathedral of Pisa to be looking at the swinging of a lamp which was hanging from the roof, he noticed that as the motion gradually died away and the extent of each oscillation became less, the time occupied by each oscillation remained sensibly the same, a result which he verified more precisely by comparison with the beating of his pulse. Further thought and trial shewed him that this property was not peculiar to cathedral lamps, but that any weight hung by a string (or any other form of pendulum) swung to and fro in a time which depended only on the length of the string and other characteristics of the pendulum itself, and not to any appreciable extent on the way in which it was set in motion or on the extent of each oscillation. He devised accordingly an instrument the oscillations of which could be used while they lasted as a measure of time, and which was in practice found very useful by doctors for measuring the rate of a patient's pulse.

115. Before very long it became evident that Galilei had no special taste for medicine, a study selected for him chiefly as leading to a reasonably lucrative professional career, and that his real bent was for mathematics and its applications to experimental science. He had received little or no formal teaching in mathematics before his second year at the University, in the course of which he happened to overhear a lesson on Euclid's geometry, given at the Grand Duke's court, and was so fascinated that he continued to attend the course, at first surreptitiously, afterwards openly; his interest in the subject was thereby so much stimulated, and his aptitude for it was so marked, that he obtained his father's consent to abandon medicine in favour of mathematics.

In 1585, however, poverty compelled him to quit the University without completing the regular course and obtaining a degree, and the next four years were spent chiefly at home, where he continued to read and to think on scientific subjects. In the year 1586 he wrote his first known scientific essay,* which was circulated in manuscript, and only printed during the present century.

116. In 1589 he was appointed for three years to a professorship of mathematics (including astronomy) at Pisa. A miserable stipend, equivalent to about five shillings a week, was attached to the post, but this he was to some extent able to supplement by taking private pupils.

In his new position Galilei had scope for his remarkable power of exposition, but far from being content with giving lectures on traditional lines he also carried out a series of scientific investigations, important both in themselves and on account of the novelty in the method of investigation employed.

It will be convenient to discuss more fully at the end of this chapter Galilei's contributions to mechanics and to scientific method, and merely to refer here briefly to his first experiments on falling bodies, which were made at this time. Some were performed by dropping various bodies from the top of the leaning tower of Pisa, and others by rolling balls down grooves arranged at different inclinations. It is difficult to us nowadays, when scientific experiments are so common, to realise the novelty and importance at the end of the 16th century of such simple experiments. The mediaeval tradition of carrying out scientific investigation largely by the interpretation of texts in Aristotle, Galen, or other great writers of the past, and by the deduction of results from general principles which were to be found in these writers without any fresh appeal to observation, still prevailed almost undisturbed at Pisa, as elsewhere. It was in particular commonly asserted, on the authority of Aristotle, that, the cause of the fall of a heavy body being its weight, a heavier body must fall faster than a lighter one and in proportion to its greater weight. It may perhaps be doubted whether any one before Galilei's time had clear enough ideas on the subject to be able to give a definite answer to such a question as how much farther a ten-pound weight would fall in a second than a one-pound

* On an instrument which he had invented, called the *hydrostatic balance.*

weight; but if so he would probably have said that it would fall ten times as far, or else that it would require ten times as long to fall the same distance. To actually try the experiment, to vary its conditions, so as to remove as many accidental causes of error as possible, to increase in some way the time of the fall so as to enable it to be measured with more accuracy, these ideas, put into practice by Galilei, were entirely foreign to the prevailing habits of scientific thought, and were indeed regarded by most of his colleagues as undesirable if not dangerous innovations. A few simple experiments were enough to prove the complete falsity of the current beliefs in this matter, and to establish that in general bodies of different weights fell nearly the same distance in the same time, the difference being not more than could reasonably be ascribed to the resistance offered by the air.

These and other results were embodied in a tract, which, like most of Galilei's earlier writings, was only circulated in manuscript, the substance of it being first printed in the great treatise on mechanics which he published towards the end of his life (§ 133).

These innovations, coupled with the slight respect that he was in the habit of paying to those who differed from him, evidently made Galilei far from popular with his colleagues at Pisa, and either on this account, or on account of domestic troubles consequent on the death of his father (1591), he resigned his professorship shortly before the expiration of his term of office, and returned to his mother's home at Florence.

117. After a few months spent at Florence he was appointed, by the influence of a Venetian friend, to a professorship of mathematics at Padua, which was then in the territory of the Venetian republic (1592). The appointment was in the first instance for a period of six years, and the salary much larger than at Pisa. During the first few years of Galilei's career at Padua his activity seems to have been very great and very varied; in addition to giving his regular lectures, to audiences which rapidly increased, he wrote tracts, for the most part not printed at the time, on astronomy, on mechanics, and on fortification, and invented a variety of scientific instruments.

No record exists of the exact time at which he first adopted the astronomical views of Coppernicus, but he himself stated that in 1597 he had adopted them some years before, and had collected arguments in their support.

In the following year his professorship was renewed for six years with an increased stipend, a renewal which was subsequently made for six years more, and finally for life, the stipend being increased on each occasion.

Galilei's first contribution to astronomical discovery was made in 1604, when a star appeared suddenly in the constellation Serpentarius, and was shewn by him to be at any rate more distant than the planets, a result confirming Tycho's conclusions (chapter v., § 100) that changes take place in the celestial regions even beyond the planets, and are by no means confined—as was commonly believed—to the earth and its immediate surroundings.

118. By this time Galilei had become famous throughout Italy, not only as a brilliant lecturer, but also as a learned and original man of science. The discoveries which first gave him a European reputation were, however, the series of telescopic observations made in 1609 and the following years.

Roger Bacon (chapter III., § 67) had claimed to have devised a combination of lenses enabling distant objects to be seen as if they were near; a similar invention was probably made by our countryman *Leonard Digges* (who died about 1571), and was described also by the Italian *Porta* in 1558. If such an instrument was actually made by any one of the three, which is not certain, the discovery at any rate attracted no attention and was again lost. The effective discovery of the telescope was made in Holland in 1608 by *Hans Lippersheim* (?–1619), a spectacle-maker of Middleburg, and almost simultaneously by two other Dutchmen, but whether independently or not it is impossible to say. Early in the following year the report of the invention reached Galilei, who, though without any detailed information as to the structure of the instrument, succeeded after a few trials in arranging two lenses—one convex and one concave—in a tube in such a way as to enlarge the apparent size of an object looked at; his first instrument made objects appear three times nearer, consequently three times greater (in breadth and height), and he was soon able to make telescopes which in the same way magnified thirty-fold.

That the new instrument might be applied to celestial as well as to terrestrial objects was a fairly obvious idea, which was acted on almost at once by the English mathematician *Thomas Harriot* (1560–1621), by *Simon Marius* (1570–1624) in Germany, and by Galilei. That the credit of first using the telescope for astronomical purposes is almost invariably attributed to Galilei, though his first observations were in all probability slightly later in date than those of Harriot and Marius, is to a great extent justified by the persistent way in which he examined object after object, whenever there seemed any reasonable prospect of results following, by the energy and acuteness with which he followed up each clue, by the independence of mind with which he interpreted his observations, and above all by the insight with which he realised their astronomical importance.

119. His first series of telescopic discoveries were published early in 1610 in a little book called *Sidereus Nuncius*, or *The Sidereal Messenger*. His first observations at once threw a flood of light on the nature of our nearest celestial neighbour, the moon. It was commonly believed that the moon, like the other celestial bodies, was perfectly smooth and spherical, and the cause of the familiar dark markings on the surface was quite unknown.*

Galilei discovered at once a number of smaller markings, both bright and dark (fig. 53), and recognised many of the latter as shadows of lunar mountains cast by the sun; and further identified bright spots seen near the boundary of the illuminated and dark portions of the moon as mountain-tops just catching the light of the rising or setting sun, while the surrounding lunar area was still in darkness. Moreover, with characteristic ingenuity and love of precision, he calculated from observations of this nature the height of some of the more conspicuous lunar moun-

* A fair idea of mediaeval views on the subject may be derived from one of the most tedious Cantos in Dante's great poem (*Paradiso*, II.), in which the poet and Beatrice expound two different "explanations" of the spots on the moon.

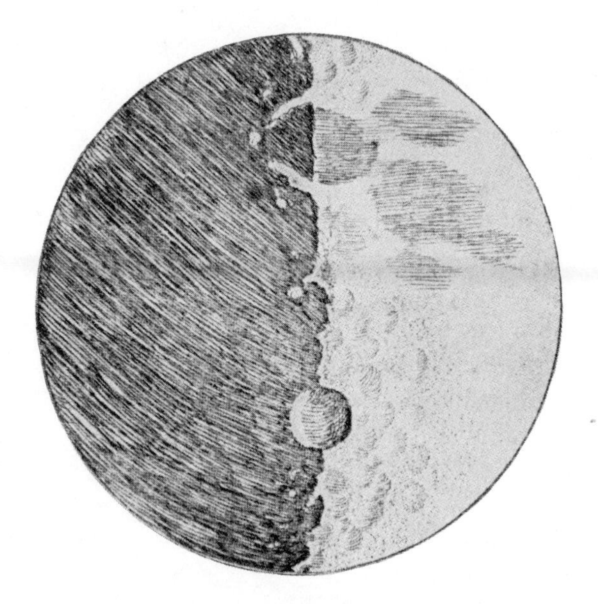

FIG. 53.—One of Galilei's drawings of the moon. From the *Sidereus Nuncius*.

[*To face p.* 150.

tains, the largest being estimated by him to be about four miles high, a result agreeing closely with modern estimates of the greatest height on the moon. The large dark spots he explained (erroneously) as possibly caused by water, though he evidently had less confidence in the correctness of the explanation than some of his immediate scientific successors, by whom the name of **seas** was given to these spots (chapter VIII., § 153). He noticed also the absence of clouds. Apart however from details, the really significant results of his observations were that the moon was in many important respects similar to the earth, that the traditional belief in its perfectly spherical form had to be abandoned, and that so far the received doctrine of the sharp distinction to be drawn between things celestial and things terrestrial was shewn to be without justification; the importance of this in connection with the Coppernican view that the earth, instead of being unique, was one of six planets revolving round the sun, needs no comment.

One of Galilei's numerous scientific opponents * attempted to explain away the apparent contradiction between the old theory and the new observations by the ingenious suggestion that the apparent valleys in the moon were in reality filled with some *invisible* crystalline material, so that the moon was in fact perfectly spherical. To this Galilei replied that the idea was so excellent that he wished to extend its application, and accordingly maintained that the moon had on it mountains of this same invisible substance, at least ten times as high as any which he had observed.

120. The telescope revealed also the existence of an immense number of stars too faint to be seen by the unaided eye; Galilei saw, for example, 36 stars in the Pleiades, which to an ordinary eye consist of six only. Portions of the Milky Way and various nebulous patches of light were also discovered to consist of multitudes of faint stars clustered together; in the cluster Præsepe (in the Crab), for example, he counted 40 stars.

121. By far the most striking discovery announced in the *Sidereal Messenger* was that of the bodies now known as

* *Ludovico delle Colombe* in a tract *Contra Il Moto della Terra*, which is reprinted in the national edition of Galilei's works, Vol. III.

the moons or satellites of Jupiter. On January 7th, 1610, Galilei turned his telescope on to Jupiter, and noticed three faint stars which caught his attention on account of their closeness to the planet and their arrangement nearly in a straight line with it. He looked again next night, and noticed that they had changed their positions relatively to Jupiter, but that the change did not seem to be such as could result from Jupiter's own motion, if the new bodies were fixed stars. Two nights later he was able to confirm this conclusion, and to infer that the new bodies were not fixed stars, but moving bodies which accompanied Jupiter in his movements. A fourth body was noticed on January 13th, and the motions of all four were soon recognised by Galilei as being motions of revolution round Jupiter as a centre. With characteristic thoroughness he

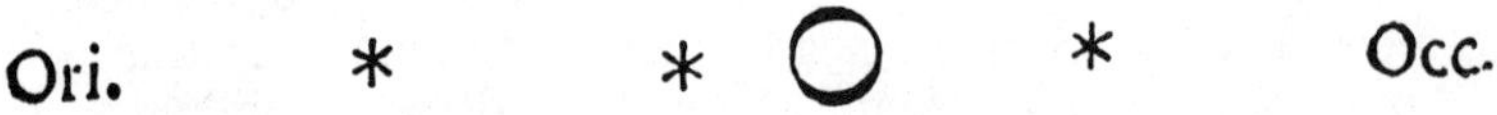

Ori. * * ○ * Occ.

Fig. 54.—Jupiter and its satellites as seen on Jan. 7, 1610. From the *Sidereus Nuncius*.

watched the motions of the new bodies night after night, and by the date of the publication of his book had already estimated with very fair accuracy their periods of revolution round Jupiter, which ranged between about 42 hours and 17 days; and he continued to watch their motions for years.

The new bodies were at first called by their discoverer Medicean planets, in honour of his patron Cosmo de Medici, the Grand Duke of Tuscany; but it was evident that bodies revolving round a planet, as the planets themselves revolved round the sun, formed a new class of bodies distinct from the known planets, and the name of **satellite**, suggested by Kepler as applicable to the new bodies as well as to the moon, has been generally accepted.

The discovery of Jupiter's satellites shewed the falsity of the old doctrine that the earth was the only centre of motion; it tended, moreover, seriously to discredit the infallibility of Aristotle and Ptolemy, who had clearly no knowledge of the existence of such bodies; and again those who had difficulty in believing that Venus and

Mercury could revolve round an apparently moving body, the sun, could not but have their doubts shaken when shewn the new satellites evidently performing a motion of just this character; and—most important consequence of all—the very real mechanical difficulty involved in the Coppernican conception of the moon revolving round the moving earth and not dropping behind was at any rate shewn not to be insuperable, as Jupiter's satellites succeeded in performing a precisely similar feat.

The same reasons which rendered Galilei's telescopic discoveries of scientific importance made them also objectionable to the supporters of the old views, and they were accordingly attacked in a number of pamphlets, some of which are still extant and possess a certain amount of interest. One *Martin Horky*, for example, a young German who had studied under Kepler, published a pamphlet in which, *after* proving to his own satisfaction that the satellites of Jupiter did not exist, he discussed at some length what they were, what they were like, and why they existed. Another writer gravely argued that because the human body had seven openings in it—the eyes, ears, nostrils, and mouth—therefore by analogy there must be seven planets (the sun and moon being included) and no more. However, confirmation by other observers was soon obtained and the pendulum even began to swing in the opposite direction, a number of new satellites of Jupiter being announced by various observers. None of these, however, turned out to be genuine, and Galilei's four remained the only known satellites of Jupiter till a few years ago (chapter XIII., § 295).

122. The reputation acquired by Galilei by the publication of the *Messenger* enabled him to bring to a satisfactory issue negotiations which he had for some time been carrying on with the Tuscan court. Though he had been well treated by the Venetians, he had begun to feel the burden of regular teaching somewhat irksome, and was anxious to devote more time to research and to writing. A republic could hardly be expected to provide him with such a sinecure as he wanted, and he accordingly accepted in the summer of 1610 an appointment as professor at Pisa, and also as "First Philosopher and Mathematician" to the Grand

Duke of Tuscany, with a handsome salary and no definite duties attached to either office.

123. Shortly before leaving Padua he turned his telescope on to Saturn, and observed that the planet appeared to consist of three parts, as shewn in the first drawing of fig. 67 (chapter VIII., § 154). On subsequent occasions, however, he failed to see more than the central body, and the appearances of Saturn continued to present perplexing variations, till the mystery was solved by Huygens in 1655 (chapter VIII., § 154).

The first discovery made at Florence (October 1610) was that Venus, which to the naked eye appears to vary very much in brilliancy but not in shape, was in reality at times crescent-shaped like the new moon and passed through phases similar to some of those of the moon. This shewed that Venus was, like the moon, a dark body in itself, deriving its light from the sun; so that its similarity to the earth was thereby made more evident.

124. The discovery of dark spots on the sun completed this series of telescopic discoveries. According to his own statement Galilei first saw them towards the end of 1610,* but apparently paid no particular attention to them at the time; and, although he shewed them as a matter of curiosity to various friends, he made no formal announcement of the discovery till May 1612, by which time the same discovery had been made independently by Harriot (§ 118) in England, by *John Fabricius* (1587-? 1615) in Holland, and by the Jesuit *Christopher Scheiner* (1575-1650) in Germany, and had been published by Fabricius (June 1611). As a matter of fact dark spots had been seen with the naked eye long before, but had been generally supposed to be caused by the passage of Mercury in front of the sun. The presence on the sun of such blemishes as black spots, the "mutability" involved in their changes in form and position, and their formation and subsequent disappearance, were all distasteful to the supporters of the old views,

* In a letter of May 4th, 1612, he says that he has seen them for eighteen months; in the *Dialogue on the Two Systems* (III., p. 312, in Salusbury's translation) he says that he saw them while he still lectured at Padua, *i.e.* presumably by September 1610, as he moved to Florence in that month.

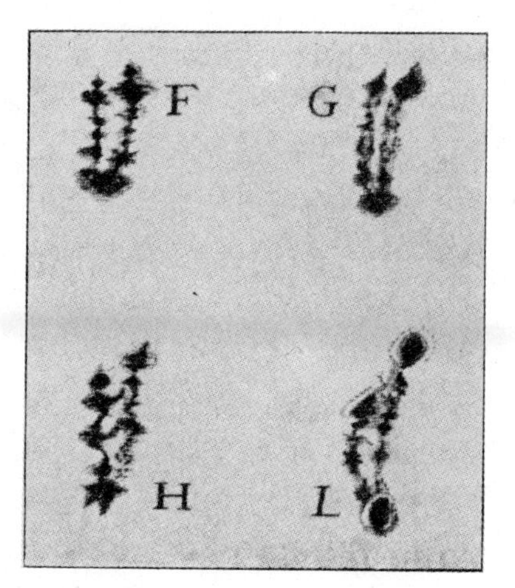

FIG. 55.—Sun-spots. From Galilei's *Macchie Solari*.

[*To face p.* 154.

according to which celestial bodies were perfect and unchangeable. The fact, noticed by all the early observers, that the spots appeared to move across the face of the sun from the eastern to the western side (*i.e.* roughly from left to right, as seen at midday by an observer in our latitudes), gave at first sight countenance to the view, championed by Scheiner among others, that the spots might really be small planets revolving round the sun, and appearing as dark objects whenever they passed between the sun and the observer. In three letters to his friend Welser, a merchant prince of Augsburg, written in 1612 and published in the following year,* Galilei, while giving a full account of his observations, gave a crushing refutation of this view; proved that the spots must be on or close to the surface of the sun, and that the motions observed were exactly such as would result if the spots were attached to the sun, and it revolved on an axis in a period of about a month; and further, while disclaiming any wish to speak confidently, called attention to several of their points of resemblance to clouds.

One of his arguments against Scheiner's views is so simple and at the same time so convincing, that it may be worth while to reproduce it as an illustration of Galilei's method, though the controversy itself is quite dead.

Galilei noticed, namely, that while a spot took about fourteen days to cross from one side of the sun to the other, and this time was the same whether the spot passed through the centre of the sun's disc, or along a shorter path at some distance from it, its rate of motion was by no means uniform, but that the spot's motion always appeared much slower when near the edge of the sun than when near the centre. This he recognised as an effect of foreshortening, which would result if, and only if, the spot were near the sun.

If, for example, in the figure, the circle represent a section of the sun by a plane through the observer at O, and A, B, C, D, E be points taken at equal distances along the surface of the sun, so as to represent the positions of an object on the sun at equal intervals of time, on the assumption that the sun revolves uniformly, then the

* *Historia e Dimostrazioni intorno alle Macchie Solari.*

apparent motion from A to B, as seen by the observer at O, is measured by the angle A O B, and is obviously much less than that from D to E, measured by the angle D O E, and consequently an object attached to the sun must appear to move more slowly from A to B, *i.e.* near the sun's edge, than from D to E, near the centre. On the other hand, if the spot be a body revolving round the sun at some distance from it, *e.g.* along the dotted circle *c d e*, then if *c*, *d*, *e* be taken at equal distances from one another, the apparent motion from *c* to *d*, measured again by the angle *c* O *d*, is only very slightly less than that from *d* to *e*, measured by the angle *d* O *e*. Moreover, it required only a simple calculation, performed by Galilei in several cases,

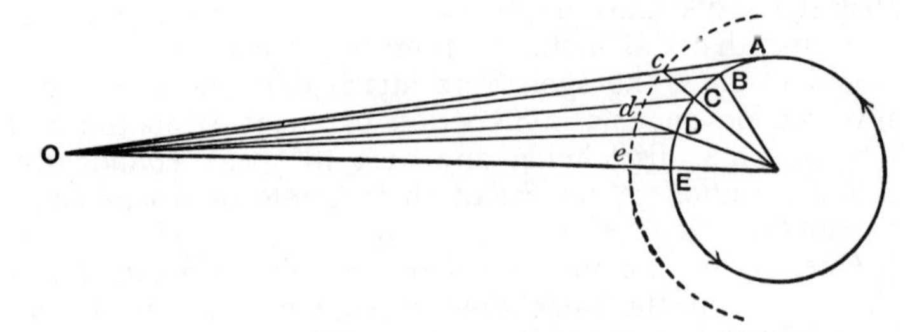

FIG. 56.—Galilei's proof that sun-spots are not planets.

to express these results in a numerical shape, and so to infer from the actual observations that the spots could not be more than a very moderate distance from the sun. The only escape from this conclusion was by the assumption that the spots, if they were bodies revolving round the sun, moved irregularly, in such a way as always to be moving fastest when they happened to be between the centre of the sun and the earth, whatever the earth's position might be at the time, a procedure for which, on the one hand, no sort of reason could be given, and which, on the other, was entirely out of harmony with the uniformity to which mediæval astronomy clung so firmly.

The rotation of the sun about an axis, thus established, might evidently have been used as an argument in support of the view that the earth also had such a motion, but, as far as I am aware, neither Galilei nor any contemporary noticed the analogy. Among other facts relating to the spots observed by Galilei were the greater darkness of the central parts, some of his drawings (see fig. 55) shewing, like most modern drawings, a fairly well-marked line of division between the central part (or **umbra**) and the less dark fringe (or **penumbra**) surrounding it; he noticed also that spots frequently appeared in groups, that the members of a group changed their positions relatively to one another, that individual spots changed their size and shape considerably during their lifetime, and that spots were usually most plentiful in two regions on each side of the sun's equator, corresponding roughly to the tropics on our own globe, and were never seen far beyond these limits.

Similar observations were made by other telescopists, and to Scheiner belongs the credit of fixing, with considerably more accuracy than Galilei, the position of the sun's axis and equator and the time of its rotation.

125. The controversy with Scheiner as to the nature of spots unfortunately developed into a personal quarrel as to their respective claims to the discovery of spots, a controversy which made Scheiner his bitter enemy, and probably contributed not a little to the hostility with which Galilei was henceforward regarded by the Jesuits. Galilei's uncompromising championship of the new scientific ideas, the slight respect which he shewed for established and traditional authority, and the biting sarcasms with which he was in the habit of greeting his opponents, had won for him a large number of enemies in scientific and philosophic circles, particularly among the large party who spoke in the name of Aristotle, although, as Galilei was never tired of reminding them, their methods of thought and their conclusions would in all probability have been rejected by the great Greek philosopher if he had been alive.

It was probably in part owing to his consciousness of a growing hostility to his views, both in scientific and in ecclesiastical circles, that Galilei paid a short visit to Rome in 1611, when he met with a most honourable reception and was treated with great friendliness by several cardinals and other persons in high places.

Unfortunately he soon began to be drawn into a controversy as to the relative validity in scientific matters of observation and reasoning on the one hand, and of the authority of the Church and the Bible on the other, a controversy which began to take shape about this time and which, though its battle-field has shifted from science to science, has lasted almost without interruption till modern times.

In 1611 was published a tract maintaining Jupiter's satellites to be unscriptural. In 1612 Galilei consulted Cardinal Conti as to the astronomical teaching of the Bible, and obtained from him the opinion that the Bible appeared to discountenance both the Aristotelian doctrine of the immutability of the heavens and the Coppernican doctrine of the motion of the earth. A tract of Galilei's on floating bodies, published in 1612, roused fresh opposition, but on the other hand Cardinal Barberini (who afterwards, as Urban VIII., took a leading part in his persecution) specially thanked him for a presentation copy of the book on sun-spots, in which Galilei, for the first time, clearly proclaimed in public his adherence to the Coppernican system. In the same year (1613) his friend and follower, Father Castelli, was appointed professor of mathematics at Pisa, with special instructions not to lecture on the motion of the earth. Within a few months Castelli was drawn into a discussion on the relations of the Bible to astronomy, at the house of the Grand Duchess, and quoted Galilei in support of his views; this caused Galilei to express his opinions at some length in a letter to Castelli, which was circulated in manuscript at the court. To this a Dominican preacher, Caccini, replied a few months afterwards by a violent sermon on the text, "Ye Galileans, why stand ye gazing up into heaven?"* and in 1615 Galilei was secretly denounced to the Inquisition on the strength of the letter to Castelli and other evidence. In the same year he expanded the letter to Castelli into a more elaborate treatise, in the form of a *Letter to the Grand Duchess Christine*, which was circulated in manuscript, but not printed till 1636. The discussion of the bearing of particular passages of the Bible (*e.g.* the account of the miracle of Joshua) on the Ptolemaic and Coppernican

* Acts i. 11. The pun is not quite so bad in its Latin form: *Viri Galilaei*, etc.

systems has now lost most of its interest; it may, however, be worth noticing that on the more general question Galilei quotes with approval the saying of Cardinal Baronius, "That the intention of the Holy Ghost is to teach us not how the heavens go, but how to go to heaven,"* and the following passage gives a good idea of the general tenor of his argument:—

"Methinks, that in the Discussion of Natural Problemes we ought not to begin at the authority of places of Scripture; but at Sensible Experiments and Necessary Demonstrations. For . . . Nature being inexorable and immutable, and never passing the bounds of the Laws assigned her, as one that nothing careth, whether her abstruse reasons and methods of operating be or be not exposed to the capacity of men; I conceive that that concerning Natural Effects, which either sensible experience sets before our eyes, or Necessary Demonstrations do prove unto us, ought not, upon any account, to be called into question, much less condemned upon the testimony of Texts of Scripture, which may under their words, couch senses seemingly contrary thereto." †

126. Meanwhile his enemies had become so active that Galilei thought it well to go to Rome at the end of 1615 to defend his cause. Early in the next year a body of theologians known as the Qualifiers of the Holy Office (Inquisition), who had been instructed to examine certain Coppernican doctrines, reported:—

"That the doctrine that the sun was the centre of the world and immoveable was false and absurd, formally heretical and contrary to Scripture, whereas the doctrine that the earth was not the centre of the world but moved, and has further a daily motion, was philosophically false and absurd and theologically at least erroneous."

In consequence of this report it was decided to censure Galilei, and the Pope accordingly instructed Cardinal Bellarmine "to summon Galilei and admonish him to

* *Spiritui sancto mentem fuisse nos docere, quo modo ad Coelum eatur, non autem, quomodo Coelum gradiatur.*

† From the translation by Salusbury, in Vol. I. of his *Mathematical Collections.*

abandon the said opinion," which the Cardinal did.* Immediately afterwards a decree was issued condemning the opinions in question and placing on the well-known *Index of Prohibited Books* three books containing Coppernican views, of which the *De Revolutionibus* of Coppernicus and another were only suspended "until they should be corrected," while the third was altogether prohibited. The necessary corrections to the *De Revolutionibus* were officially published in 1620, and consisted only of a few alterations which tended to make the essential principles of the book appear as mere mathematical hypotheses, convenient for calculation. Galilei seems to have been on the whole well satisfied with the issue of the inquiry, as far as he was personally concerned, and after obtaining from Cardinal Bellarmine a certificate that he had neither abjured his opinions nor done penance for them, stayed on in Rome for some months to shew that he was in good repute there.

127. During the next few years Galilei, who was now more than fifty, suffered a good deal from ill-health and was comparatively inactive. He carried on, however, a correspondence with the Spanish court on a method of ascertaining the longitude at sea by means of Jupiter's satellites. The essential problem in finding the longitude is to obtain the time as given by the sun at the required place and also that at some place the longitude of which is known. If, for example, midday at Rome occurs an hour earlier than in London, the sun takes an hour to travel from the meridian of Rome to that of London, and the longitude of Rome is 15° east of that of London. At sea it is easy to ascertain the local time, *e.g.* by observing when the sun is highest in the sky, but the great difficulty, felt in Galilei's time and long afterwards (chapter x., §§ 197, 226), was that of ascertaining the time at some standard place. Clocks were then, and long afterwards, not to be relied upon to keep time accurately during

* The only point of any importance in connection with Galilei's relations with the Inquisition on which there seems to be room for any serious doubt is as to the stringency of this warning. It is probable that Galilei was at the same time specifically forbidden to "hold, teach, or defend in any way, whether verbally or in writing," the obnoxious doctrine.

a long ocean voyage, and some astronomical means of determining the time was accordingly wanted. Galilei's idea was that if the movements of Jupiter's satellites, and in particular the eclipses which constantly occurred when a satellite passed into Jupiter's shadow, could be predicted, then a table could be prepared giving the times, according to some standard place, say Rome, at which the eclipses would occur, and a sailor by observing the local time of an eclipse and comparing it with the time given in the table could ascertain by how much his longitude differed from that of Rome. It is, however, doubtful whether the movements of Jupiter's satellites could at that time be predicted accurately enough to make the method practically useful, and in any case the negotiations came to nothing.

In 1618 three comets appeared, and Galilei was soon drawn into a controversy on the subject with a Jesuit of the name of Grassi. The controversy was marked by the personal bitterness which was customary, and soon developed so as to include larger questions of philosophy and astronomy. Galilei's final contribution to it was published in 1623 under the title *Il Saggiatore* (The Assayer), which dealt incidentally with the Coppernican theory, though only in the indirect way which the edict of 1616 rendered necessary. In a characteristic passage, for example, Galilei says:—

> "Since the motion attributed to the earth, which I, as a pious and Catholic person, consider most false, and not to exist, accommodates itself so well to explain so many and such different phenomena, I shall not feel sure . . . that, false as it is, it may not just as deludingly correspond with the phenomena of comets";

and again, in speaking of the rival systems of Coppernicus and Tycho, he says:—

> "Then as to the Copernican hypothesis, if by the good fortune of us Catholics we had not been freed from error and our blindness illuminated by the Highest Wisdom, I do not believe that such grace and good fortune could have been obtained by means of the reasons and observations given by Tycho."

Although in scientific importance the *Saggiatore* ranks far below many others of Galilei's writings, it had a great reputation as a piece of brilliant controversial writing, and notwithstanding its thinly veiled Coppernicanism, the new Pope, Urban VIII., to whom it was dedicated, was so much pleased with it that he had it read aloud to him at meals. The book must, however, have strengthened the hands of Galilei's enemies, and it was probably with a view to counteracting their influence that he went to Rome next year, to pay his respects to Urban and congratulate him on his recent elevation. The visit was in almost every way a success; Urban granted to him several friendly interviews, promised a pension for his son, gave him several presents, and finally dismissed him with a letter of special recommendation to the new Grand Duke of Tuscany, who had shewn some signs of being less friendly to Galilei than his father. On the other hand, however, the Pope refused to listen to Galilei's request that the decree of 1616 should be withdrawn.

128. Galilei now set seriously to work on the great astronomical treatise, the *Dialogue on the Two Chief Systems of the World, the Ptolemaic and Coppernican*, which he had had in mind as long ago as 1610, and in which he proposed to embody most of his astronomical work and to collect all the available evidence bearing on the Coppernican controversy. The form of a dialogue was chosen, partly for literary reasons, and still more because it enabled him to present the Coppernican case as strongly as he wished through the mouths of some of the speakers, without necessarily identifying his own opinions with theirs. The manuscript was almost completed in 1629, and in the following year Galilei went to Rome to obtain the necessary licence for printing it. The censor had some alterations made and then gave the desired permission for printing at Rome, on condition that the book was submitted to him again before being finally printed off. Soon after Galilei's return to Florence the plague broke out, and quarantine difficulties rendered it almost necessary that the book should be printed at Florence instead of at Rome. This required a fresh licence, and the difficulty experienced in obtaining it shewed that the Roman censor was getting

more and more doubtful about the book. Ultimately, however, the introduction and conclusion having been sent to Rome for approval and probably to some extent re-written there, and the whole work having been approved by the Florentine censor, the book was printed and the first copies were ready early in 1632, bearing both the Roman and the Florentine *imprimatur*.

129. The Dialogue extends over four successive days, and is carried on by three speakers, of whom Salviati is a Coppernican and Simplicio an Aristotelian philosopher, while Sagredo is avowedly neutral, but on almost every occasion either agrees with Salviati at once or is easily convinced by him, and frequently joins in casting ridicule upon the arguments of the unfortunate Simplicio. Though many of the arguments have now lost their immediate interest, and the book is unduly long, it is still very readable, and the specimens of scholastic reasoning put into the mouth of Simplicio and the refutation of them by the other speakers strike the modern reader as excellent fooling.

Many of the arguments used had been published by Galilei in earlier books, but gain impressiveness and cogency by being collected and systematically arranged. The Aristotelian dogma of the immutability of the celestial bodies is once more belaboured, and shewn to be not only inconsistent with observations of the moon, the sun, comets, and new stars, but to be in reality incapable of being stated in a form free from obscurity and self-contradiction. The evidence in favour of the earth's motion derived from the existence of Jupiter's satellites and from the undoubted phases of Venus, from the suspected phases of Mercury and from the variations in the apparent size of Mars, are once more insisted on. The greater simplicity of the Coppernican explanation of the daily motion of the celestial sphere and of the motion of the planets is forcibly urged and illustrated in detail. It is pointed out that on the Coppernican hypothesis all motions of revolution or rotation take place in the same direction (from west to east), whereas the Ptolemaic hypothesis requires some to be in one direction, some in another. Moreover the apparent daily motion of the stars, which appears simple enough if the stars are regarded as rigidly attached to a material sphere, is shewn in a quite different aspect if, as even Simplicio admits, no such sphere exists, and each star moves in some sense independently. A star near the pole must then be supposed to move far more slowly than one near the equator, since it describes a much smaller circle in the same time; and further—an argument very characteristic of Galilei's ingenuity in drawing conclusions from known facts—owing to the precession of the equinoxes (chapter II., § 42, and IV., § 84) and the consequent change of the position of the pole among the stars, some of those stars which in Ptolemy's time were describing very small circles, and therefore moving slowly, must now be describing large ones at a greater speed, and *vice versa*. An extremely complicated adjustment of motions becomes therefore necessary to account for observations which Coppernicus explained adequately by the rotation of the earth and a simple displacement of its axis of rotation.

Salviati deals also with the standing difficulty that the annual motion of the earth ought to cause a corresponding apparent motion of the stars, and that if the stars be assumed so far off that this motion is imperceptible, then some of the stars themselves must be at least as large as the earth's orbit round the sun. Salviati points out that the apparent or angular magnitudes of the fixed stars, avowedly difficult to determine, are in reality almost entirely illusory, being due in great part to an optical effect known as **irradiation,** in virtue of which a bright object always tends to appear enlarged; * and that there is in consequence no reason to suppose the stars nearly as large as they might otherwise be thought to be. It is suggested also that the most promising way of detecting the annual motion of stars resulting from the motion of the earth would be by observing the relative displacement of two stars close together in the sky (and therefore nearly in the same direction), of which one might be presumed from its greater

* This is illustrated by the well-known optical illusion whereby a white circle on a black background appears larger than an equal black one on a white background. The apparent size of the hot filament in a modern incandescent electric lamp is another good illustration.

brightness to be nearer than the other. It is, for example, evident that if, in the figure, E, E′ represent two positions of the earth in its path round the sun, and A, B two stars at different distances, but nearly in the same direction, then to the observer at E the star A appears to the left of B, whereas six months afterwards, when the observer is at E′, A appears to the right of B. Such a motion of one star with respect to another close to it would be much more easily observed than an alteration of the same amount in the distance of the star from some standard point such as the pole. Salviati points out that accurate observations of

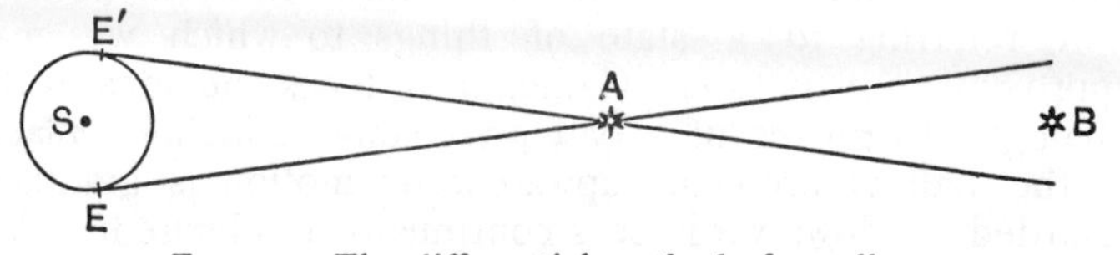

FIG. 57.—The differential method of parallax.

this kind had not been made, and that the telescope might be of assistance for the purpose. This method, known as the **double-star** or **differential method of parallax,** was in fact the first to lead—two centuries later—to a successful detection of the motion in question (chapter XIII., § 278).

130. Entirely new ground is broken in the *Dialogue* when Galilei's discoveries of the laws of motion of bodies are applied to the problem of the earth's motion. His great discovery, which threw an entirely new light on the mechanics of the solar system, was substantially the law afterwards given by Newton as the first of his three laws of motion, in the form: *Every body continues in its state of rest or of uniform motion in a straight line, except in so far as it is compelled by force applied to it to change that state.* Putting aside for the present any discussion of *force*, a conception first made really definite by Newton, and only imperfectly grasped by Galilei, we may interpret this law as meaning that a body has no more inherent tendency to diminish its motion or to stop than it has to increase its motion or to start, and that any alteration in either the speed or the direction of a body's motion is to be explained by the action on it of some other body, or at any rate by

some other assignable cause. Thus a stone thrown along a road comes to rest on account of the friction between it and the ground, a ball thrown up into the air ascends more and more slowly and then falls to the ground on account of that attraction of the earth on it which we call its weight. As it is impossible to entirely isolate a body from all others, we cannot experimentally realise the state of things in which a body goes on moving indefinitely in the same direction and at the same rate; it may, however, be shewn that the more we remove a body from the influence of others, the less alteration is there in its motion. The law is therefore, like most scientific laws, an abstraction referring to a state of things to which we may approximate in nature. Galilei introduces the idea in the Dialogue by means of a ball on a smooth inclined plane. If the ball is projected upwards, its motion is gradually retarded; if downwards, it is continually accelerated. This is true if the plane is fairly smooth—like a well-planed plank—and the inclination of the plane not very small. If we imagine the experiment performed on an ideal plane, which is supposed *perfectly* smooth, we should expect the same results to follow, however small the inclination of the plane. Consequently, if the plane were quite level, so that there is no distinction between up and down, we should expect the motion to be neither retarded nor accelerated, but to continue without alteration. Other more familiar examples are also given of the tendency of a body, when once in motion, to continue in motion, as in the case of a rider whose horse suddenly stops, or of bodies in the cabin of a moving ship which have no tendency to lose the motion imparted to them by the ship, so that, *e.g.*, a body falls down to all appearances exactly as if the rest of the cabin were at rest, and therefore, in reality, while falling retains the forward motion which it shares with the ship and its contents. Salviati states also that—contrary to general belief—a stone dropped from the mast-head of a ship in motion falls at the foot of the mast, not behind it, but there is no reference to the experiment having been actually performed.

This mechanical principle being once established, it becomes easy to deal with several common objections to the supposed motion of the earth. The case of a stone dropped from the top of a tower, which if the earth be in reality moving rapidly from west to east might be expected to fall to the west in its descent, is easily shewn to be exactly parallel to the case of a stone dropped from the mast-head of a ship in motion. The motion towards the east, which the stone when resting on the tower shares with the tower and the earth, is not destroyed in its descent, and it is therefore entirely in accordance with the Coppernican theory that the stone should fall as it does at the foot of the tower.* Similarly, the fact that the clouds, the atmosphere in general, birds flying in it, and loose objects on the surface of the earth, shew no tendency to be left behind as the earth moves rapidly eastward, but are apparently unaffected by the motion of the earth, is shewn to be exactly parallel to the fact that the flies in a ship's cabin and the loose objects there are in no way affected by the uniform onward motion of the ship (though the irregular motions of pitching and rolling do affect them). The stock objection that a cannon-ball shot westward should, on the Coppernican hypothesis, carry farther than one shot eastward under like conditions, is met in the same way; but it is further pointed out that, owing to the imperfection of gunnery practice, the experiment could not really be tried accurately enough to yield any decisive result.

The most unsatisfactory part of the *Dialogue* is the fourth day's discussion, on the tides, of which Galilei suggests with great confidence an explanation based merely on the motion of the earth, while rejecting with scorn the suggestion of Kepler and others—correct as far as it went—that they were caused by some influence emanating from the moon. It is hardly to be wondered at that the rudimentary mechanical and mathematical knowledge at Galilei's command should not have enabled him to deal

* Actually, since the top of the tower is describing a slightly larger circle than its foot, the stone is at first moving eastward slightly faster than the foot of the tower, and therefore should reach the ground slightly to the *east* of it. This displacement is, however, very minute, and can only be detected by more delicate experiments than any devised by Galilei.

correctly with a problem of which the vastly more powerful resources of modern science can only give an imperfect solution (cf. chapter XI., § 248, and chapter XIII., § 292).

131. The book as a whole was in effect, though not in form, a powerful—indeed unanswerable—plea for Coppernicanism. Galilei tried to safeguard his position, partly by the use of dialogue, and partly by the very remarkable introduction, which was not only read and approved by the licensing authorities, but was in all probability in part the composition of the Roman censor and of the Pope. It reads to us like a piece of elaborate and thinly veiled irony, and it throws a curious light on the intelligence or on the seriousness of the Pope and the censor, that they should have thus approved it :—

"Judicious reader, there was published some years since in *Rome* a salutiferous Edict, that, for the obviating of the dangerous Scandals of the present Age, imposed a reasonable Silence upon the Pythagorean Opinion of the Mobility of the Earth. There want not such as unadvisedly affirm, that the Decree was not the production of a sober Scrutiny, but of an illformed passion; and one may hear some mutter that Consultors altogether ignorant of Astronomical observations ought not to clipp the wings of speculative wits with rash prohibitions. My zeale cannot keep silence when I hear these inconsiderate complaints. I thought fit, as being thoroughly acquainted with that prudent Determination, to appear openly upon the Theatre of the World as a Witness of the naked Truth. I was at that time in *Rome*, and had not only the audiences, but applauds of the most Eminent Prelates of that Court; nor was that Decree published without Previous Notice given me thereof. Therefore it is my resolution in the present case to give Foreign Nations to see, that this point is as well understood in *Italy*, and particularly in *Rome*, as Transalpine Diligence can imagine it to be: and collecting together all the proper speculations that concerne the *Copernican Systeme* to let them know, that the notice of all preceded the Censure of the *Roman Court*; and that there proceed from this Climate not only Doctrines for the health of the Soul, but also ingenious Discoveries for the recreating of the Mind. . . . I hope that by these considerations the world will know, that if other Nations have Navigated more than we, we have not studied less than they; and that our returning to assert the Earth's stability, and to take the contrary only for a Mathematical *Capriccio*, proceeds not from inadvertency of what others have thought thereof, but (had one no other inducements), from these reasons that Piety, Religion, the Knowledge of the Divine Omnipotency, and a consciousness of the incapacity of man's understanding dictate unto us." *

132. Naturally Galilei's many enemies were not long in penetrating these thin disguises, and the immense success of the book only intensified the opposition which it excited; the Pope appears to have been persuaded that Simplicio—the butt of the whole dialogue—was intended for himself, a supposed insult which bitterly wounded his vanity; and it was soon evident that the publication of the book could not be allowed to pass without notice. In June 1632 a special commission was appointed to inquire into the matter—an unusual procedure, probably meant as a mark of consideration for Galilei—and two months later the further issue of copies of the book was prohibited, and in September a papal mandate was issued requiring Galilei to appear personally before the Inquisition. He was evidently frightened by the summons, and tried to avoid compliance through the good offices of the Tuscan court and by pleading his age and infirmities, but after considerable delay, at the end of which the Pope issued instructions to bring him if necessary by force and in chains, he had to submit, and set off for Rome early in 1633. Here he was treated with unusual consideration, for whereas in general even the most eminent offenders under trial by the Inquisition were confined in its prisons, he was allowed to live with his friend Niccolini, the Tuscan ambassador, throughout the trial, with the exception of a period of about three weeks, which he spent within the buildings of the Inquisition, in comfortable rooms belonging to one of the officials, with permission to correspond with his friends, to take exercise in the garden, and other privileges. At his first hearing before the Inquisition, his reply to the charge of having violated the decree of 1616 (§ 126) was that he had not understood that the decree or the admonition given to him forbade the teaching of the Coppernican theory as a mere "hypothesis," and that his book had not upheld the doctrine in any other way. Between his first and second hearing the Commission, which had been

* From the translation by Salusbury, in Vol. I. of his *Mathematical Collections*.

examining his book, reported that it did distinctly defend and maintain the obnoxious doctrines, and Galilei, having been meanwhile privately advised by the Commissary-General of the Inquisition to adopt a more submissive attitude, admitted at the next hearing that on reading his book again he recognised that parts of it gave the arguments for Coppernicanism more strongly than he had at first thought. The pitiable state to which he had been reduced was shewn by the offer which he now made to write a continuation to the Dialogue which should as far as possible refute his own Coppernican arguments. At the final hearing on June 21st he was examined under threat of torture,* and on the next day he was brought up for sentence. He was convicted "of believing and holding the doctrines—false and contrary to the Holy and Divine Scriptures—that the sun is the centre of the world, and that it does not move from east to west, and that the earth does move and is not the centre of the world; also that an opinion can be held and supported as probable after it has been declared and decreed contrary to the Holy Scriptures." In punishment, he was required to "abjure, curse, and detest the aforesaid errors," the abjuration being at once read by him on his knees; and was further condemned to the "formal prison of the Holy Office" during the pleasure of his judges, and required to repeat the seven penitential psalms once a week for three years. On the following day the Pope changed the sentence of imprisonment into confinement at a country-house near Rome belonging to the Grand Duke, and Galilei moved there on June 24th.† On petitioning to be allowed to return to Florence, he was at first allowed to go as far as Siena, and at the end of the year was permitted to retire to his country-house at Arcetri near Florence, on condition of not leaving it for the future without permission, while his intercourse with scientific and other friends was jealously watched.

* The official minute is: *Et ei dicto quod dicat veritatem, alias devenietur ad torturam.*

† The three days June 21–24 are the only ones which Galilei *could* have spent in an actual prison, and there seems no reason to suppose that they were spent elsewhere than in the comfortable rooms in which it is known that he lived during most of April.

GALILEI.

[*To face p.* 171.

The story of the trial reflects little credit either on Galilei or on his persecutors. For the latter, it may be urged that they acted with unusual leniency considering the customs of the time; and it is probable that many of those who were concerned in the trial were anxious to do as little injury to Galilei as possible, but were practically forced by the party personally hostile to him to take some notice of the obvious violation of the decree of 1616. It is easy to condemn Galilei for cowardice, but it must be borne in mind, on the one hand, that he was at the time nearly seventy, and much shaken in health, and, on the other, that the Roman Inquisition, if not as cruel as the Spanish, was a very real power in the early 17th century; during Galilei's life-time (1600) Giordano Bruno had been burnt alive at Rome for writings which, in addition to containing religious and political heresies, supported the Coppernican astronomy and opposed the traditional Aristotelian philosophy. Moreover, it would be unfair to regard his submission as due merely to considerations of personal safety, for—apart from the question whether his beloved science would have gained anything by his death or permanent imprisonment—there can be no doubt that Galilei was a perfectly sincere member of his Church, and although he did his best to convince individual officers of the Church of the correctness of his views, and to minimise the condemnation of them passed in 1616, yet he was probably prepared, when he found that the condemnation was seriously meant by the Pope, the Holy Office, and others, to believe that in some senses at least his views must be wrong, although, as a matter of observation and pure reason, he was unable to see how or why. In fact, like many other excellent people, he kept water-tight compartments in his mind, respect for the Church being in one and scientific investigation in another.

Copies of the sentence on Galilei and of his abjuration were at once circulated in Italy and in Roman Catholic circles elsewhere, and a decree of the Congregation of the Index was also issued adding the *Dialogue* to the three Coppernican books condemned in 1616, and to Kepler's *Epitome* of the Coppernican Astronomy (chapter VII., § 145), which had been put on the *Index* shortly afterwards. It may be of interest to note that these five books still remained in the edition of the *Index of Prohibited Books* which was issued in 1819 (with appendices dated as late as 1821), but disappeared from the next edition, that of 1835.

133. The rest of Galilei's life may be described very briefly. With the exception of a few months, during which he was allowed to be at Florence for the sake of medical treatment, he remained continuously at Arcetri, evidently pretty closely watched by the agents of the Holy Office, much restricted in his intercourse with his friends, and prevented from carrying on his studies in the directions which he liked best. He was moreover very infirm, and he was afflicted by domestic troubles, especially by the death in 1634 of his favourite child, a nun in a neighbouring convent. But his spirit was not broken, and he went on with several important pieces of work, which he had begun earlier in his career. He carried a little further the study of his beloved Medicean Planets and of the method of finding longitude based on their movements (§ 127), and negotiated on the subject with the Dutch government. He made also a further discovery relating to the moon, of sufficient importance to deserve a few words of explanation.

It had long been well known that as the moon describes her monthly path round the earth we see the same markings substantially in the same positions on the disc, so that substantially the same face of the moon is turned towards the earth. It occurred to Galilei to inquire whether this was accurately the case, or whether, on the contrary, some change in the moon's disc could be observed. He saw that if, as seemed likely, the line joining the centres of the earth and moon always passed through the same point on the moon's surface, nevertheless certain alterations in an observer's position on the earth would enable him to see different portions of the moon's surface from time to time. The simplest of these alterations is due to the daily motion of the earth. Let us suppose for simplicity that the observer is on the earth's equator, and that the moon is at the time in the celestial equator. Let the larger circle in fig. 58 represent the earth's equator, and the smaller circle the section of the moon by the plane of the equator. Then in about 12 hours the earth's rotation carries the

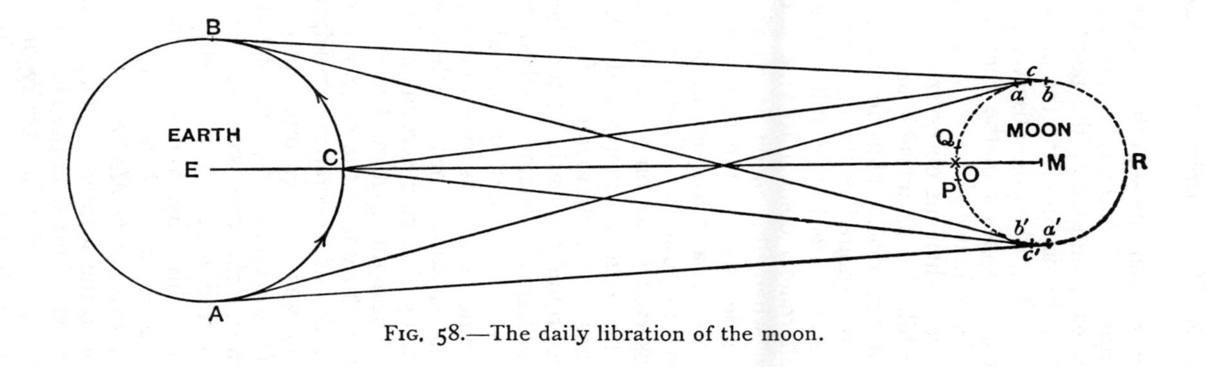

Fig. 58.—The daily libration of the moon.

observer from A, where he sees the moon rising, to B, where he sees it setting. When he is at C, on the line joining the centres of the earth and moon, the point O appears to be in the centre of the moon's disc, and the portion *c* O *c'* is visible, *c* R *c'* invisible. But when the observer is at A, the point P, on the right of O, appears in the centre, and the portion *a* P *a'* is visible, so that *c' a'* is now visible and *a c* invisible. In the same way, when the observer is at B, he can see the portion *c b*, while *b' c'* is invisible and Q appears to be in the centre of the disc. Thus in the course of the day the portion *a* O *b'* (dotted in the figure) is constantly visible and *b* R *a'* (also dotted) constantly invisible, while *a c b* and *a' c' b'* alternately come into view and disappear. In other words, when the moon is rising we see a little more of the side which is the then uppermost, and when she is setting we see a little more of the other side which is uppermost in this position. A similar explanation applies when the observer is not on the earth's equator, but the geometry is slightly more complicated. In the same way, as the moon passes from south to north of the equator and back as she revolves round the earth, we see alternately more and less of the northern and southern half of the moon. This set of changes—the simplest of several somewhat similar ones which are now known as **librations** of the moon—being thus thought of as likely to occur, Galilei set to work to test their existence by observing certain markings of the moon usually visible near the edge, and at once detected alterations in their distance from the edge, which were in general accordance with his theoretical anticipations. A more precise inquiry was however interrupted by failing sight, culminating (at the end of 1636) in total blindness.

But the most important work of these years was the completion of the great book, in which he summed up and completed his discoveries in mechanics, *Mathematical Discourses and Demonstrations concerning Two New Sciences, relating to Mechanics and to Local Motion.* It was written in the form of a dialogue between the same three speakers who figured in the Dialogue on the Systems, but is distinctly inferior in literary merit to the earlier work. We have here no concern with a large part of the book, which deals with the conditions under which bodies are kept at rest by forces applied to them (statics), and certain problems relating to the resistance of bodies to fracture and to bending, though in both of these subjects Galilei broke new ground. More important astronomically—and probably intrinsically also—is what he calls the science of local motion,* which deals with the motion of bodies. He builds up on the basis of his early experiments (§ 116) a theory of falling bodies, in which occurs for the first time the important idea of **uniformly accelerated motion**, or **uniform acceleration**, *i.e.* motion in which the moving body receives in every equal interval of time an equal increase of velocity. He shews that the motion of a falling body is—except in so far as it is disturbed by the air—of this nature, and that, as already stated, the motion is the same for all bodies, although his numerical estimate is not at all accurate.† From this fundamental law he works out a number of mathematical deductions, connecting the space fallen through, the velocity, and the time elapsed, both for the case of a body falling freely and for one falling down an inclined plane. He gives also a correct elementary theory of projectiles, in the course of which he enunciates more completely than before the law of inertia already referred to (§ 130), although Galilei's form is still much less general than Newton's:—

Conceive a body projected or thrown along a horizontal plane, all impediments being removed. Now it is clear by what we have said before at length that its motion will be uniform and perpetual along the said plane, if the plane extend indefinitely.

In connection with projectiles, Galilei also appears to realise that a body may be conceived as having motions in two different directions simultaneously, and that each may be treated as independent of the other, so that, for example, if a bullet is shot horizontally out of a gun, its downward motion, due to its weight, is unaffected

* Equivalent to portions of the subject now called *dynamics* or (more correctly) *kinematics* and *kinetics*.

† He estimates that a body falls in a second a distance of 4 "bracchia," equivalent to about 8 feet, the true distance being slightly over 16.

by its horizontal motion, and consequently it reaches the ground at the same time as a bullet simply allowed to drop; but Galilei gives no general statement of this principle, which was afterwards embodied by Newton in his Second Law of Motion.

The treatise on the *Two New Sciences* was finished in 1636, and, since no book of Galilei's could be printed in Italy, it was published after some little delay at Leyden in 1638. In the same year his eyesight, which he had to some extent recovered after his first attack of blindness, failed completely, and four years later (January 8th, 1642) the end came.

134. Galilei's chief scientific discoveries have already been noticed. The telescopic discoveries, on which much of his popular reputation rests, have probably attracted more than their fair share of attention; many of them were made almost simultaneously by others, and the rest, being almost inevitable results of the invention of the telescope, could not have been delayed long. But the skilful use which Galilei made of them as arguments for the Coppernican system, the no less important support which his dynamical discoveries gave to the same cause, the lucidity and dialectic brilliance with which he marshalled the arguments in favour of his views and demolished those of his opponents, together with the sensational incidents of his persecution, formed conjointly a contribution to the Coppernican controversy which was in effect decisive. Astronomical text-books still continued to give side by side accounts of the Ptolemaic and of the Coppernican systems, and the authors, at any rate if they were good Roman Catholics, usually expressed, in some more or less perfunctory way, their adherence to the former, but there was no real life left in the traditional astronomy; new advances in astronomical theory were all on Coppernican lines, and in the extensive scientific correspondence of Newton and his contemporaries the truth of the Coppernican system scarcely ever appears as a subject for discussion.

Galilei's dynamical discoveries, which are only in part of astronomical importance, are in many respects his most remarkable contribution to science. For whereas in astronomy he was building on foundations laid by previous generations, in dynamics it was no question of improving or developing an existing science, but of creating a new one. From his predecessors he inherited nothing but erroneous traditions and obscure ideas; and when these had been discarded, he had to arrive at clear fundamental notions, to devise experiments and make observations, to interpret his experimental results, and to follow out the mathematical consequences of the simple laws first arrived at. The positive results obtained may not appear numerous, if viewed from the standpoint of our modern knowledge, but they sufficed to constitute a secure basis for the superstructure which later investigators added.

It is customary to associate with our countryman Francis Bacon (1561–1627) the reform in methods of scientific discovery which took place during the seventeenth century, and to which much of the rapid progress in the natural sciences made since that time must be attributed. The value of Bacon's theory of scientific discovery is very differently estimated by different critics, but there can be no question of the singular ill-success which attended his attempts to apply it in particular cases, and it may fairly be questioned whether the scientific methods constantly referred to incidentally by Galilei, and brilliantly exemplified by his practice, do not really contain a large part of what is valuable in the Baconian philosophy of science, while at the same time avoiding some of its errors. Reference has already been made on several occasions to Galilei's protests against the current method of dealing with scientific questions by the interpretation of passages in Aristotle, Ptolemy, or other writers; and to his constant insistence on the necessity of appealing directly to actual observation of facts. But while thus agreeing with Bacon in these essential points, he differed from him in the recognition of the importance, both of deducing new results from established ones by mathematical or other processes of exact reasoning, and of using such deductions, when compared with fresh experimental results, as a means of verifying hypotheses provisionally adopted. This method of proof, which lies at the base of nearly all important scientific discovery, can hardly be described better than by

Galilei's own statement of it, as applied to a particular case :—

"Let us therefore take this at present as a *Postulatum*, the truth whereof we shall afterwards find established, when we shall see other conclusions built upon this *Hypothesis*, to answer and most exactly to agree with Experience." *

* *Two New Sciences*, translated by Weston, p. 255.

CHAPTER VII.

KEPLER.

"His celebrated laws were the outcome of a lifetime of speculation, for the most part vain and groundless. . . . But Kepler's name was destined to be immortal, on account of the patience with which he submitted his hypotheses to comparison with observation, the candour with which he acknowledged failure after failure, and the perseverance and ingenuity with which he renewed his attack upon the riddles of nature."

JEVONS.

135. JOHN KEPLER, or Keppler,* was born in 1571, seven years after Galilei, at Weil in Würtemberg; his parents were in reduced circumstances, though his father had some claims to noble descent. Though Weil itself was predominantly Roman Catholic, the Keplers were Protestants, a fact which frequently stood in Kepler's way at various stages of his career. But the father could have been by no means zealous in his faith, for he enlisted in the army of the notorious Duke of Alva when it was engaged in trying to suppress the revolt of the Netherlands against Spanish persecution.

John Kepler's childhood was marked by more than the usual number of illnesses, and his bodily weaknesses, combined with a promise of great intellectual ability, seemed to point to the Church as a suitable career for him. After attending various elementary schools with great irregularity —due partly to ill-health, partly to the requirements of

* The astronomer appears to have used both spellings of his name almost indifferently. For example, the title-page of his most important book, the *Commentaries on the Motions of Mars* (§ 141), has the form Kepler, while the dedication of the same book is signed Keppler.

manual work at home—he was sent in 1584 at the public expense to the monastic school at Adelberg, and two years later to the more advanced school or college of the same kind at Maulbronn, which was connected with the University of Tübingen, then one of the great centres of Protestant theology.

In 1588 he obtained the B.A. degree, and in the following year entered the philosophical faculty at Tübingen.

There he came under the influence of Maestlin, the professor of mathematics, by whom he was in private taught the principles of the Coppernican system, though the professorial lectures were still on the traditional lines.

In 1591 Kepler graduated as M.A., being second out of fourteen candidates, and then devoted himself chiefly to the study of theology.

136. In 1594, however, the Protestant Estates of Styria applied to Tübingen for a lecturer on mathematics (including astronomy) for the high school of Gratz, and the appointment was offered to Kepler. Having no special knowledge of the subject and as yet no taste for it, he naturally hesitated about accepting the offer, but finally decided to do so, expressly stipulating, however, that he should not thereby forfeit his claims to ecclesiastical preferment in Würtemberg. The demand for higher mathematics at Gratz seems to have been slight; during his first year Kepler's mathematical lectures were attended by very few students, and in the following year by none, so that to prevent his salary from being wasted he was set to teach the elements of various other subjects. It was moreover one of his duties to prepare an annual almanack or calendar, which was expected to contain not merely the usual elementary astronomical information such as we are accustomed to in the calendars of to-day, but also astrological information of a more interesting character, such as predictions of the weather and of remarkable events, guidance as to unlucky and lucky times, and the like. Kepler's first calendar, for the year 1595, contained some happy weather-prophecies, and he acquired accordingly a considerable popular reputation as a prophet and astrologer, which remained throughout his life.

Meanwhile his official duties evidently left him a good

deal of leisure, which he spent with characteristic energy in acquiring as thorough a knowledge as possible of astronomy, and in speculating on the subject.

According to his own statement, "there were three things in particular, *viz.* the number, the size, and the motion of the heavenly bodies, as to which he searched zealously for reasons why they were as they were and not otherwise"; and the results of a long course of wild speculation on the subject led him at last to a result with which he was immensely pleased—a numerical relation connecting the distances of the several planets from the sun with certain geometrical bodies known as the regular solids (of which the cube is the best known), a relation which is not very accurate numerically, and is of absolutely no significance or importance.* This discovery, together with a detailed account of the steps which led to it, as well as of a number of other steps which led nowhere, was published in 1596 in a book a portion of the title of which may be translated as *The Forerunner of Dissertations on the Universe, containing the Mystery of the Universe*, commonly referred to as the *Mysterium Cosmographicum*. The contents were probably much more attractive and seemed more valuable to Kepler's contemporaries than to us, but even to those who were least inclined to attach weight to its conclusions, the book shewed evidence of considerable astronomical knowledge and very great ingenuity; and both Tycho Brahe and Galilei, to whom copies were sent, recognised in the author a rising astronomer likely to do good work.

137. In 1597 Kepler married. In the following year the religious troubles, which had for some years been steadily growing, were increased by the action of the Archduke Ferdinand of Austria (afterwards the Emperor Ferdinand II.), who on his return from a pilgrimage to Loretto started a

* The regular solids being taken in the order: cube, tetrahedron, dodecahedron, icosahedron, octohedron, and of such magnitude that a sphere can be circumscribed to each and at the same time inscribed in the preceding solid of the series, then the radii of the six spheres so obtained were shewn by Kepler to be approximately proportional to the distances from the sun of the six planets Saturn, Jupiter, Mars, Earth, Venus, and Mercury.

vigorous persecution of Protestants in his dominions, one step in which was an order that all Protestant ministers and teachers in Styria should quit the country at once (1598). Kepler accordingly fled to Hungary, but returned after a few weeks by special permission of the Archduke, given apparently on the advice of the Jesuit party, who had hopes of converting the astronomer. Kepler's hearers had, however, mostly been scattered by the persecution, it became difficult to ensure regular payment of his stipend, and the rising tide of Catholicism made his position increasingly insecure. Tycho's overtures were accordingly welcome, and in 1600 he paid a visit to him, as already described (chapter v., § 108), at Benatek and Prague. He returned to Gratz in the autumn, still uncertain whether to accept Tycho's offer or not, but being then definitely dismissed from his position at Gratz on account of his Protestant opinions, he returned finally to Prague at the end of the year.

138. Soon after Tycho's death Kepler was appointed his successor as mathematician to the Emperor Rudolph (1602), but at only half his predecessor's salary, and even this was paid with great irregularity, so that complaints as to arrears and constant pecuniary difficulties played an important part in his future life, as they had done during the later years at Gratz. Tycho's instruments never passed into his possession, but as he had little taste or skill for observing, the loss was probably not great; fortunately, after some difficulties with the heirs, he secured control of the greater part of Tycho's incomparable series of observations, the working up of which into an improved theory of the solar system was the main occupation of the next 25 years of his life. Before, however, he had achieved any substantial result in this direction, he published several minor works—for example, two pamphlets on a new star which appeared in 1604, and a treatise on the applications of optics to astronomy (published in 1604 with a title beginning *Ad Vitellionem Paralipomena quibus Astronomiae Pars Optica Traditur . . .*), the most interesting and important part of which was a considerable improvement in the theory of astronomical refraction (chapter II., § 46, and chapter v., § 110). A later optical treatise (the *Dioptrice* of 1611) contained a

KEPLER. [*To face p.* 183.

suggestion for the construction of a telescope by the use of two convex lenses, which is the form now most commonly adopted, and is a notable improvement on Galilei's instrument (chapter VI., § 118), one of the lenses of which is concave; but Kepler does not seem himself to have had enough mechanical skill to actually construct a telescope on this plan, or to have had access to workmen capable of doing so for him; and it is probable that Galilei's enemy Scheiner (chapter VI., §§ 124, 125) was the first person to use (about 1613) an instrument of this kind.

139. It has already been mentioned (chapter V., § 108) that when Tycho was dividing the work of his observatory among his assistants he assigned to Kepler the study of the planet Mars, probably as presenting more difficulties than the subjects assigned to the others. It had been known since the time of Coppernicus that the planets, including the earth, revolved round the sun in paths that were at any rate not very different from circles, and that the deviations from uniform circular motion could be represented roughly by systems of eccentrics and epicycles. The deviations from uniform circular motion were, however, notably different in amount in different planets, being, for example, very small in the case of Venus, relatively large in the case of Mars, and larger still in that of Mercury. The *Prussian Tables* calculated by Reinhold on a Coppernican basis (chapter V., § 94) were soon found to represent the actual motions very imperfectly, errors of 4° and 5° having been noted by Tycho and Kepler, so that the principles on which the tables were calculated were evidently at fault.

The solution of the problem was clearly more likely to be found by the study of a planet in which the deviations from circular motion were as great as possible. In the case of Mercury satisfactory observations were scarce, whereas in the case of Mars there was an abundant series recorded by Tycho, and hence it was true insight on Tycho's part to assign to his ablest assistant this particular planet, and on Kepler's to continue the research with unwearied patience. The particular system of epicycles used by Coppernicus (chapter IV., § 87) having proved defective, Kepler set to work to devise other geometrical schemes, the results of which could be compared with observation. The places of Mars as seen on the sky being a combined result of the motions of Mars and of the earth in their respective orbits round the sun, the irregularities of the two orbits were apparently inextricably mixed up, and a great simplification was accordingly effected when Kepler succeeded, by an ingenious combination of observations taken at suitable times, in disentangling the irregularities due to the earth from those due to the motion of Mars itself, and thus rendering it possible to concentrate his attention on the latter. His fertile imagination suggested hypothesis after hypothesis, combination after combination of eccentric, epicycle, and equant; he calculated the results of each and compared them rigorously with observation; and at one stage he arrived at a geometrical scheme which was capable of representing the observations with errors not exceeding 8′.* A man of less intellectual honesty, or less convinced of the necessity of subordinating theory to fact when the two conflict, might have rested content with this degree of accuracy, or might have supposed Tycho's refractory observations to be in error. Kepler, however, thought otherwise:—

"Since the divine goodness has given to us in Tycho Brahe a most careful observer, from whose observations the error of 8′ is shewn in this calculation, . . . it is right that we should with gratitude recognise and make use of this gift of God. . . . For if I could have treated 8′ of longitude as negligible I should have already corrected sufficiently the hypothesis . . . discovered in chapter XVI. But as they could not be neglected, these 8′ alone have led the way towards the complete reformation of astronomy, and have been made the subject-matter of a great part of this work." †

140. He accordingly started afresh, and after trying a variety of other combinations of circles decided that the path of Mars must be an oval of some kind. At first he was inclined to believe in an egg-shaped oval, larger at one end than at the other, but soon had to abandon this idea. Finally

* Two stars 4′ apart only just appear distinct to the naked eye of a person with average keenness of sight.

† *Commentaries on the Motions of Mars*, Part II., end of chapter XIX.

he tried the simplest known oval curve, the **ellipse**,* and found to his delight that it satisfied the conditions of the problem, if the sun were taken to be at a focus of the ellipse described by Mars.

It was further necessary to formulate the law of variation of the rate of motion of the planet in different parts of its orbit. Here again Kepler tried a number of hypotheses, in the course of which he fairly lost his way in the intricacies of the mathematical questions involved, but fortunately arrived, after a dubious process of compensation of errors, at a simple law which agreed with observation. He found that the planet moved fast when near the sun and slowly when distant from it, in such a way that the area described or swept out in any time by the line joining the sun to Mars was always proportional to the time. Thus in fig. 60† the motion of Mars is most rapid at the point A nearest to the focus S where the sun is, least rapid at A′, and the

* An ellipse is one of several curves, known as **conic sections**, which can be formed by taking a section of a cone, and may also be defined as a curve the sum of the distances of any point on which from two fixed points inside it, known as the **foci**, is always the same. Thus if, in the figure, S and H are the foci, and P, Q are *any* two

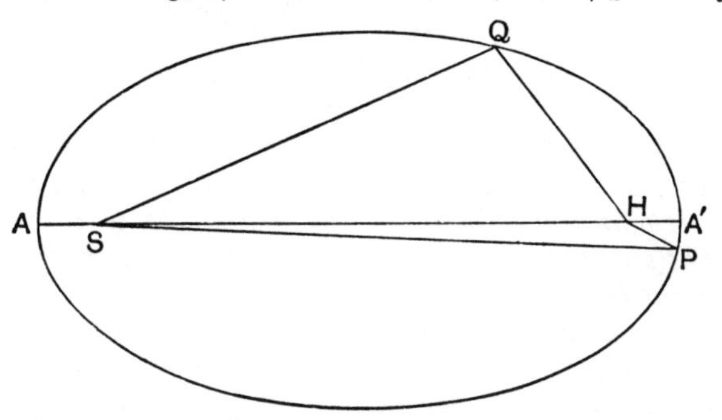

FIG. 59.—An ellipse.

points on the curve, then the distances S P, H P added together are equal to the distances S Q, Q H added together, and each sum is equal to the length A A′ of the ellipse. The ratio of the distance S H to the length A A′ is known as the **eccentricity**, and is a convenient measure of the extent to which the ellipse differs from a circle.

† The ellipse is *more* elongated than the actual path of Mars, an accurate drawing of which would be undistinguishable to the eye from a circle. The eccentricity is $\frac{1}{3}$ in the figure, that of Mars being $\frac{1}{10}$.

shaded and unshaded portions of the figure represent equal areas each corresponding to the motion of the planet during a month. Kepler's triumph at arriving at this result is expressed by the figure of victory in the corner of the diagram (fig. 61) which was used in establishing the last stage of his proof.

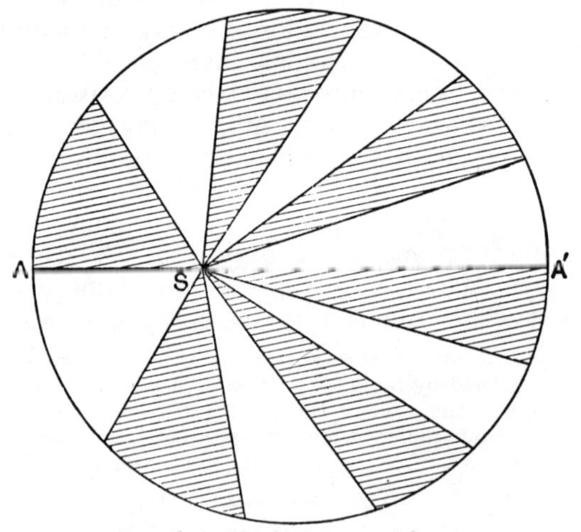

FIG. 60.—Kepler's second law.

141. Thus were established for the case of Mars the two important results generally known as Kepler's first two laws :—

1. *The planet describes an ellipse, the sun being in one focus.*

2. *The straight line joining the planet to the sun sweeps out equal areas in any two equal intervals of time.*

The full history of this investigation, with the results already stated and a number of developments and results of minor importance, together with innumerable digressions and quaint comments on the progress of the inquiry, was published in 1609 in a book of considerable length, the *Commentaries on the Motions of Mars.**

142. Although the two laws of planetary motion just given were only fully established for the case of Mars,

* *Astronomia Nova* αἰτιολογητος *seu Physica Coelestis, tradita Commentariis de Motibus Stellae Martis. Ex Observationibus G. V. Tychonis Brahe.*

Kepler stated that the earth's path also must be an oval of some kind, and was evidently already convinced—aided by his firm belief in the harmony of Nature—that all the planets moved in accordance with the same laws. This view is indicated in the dedication of the book to the Emperor Rudolph, which gives a fanciful account of the work as a

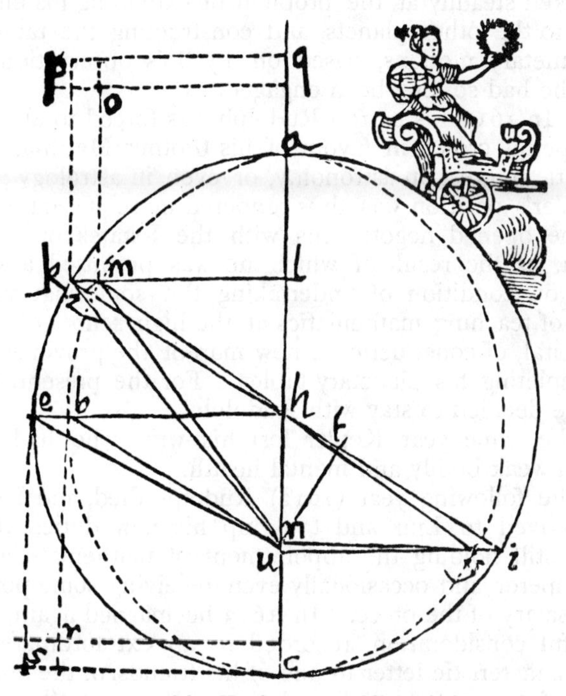

FIG. 61.—Diagram used by Kepler to establish his laws of planetary motion. From the *Commentaries on Mars*.

struggle against the rebellious War-God Mars, as the result of which he is finally brought captive to the feet of the Emperor and undertakes to live for the future as a loyal subject. As, however, he has many relations in the ethereal spaces—his father Jupiter, his grandfather Saturn, his dear sister Venus, his faithful brother Mercury—and he yearns for them and they for him *on account of the similarity of their habits*, he entreats the Emperor to send out an expedition as soon as possible to capture them also, and with that object to provide Kepler with the "sinews of war" in order that he may equip a suitable army.

Although the money thus delicately asked for was only supplied very irregularly, Kepler kept steadily in view the expedition for which it was to be used, or, in plainer words, he worked steadily at the problem of extending his elliptic theory to the other planets, and constructing the tables of the planetary motions, based on Tycho's observations, at which he had so long been engaged.

143. In 1611 his patron Rudolph was forced to abdicate the imperial crown in favour of his brother Matthias, who had little interest in astronomy, or even in astrology; and as Kepler's position was thus rendered more insecure than ever, he opened negotiations with the Estates of Upper Austria, as the result of which he was promised a small salary, on condition of undertaking the somewhat varied duties of teaching mathematics at the high school of Linz, the capital, of constructing a new map of the province, and of completing his planetary tables. For the present, however, he decided to stay with Rudolph.

In the same year Kepler lost his wife, who had long been in weak bodily and mental health.

In the following year (1612) Rudolph died, and Kepler then moved to Linz and took up his new duties there, though still holding the appointment of mathematician to the Emperor and occasionally even receiving some portion of the salary of the office. In 1613 he married again, after a careful consideration, recorded in an extraordinary but very characteristic letter to one of his friends, of the relative merits of eleven ladies whom he regarded as possible; and the provision of a proper supply of wine for his new household led to the publication of a pamphlet, of some mathematical interest, dealing with the proper way of measuring the contents of a cask with curved sides.*

144. In the years 1618–1621, although in some ways the most disturbed years of his life, he published three books of importance—an *Epitome of the Copernican Astronomy*, the *Harmony of the World*,† and a treatise on *Comets*.

* It contains the germs of the method of infinitesimals.

† *Harmonices Mundi Libri V.*

The second and most important of these, published in 1619, though the leading idea in it was discovered early in 1618, was regarded by Kepler as a development of his early *Mysterium Cosmographicum* (§ 136). His speculative and mystic temperament led him constantly to search for relations between the various numerical quantities occurring in the solar system; by a happy inspiration he thought of trying to get a relation connecting the sizes of the orbits of the various planets with their times of revolution round the sun, and after a number of unsuccessful attempts discovered a simple and important relation, commonly known as Kepler's third law:—

The squares of the times of revolution of any two planets (including the earth) about the sun are proportional to the cubes of their mean distances from the sun.

If, for example, we express the times of revolution of the various planets in terms of any one, which may be conveniently taken to be that of the earth, namely a year, and in the same way express the distances in terms of the distance of the earth from the sun as a unit, then the times of revolution of the several planets taken in the order Mercury, Venus, Earth, Mars, Jupiter, Saturn are approximately ·24, ·615, 1, 1·88, 11·86, 29·457, and their distances from the sun are respectively ·387, ·723, 1, 1·524, 5·203, 9·539; if now we take the squares of the first series of numbers (the square of a number being the number multiplied by itself) and the cubes of the second series (the cube of a number being the number multiplied by itself twice, or the square multiplied again by the number), we get the two series of numbers given approximately by the table:—

	Mercury.	Venus.	Earth.	Mars.	Jupiter.	Saturn.
Square of periodic time	·058	·378	1	3·54	140·7	867·7
Cube of mean distance	·058	·378	1	3·54	140·8	867·9

Here it will be seen that the two series of numbers, in the

upper and lower row respectively, agree completely for as many decimal places as are given, except in the cases of the two outer planets, where the lower numbers are slightly in excess of the upper. For this discrepancy Newton afterwards assigned a reason (chapter IX., § 186), but with the somewhat imperfect knowledge of the times of revolution and distances which Kepler possessed the discrepancy was barely capable of detection, and he was therefore justified—from his standpoint—in speaking of the law as "precise."*

It should be noticed further that Kepler's law requires no knowledge of the actual distances of the several planets

FIG. 62.—The "music of the spheres," according to Kepler. From the *Harmony of the World*.

from the sun, but only of their relative distances, *i.e.* the number of times farther off from the sun or nearer to the sun any planet is than any other. In other words, it is necessary to have or to be able to construct a map of the solar system correct in its *proportions*, but it is quite unnecessary for this purpose to know the *scale* of the map.

Although the *Harmony of the World* is a large book, there is scarcely anything of value in it except what has already been given. A good deal of space is occupied with repetitions of the earlier speculations contained in the

* There may be some interest in Kepler's own statement of the law: "Res est certissima exactissimaque, quod proportionis quae est inter binorum quorumque planetarum tempora periodica, sit praecise sesquialtera proportionis mediarum distantiarum, id est orbium ipsorum."—*Harmony of the World*, Book V., chapter III.

Mysterium Cosmographicum, and most of the rest is filled with worthless analogies between the proportions of the solar system and the relations between various musical scales.

He is bold enough to write down in black and white the "music of the spheres" (in the form shewn in fig. 62), while the nonsense which he was capable of writing may be further illustrated by the remark which occurs in the same part of the book: "The Earth sings the notes M I, F A, M I, so that you may guess from them that in this abode of ours MIsery (*miseria*) and FAmine (*fames*) prevail."

145. The *Epitome of the Copernican Astronomy*, which appeared in parts in 1618, 1620, and 1621, although there are no very striking discoveries in it, is one of the most attractive of Kepler's books, being singularly free from the extravagances which usually render his writings so tedious. It contains within moderately short compass, in the form of question and answer, an account of astronomy as known at the time, expounded from the Coppernican standpoint, and embodies both Kepler's own and Galilei's latest discoveries. Such a text-book supplied a decided want, and that this was recognised by enemies as well as by friends was shewn by its prompt appearance in the Roman *Index of Prohibited Books* (cf. chapter VI., §§ 126, 132). The *Epitome* contains the first clear statement that the two fundamental laws of planetary motion established for the case of Mars (§ 141) were true also for the other planets (no satisfactory proof being, however, given), and that they applied also to the motion of the moon round the earth, though in this case there were further irregularities which complicated matters. The theory of the moon is worked out in considerable detail, both evection (chapter II., § 48) and variation (chapter III., § 60; chapter V., § 111) being fully dealt with, though the "annual equation" which Tycho had just begun to recognise at the end of his life (chapter V., § 111) is not discussed. Another interesting development of his own discoveries is the recognition that his third law of planetary motion applied also to the movements of the four satellites round Jupiter, as recorded by Galilei and Simon Marius (chapter VI., § 118). Kepler also introduced in the *Epitome* a considerable

improvement in the customary estimate of the distance of the earth from the sun, from which those of the other planets could at once be deduced.

If, as had been generally believed since the time of Hipparchus and Ptolemy, the distance of the sun were 1,200 times the radius of the earth, then the parallax (chapter II., §§ 43, 49) of the sun would at times be as much as 3′, and that of Mars, which in some positions is much nearer to the earth, proportionally larger. But Kepler had been unable to detect any parallax of Mars, and therefore inferred that the distances of Mars and of the sun must be greater than had been supposed. Having no exact data to go on, he produced out of his imagination and his ideas of the harmony of the solar system a distance about three times as great as the traditional one. He argued that, as the earth was the abode of measuring creatures, it was reasonable to expect that the measurements of the solar system would bear some simple relation to the dimensions of the earth. Accordingly he assumed that the volume of the sun was as many times greater than the volume of the earth as the distance of the sun was greater than the radius of the earth, and from this quaint assumption deduced the value of the distance already stated, which, though an improvement on the old value, was still only about one-seventh of the true distance.

The *Epitome* contains also a good account of eclipses both of the sun and moon, with the causes, means of predicting them, etc. The faint light (usually reddish) with which the face of the eclipsed moon often shines is correctly explained as being sunlight which has passed through the atmosphere of the earth, and has there been bent from a straight course so as to reach the moon, which the light of the sun in general is, owing to the interposition of the earth, unable to reach. Kepler mentions also a ring of light seen round the eclipsed sun in 1567, when the eclipse was probably total, not annular (chapter II., § 43), and ascribes it to some sort of luminous atmosphere round the sun, referring to a description in Plutarch of the same appearance. This seems to have been an early observation, and a rational though of course very imperfect explanation, of that remarkable solar envelope known as the **corona** which has attracted so much attention in the last half-century (chapter XIII., § 301).

146. The treatise on *Comets* (1619) contained an account of a comet seen in 1607, afterwards famous as Halley's comet (chapter X., § 200), and of three comets seen in 1618. Following Tycho, Kepler held firmly the view that comets were celestial not terrestrial bodies, and accounted for their appearance and disappearance by supposing that they moved in straight lines, and therefore after having once passed near the earth receded indefinitely into space; he does not appear to have made any serious attempt to test this theory by comparison with observation, being evidently of opinion that the path of a body which would never reappear was not a suitable object for serious study. He agreed with the observation made by Fracastor and Apian (chapter III., § 69) that comets' tails point away from the sun, and explained this by the supposition that the tail is formed by rays of the sun which penetrate the body of the comet and carry away with them some portion of its substance, a theory which, allowance being made for the change in our views as to the nature of light, is a curiously correct anticipation of modern theories of comets' tails (chapter XIII., § 304).

In a book intended to have a popular sale it was necessary to make the most of the "meaning" of the appearance of a comet, and of its influence on human affairs, and as Kepler was writing when the Thirty Years' War had just begun, while religious persecutions and wars had been going on in Europe almost without interruption during his lifetime, it was not difficult to find sensational events which had happened soon after or shortly before the appearance of the comets referred to. Kepler himself was evidently not inclined to attach much importance to such coincidences; he thought that possibly actual contact with a comet's tail might produce pestilence, but beyond that was not prepared to do more than endorse the pious if somewhat neutral opinion that one of the uses of a comet is to remind us that we are mortal. His belief that comets are very numerous is expressed in the curious form: "There are as many arguments to prove the annual motion of the earth round the sun as there are comets in the heavens."

147. Meanwhile Kepler's position at Linz had become more and more uncomfortable, owing to the rising tide of the religious and political disturbances which finally led to the outbreak of the Thirty Years' War in 1618; but notwithstanding this he had refused in 1617 an offer of a chair of mathematics at Bologna, partly through attachment to his native country and partly through a well-founded distrust of the Papal party in Italy. Three years afterwards he rejected also the overtures made by the English ambassador, with a view to securing him as an ornament to the court of James I., one of his chief grounds for refusal in this case being a doubt whether he would not suffer from being cooped up within the limits of an island. In 1619 the Emperor Matthias died, and was succeeded by Ferdinand II., who as Archduke had started the persecution of the Protestants at Gratz (§ 137) and who had few scientific interests. Kepler was, however, after some delay, confirmed in his appointment as Imperial Mathematician. In 1620 Linz was occupied by the Imperialist troops, and by 1626 the oppression of the Protestants by the Roman Catholics had gone so far that Kepler made up his mind to leave, and, after sending his family to Regensburg, went himself to Ulm.

148. At Ulm Kepler published his last great work. For more than a quarter of a century he had been steadily working out in detail, on the basis of Tycho's observations and of his own theories, the motions of the heavenly bodies, expressing the results in such convenient tabular form that the determination of the place of any body at any required time, as well as the investigation of other astronomical events such as eclipses, became merely a matter of calculation according to fixed rules; this great undertaking, in some sense the summing up of his own and of Tycho's work, was finally published in 1627 as the *Rudolphine Tables* (the name being given in honour of his former patron), and remained for something like a century the standard astronomical tables.

It had long been Kepler's intention, after finishing the tables, to write a complete treatise on astronomy, to be called the *New Almagest*; but this scheme was never fairly started, much less carried out.

149. After a number of unsuccessful attempts to secure the arrears of his salary, he was told to apply to Wallenstein, the famous Imperialist general, then established in Silesia in a semi-independent position, who was keenly interested in astrology and usually took about with him one or more representatives of the art. Kepler accordingly joined Wallenstein in 1628, and did astrology for him, in addition to writing some minor astronomical and astrological treatises. In 1630 he travelled to Regensburg, where the Diet was then sitting, to press in person his claims for various arrears of salary; but, worn out by anxiety and by the fatigues of the journey, he was seized by a fever a few days after his arrival, and died on November 15th (N.S.), 1630, in his 59th year.

The inventory of his property, made after his death, shews that he was in possession of a substantial amount, so that the effect of extreme poverty which his letters convey must have been to a considerable extent due to his over-anxious and excitable temperament.

150. In addition to the great discoveries already mentioned Kepler made a good many minor contributions to astronomy, such as new methods of finding the longitude, and various improvements in methods of calculation required for astronomical problems. He also made speculations of some interest as to possible causes underlying the known celestial motions. Whereas the Ptolemaic system required a number of motions round mere geometrical points, centres of epicycles or eccentrics, equants, etc., unoccupied by any real body, and many such motions were still required by Coppernicus, Kepler's scheme of the solar system placed a real body, the sun, at the most important point connected with the path of each planet, and dealt similarly with the moon's motion round the earth and with that of the four satellites round Jupiter. Motions of revolution came in fact to be associated not with some central *point* but with some central *body*, and it became therefore an inquiry of interest to ascertain if there were any connection between the motion and the central body. The property possessed by a magnet of attracting a piece of iron at some little distance from it suggested a possible analogy to Kepler, who had read with care and was evidently impressed by

the treatise *On the Magnet* (De Magnete) published in 1600 by our countryman *William Gilbert* of Colchester (1540–1603). He suggested that the planets might thus be regarded as connected with the sun, and therefore as sharing to some extent the sun's own motion of revolution. In other words, a certain "carrying virtue" spread out from the sun, with or like the rays of light and heat, and tried to carry the planets round with the sun.

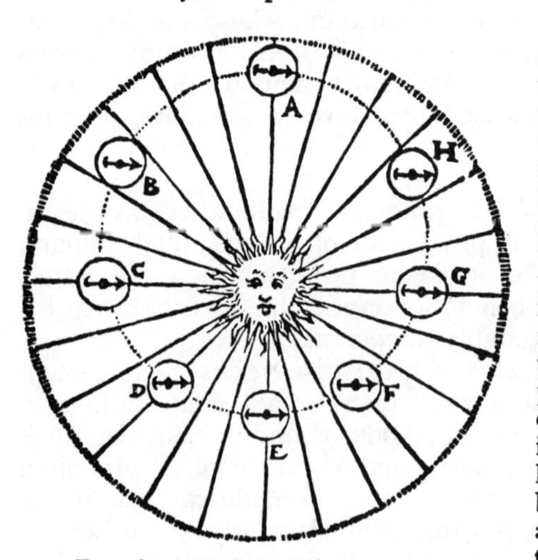

FIG. 63.—Kepler's idea of gravity. From the *Epitome*.

"There is therefore a conflict between the carrying power of the sun and the impotence or material sluggishness (*inertia*) of the planet; each enjoys some measure of victory, for the former moves the planet from its position and the latter frees the planet's body to some extent from the bonds in which it is thus held, . . . but only to be captured again by another portion of this rotatory virtue."*

The annexed diagram is given by Kepler in illustration of this rather confused and vague theory.

He believed also in a more general "gravity," which he defined † as "a mutual bodily affection between allied bodies tending towards their union or junction," and regarded the tides as due to an action of this sort between the moon and the water of the earth. But the speculative ideas thus thrown out, which it is possible to regard as anticipations of Newton's discovery of universal gravitation, were not in any way developed logically, and Kepler's mechanical ideas

* *Epitome*, Book IV., Part 2.

† Introduction to the *Commentaries on the Motions of Mars*.

were too imperfect for him to have made real progress in this direction.

151. There are few astronomers about whose merits such different opinions have been held as about Kepler. There is, it is true, a general agreement as to the great importance of his three laws of planetary motion, and as to the substantial value of the *Rudolphine Tables* and of various minor discoveries. These results, however, fill but a small part of Kepler's voluminous writings, which are encumbered with masses of wild speculation, of mystic and occult fancies, of astrology, weather prophecies, and the like, which are not only worthless from the standpoint of modern astronomy, but which—unlike many erroneous or imperfect speculations—in no way pointed towards the direction in which the science was next to make progress, and must have appeared almost as unsound to sober-minded contemporaries like Galilei as to us. Hence as one reads chapter after chapter without a lucid still less a correct idea, it is impossible to refrain from regrets that the intelligence of Kepler should have been so wasted, and it is difficult not to suspect at times that some of the valuable results which lie imbedded in this great mass of tedious speculation were arrived at by a mere accident. On the other hand, it must not be forgotten that such accidents have a habit of happening only to great men, and that if Kepler loved to give reins to his imagination he was equally impressed with the necessity of scrupulously comparing speculative results with observed facts, and of surrendering without demur the most beloved of his fancies if it was unable to stand this test. If Kepler had burnt three-quarters of what he printed, we should in all probability have formed a higher opinion of his intellectual grasp and sobriety of judgment, but we should have lost to a great extent the impression of extraordinary enthusiasm and industry, and of almost unequalled intellectual honesty, which we now get from a study of his works.

CHAPTER VIII.

FROM GALILEI TO NEWTON.

> "And now the lofty telescope, the scale
> By which they venture heaven itself t'assail,
> Was raised, and planted full against the moon."
>
> *Hudibras*

152. BETWEEN the publication of Galilei's *Two New Sciences* (1638) and that of Newton's *Principia* (1687) a period of not quite half a century elapsed; during this interval no astronomical discovery of first-rate importance was published, but steady progress was made on lines already laid down.

On the one hand, while the impetus given to exact observation by Tycho Brahe had not yet spent itself, the invention of the telescope and its gradual improvement opened out an almost indefinite field for possible discovery of new celestial objects of interest. On the other hand, the remarkable character of the three laws in which Kepler had summed up the leading characteristics of the planetary motions could hardly fail to suggest to any intelligent astronomer the question *why* these particular laws should hold, or, in other words, to stimulate the inquiry into the possibility of shewing them to be necessary consequences of some simpler and more fundamental law or laws, while Galilei's researches into the laws of motion suggested the possibility of establishing some connection between the causes underlying these celestial motions and those of ordinary terrestrial objects.

153. It has been already mentioned how closely Galilei was followed by other astronomers (if not in some cases actually anticipated) in most of his telescopic discoveries.

To his rival Christopher Scheiner (chapter VI., §§ 124, 125) belongs the credit of the discovery of bright cloud-like objects on the sun, chiefly visible near its edge, and from their brilliancy named **faculae** (little torches). Scheiner made also a very extensive series of observations of the motions and appearances of spots.

The study of the surface of the moon was carried on with great care by *John Hevel* of Danzig (1611–1687), who published in 1647 his *Selenographia*, or description of the moon, magnificently illustrated by plates engraved as well as drawn by himself. The chief features of the moon—mountains, craters, and the dark spaces then believed to be seas—were systematically described and named, for the most part after corresponding features of our own earth. Hevel's names for the chief mountain ranges, *e.g.* the *Apennines* and the *Alps*, and for the seas, *e.g.* *Mare Serenitatis* or Pacific Ocean, have lasted till to-day; but similar names given by him to single mountains and craters have disappeared, and they are now called after various distinguished men of science and philosophers, *e.g.* *Plato* and *Coppernicus*, in accordance with a system introduced by *John Baptist Riccioli* (1598–1671) in his bulky treatise on astronomy called the *New Almagest* (1651).

Hevel, who was an indefatigable worker, published two large books on comets, *Prodromus Cometicus* (1654) and *Cometographia* (1668), containing the first systematic account of all recorded comets. He constructed also a catalogue of about 1,500 stars, observed on the whole with accuracy rather greater than Tycho's, though still without the use of the telescope; he published in addition an improved set of tables of the sun, and a variety of other calculations and observations.

154. The planets were also watched with interest by a number of observers, who detected at different times bright or dark markings on Jupiter, Mars, and Venus. The two appendages of Saturn which Galilei had discovered in 1610 and had been unable to see two years later (chapter VI., § 123) were seen and described by a number of astronomers under a perplexing variety of appearances, and the mystery was only unravelled, nearly half a century after Galilei's first observation, by the greatest astronomer of this period,

Christiaan Huygens (1629–1695), a native of the Hague. Huygens possessed remarkable ability, both practical and theoretical, in several different directions, and his contributions to astronomy were only a small part of his services to science. Having acquired the art of grinding lenses with unusual accuracy, he was able to construct telescopes of much greater power than his predecessors. By the help of one of these instruments he discovered in 1655 a satellite of Saturn (*Titan*). With one of those remnants of mediaeval mysticism from which even the soberest minds of the century freed themselves with the greatest difficulty, he asserted that, as the total number of planets and satellites now reached the perfect number 12, no more remained to be discovered—a prophecy which has been abundantly falsified since (§ 160; chapter XII., §§ 253, 255; chapter XIII., §§ 289, 294, 295).

Using a still finer telescope, and aided by his acuteness in interpreting his observations, Huygens made the much more interesting discovery that the puzzling appearances seen round Saturn were due to a thin ring (fig. 64) inclined at a considerable angle (estimated by him at 31°) to the plane of the ecliptic, and therefore also to the plane in which Saturn's path round the sun lies. This result was first announced—according to the curious custom of the time—by an anagram, in the same pamphlet in which the discovery of the satellite was published, *De Saturni Luna Observatio Nova* (1656); and three years afterwards (1659) the larger *Systema Saturnium* appeared, in which the interpretation of the anagram was given, and the varying appearances seen both by himself and by earlier observers were explained with admirable lucidity and thoroughness. The ring being extremely thin is invisible either when its edge is presented to the observer or when it is presented to the sun, because in the latter position the rest of the ring catches no light. Twice in the course of Saturn's revolution round the sun (at B and D in fig. 66), *i.e.* at intervals of about 15 years, the plane of the ring passes for a short time through or very close both to the earth and to the sun, and at these two periods the ring is consequently invisible (fig. 65). Near these positions (as at Q, R, S, T) the ring appears much foreshortened, and presents the appearance of two arms projecting from the body

FIG. 64.—Saturn's ring, as drawn by Huygens. From the *Systema Saturnium.*

FIG. 65.—Saturn, with the ring seen edge-wise. From the *Systema Saturnium.*

[*To face p.* 200

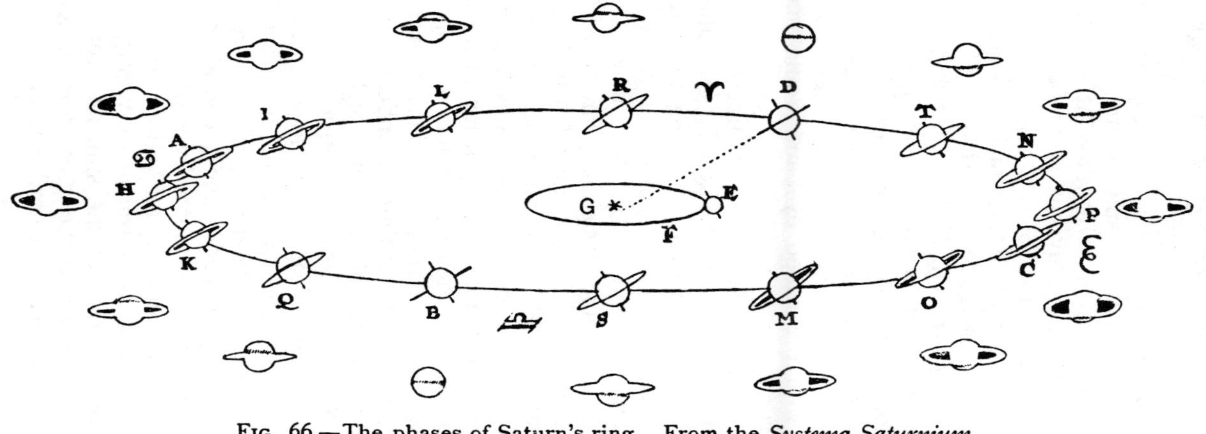

FIG. 66.—The phases of Saturn's ring From the *Systema Saturnium*.

of Saturn; farther off still the ring appears wider and the opening becomes visible; and about seven years before and after the periods of invisibility (at A and C) the ring is seen at its widest. Huygens gives for comparison with his own results a number of drawings by earlier observers (reproduced in fig. 67), from which it may be seen how near some of them were to the discovery of the ring.

155. To our countryman *William Gascoigne* (1612?–1644) is due the first recognition that the telescope could be utilised, not merely for observing generally the appearances of celestial bodies, but also as an instrument of precision, which would give the directions of stars, etc., with greater accuracy than is possible with the naked eye, and would magnify small angles in such a way as to facilitate the measurement of angular distances between neighbouring stars, of the diameters of the planets, and of similar quantities. He was unhappily killed when quite a young man at the battle of Marston Moor (1644), but his letters, published many years afterwards shew that by 1640 he was familiar with the use of telescopic "sights," for determining with accuracy the position of a star, and that he had constructed a so-called **micrometer** * with which he was able to measure angles of a few seconds. Nothing was known of his discoveries at the time, and it was left for Huygens to invent independently a micrometer of an inferior kind (1658), and for *Adrien Auzout* (?–1691) to introduce as an improvement (about 1666) an instrument almost identical with Gascoigne's.

The systematic use of telescopic sights for the regular work of an observatory was first introduced about 1667 by Auzout's friend and colleague *Jean Picard* (1620–1682).

156. With Gascoigne should be mentioned his friend *Jeremiah Horrocks* (1617?–1641), who was an enthusiastic admirer of Kepler and had made a considerable improvement in the theory of the moon, by taking the elliptic orbit as a basis and then allowing for various irregularities. He was the first observer of a transit of Venus, *i.e.* a passage of Venus over the disc of the sun, an event which took place in 1639, contrary to the prediction of Kepler in the *Rudolphine Tables*, but in accordance with the rival tables

* Substantially the *filar micrometer* of modern astronomy.

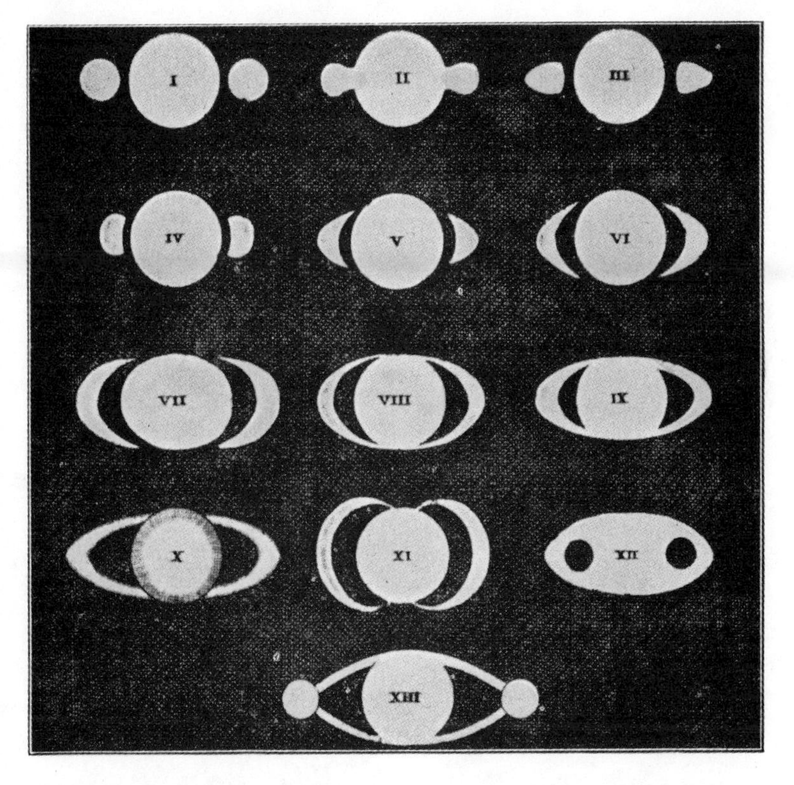

FIG. 67.—Early drawings of Saturn. From the *Systema Saturnium.*

[*To face p.* 202.

of *Philips von Lansberg* (1561–1632), which Horrocks had verified for the purpose. It was not, however, till long afterwards that Halley pointed out the importance of the transit of Venus as a means of ascertaining the distance of the sun from the earth (chapter x., § 202). It is also worth noticing that Horrocks suggested the possibility of the irregularities of the moon's motion being due to the disturbing action of the sun, and that he also had some idea of certain irregularities in the motion of Jupiter and Saturn, now known to be due to their mutual attraction (chapter x., § 204; chapter xi., § 243).

157. Another of Huygens's discoveries revolutionised the art of exact astronomical observation. This was the invention of the pendulum-clock (made 1656, patented in 1657). It has been already mentioned how the same discovery was made by Bürgi, but virtually lost (see chapter v., § 98); and how Galilei again introduced the pendulum as a time-measurer (chapter vi., § 114). Galilei's pendulum, however, could only be used for measuring very short times, as there was no mechanism to keep it in motion, and the motion soon died away. Huygens attached a pendulum to a clock driven by weights, so that the clock kept the pendulum going and the pendulum regulated the clock.* Henceforward it was possible to take reasonably accurate time-observations, and, by noticing the interval between the passage of two stars across the meridian, to deduce, from the known rate of motion of the celestial sphere, their angular distance east and west of one another, thus helping to fix the position of one with respect to the other. It was again Picard (§ 155) who first recognised the astronomical importance of this discovery, and introduced regular time-observations at the new Observatory of Paris.

158. Huygens was not content with this practical. use of the pendulum, but worked out in his treatise called *Oscillatorium Horologium* or *The Pendulum Clock* (1673) a number of important results in the theory of the pendulum, and in the allied problems connected with the motion of a body in a circle or other curve. The greater part of these

* Galilei, at the end of his life, appears to have thought of contriving a pendulum with clockwork, but there is no satisfactory evidence that he ever carried out the idea.

investigations lie outside the field of astronomy, but his formula connecting the time of oscillation of a pendulum with its length and the intensity of gravity * (or, in other words, the rate of falling of a heavy body) afforded a practical means of measuring gravity, of far greater accuracy than any direct experiments on falling bodies; and his study of circular motion, leading to the result that a body moving in a circle must be acted on by some force towards the *centre*, the magnitude of which depended in a definite way on the speed of the body and the size of the circle,† is of fundamental importance in accounting for the planetary motions by gravitation.

159. During the 17th century also the first measurements of the earth were made which were a definite advance on those of the Greeks and Arabs (chapter II., §§ 36, 45, and chapter III., § 57). *Willebrord Snell* (1591–1626), best known by his discovery of the law of refraction of light, made a series of measurements in Holland in 1617, from which the length of a degree of a meridian appeared to be about 67 miles, an estimate subsequently altered to about 69 miles by one of his pupils, who corrected some errors in the calculations, the result being then within a few hundred feet of the value now accepted. Next, *Richard Norwood* (1590 ?–1675) measured the distance from London to York, and hence obtained (1636) the length of the degree with an error of less than half a mile. Lastly, Picard in 1671 executed some measurements near Paris leading to a result only a few yards wrong. The length of a degree being known, the circumference and radius of the earth can at once be deduced.

160. Auzout and Picard were two members of a group of observational astronomers working at Paris, of whom the best known, though probably not really the greatest, was *Giovanni Domenico Cassini* (1625–1712). Born in the north of Italy, he acquired a great reputation, partly by some rather fantastic schemes for the construction of gigantic instruments, partly by the discovery of the rotation

* In modern notation: time of oscillation $= 2\pi\sqrt{l/g}$.

† *I.e.* he obtained the familiar formula v^2/r, and several equivalent forms for *centrifugal force*.

of Jupiter (1665), of Mars (1666), and possibly of Venus (1667), and also by his tables of the motions of Jupiter's moons (1668). The last caused Picard to procure for him an invitation from Louis XIV. (1669) to come to Paris and to exercise a general superintendence over the Observatory, which was then being built and was substantially completed in 1671. Cassini was an industrious observer and a voluminous writer, with a remarkable talent for impressing the scientific public as well as the Court. He possessed a strong sense of the importance both of himself and of his work, but it is more than doubtful if he had as clear ideas as Picard of the really important pieces of work which ought to be done at a public observatory, and of the way to set about them. But, notwithstanding these defects, he rendered valuable services to various departments of astronomy. He discovered four new satellites of Saturn: *Japetus* in 1671, *Rhea* in the following year, *Dione* and *Thetis* in 1684; and also noticed in 1675 a dark marking in Saturn's ring, which has subsequently been more distinctly recognised as a division of the ring into two, an inner and an outer, and is known as Cassini's division (see fig. 95 facing p. 384). He also improved to some extent the theory of the sun, calculated a fresh table of atmospheric refraction which was an improvement on Kepler's (chapter VII., § 138), and issued in 1693 a fresh set of tables of Jupiter's moons, which were much more accurate than those which he had published in 1668, and much the best existing.

161. It was probably at the suggestion of Picard or Cassini that one of their fellow astronomers, *John Richer* (?–1696), otherwise almost unknown, undertook (1671–3) a scientific expedition to Cayenne (in latitude 5° N.). Two important results were obtained. It was found that a pendulum of given length beat more slowly at Cayenne than at Paris, thus shewing that the intensity of gravity was less near the equator than in higher latitudes. This fact suggested that the earth was not a perfect sphere, and was afterwards used in connection with theoretical investigations of the problem of the earth's shape (cf. chapter IX., § 187). Again, Richer's observations of the position of Mars in the sky, combined with observations taken at the same time by Cassini, Picard,

and others in France, led to a reasonably accurate estimate of the distance of Mars and hence of that of the sun. Mars was at the time in opposition (chapter II., § 43), so that it was nearer to the earth than at other times (as shewn in fig. 68), and therefore favourably situated for such observations. The principle of the method is extremely simple and substantially identical with that long used in the case of the moon (chapter II., § 49). One observer is, say, at Paris (P, in fig. 69), and observes the direction in which Mars appears, *i.e.* the direction of the line P M; the other at Cayenne (C) observes similarly the direction of the line C M. The line C P, joining Paris and Cayenne, is known geographically; the shape of the triangle C P M and

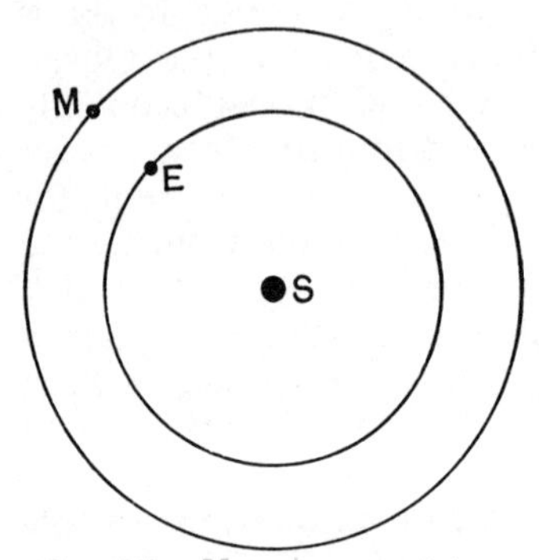

FIG. 68.—Mars in opposition.

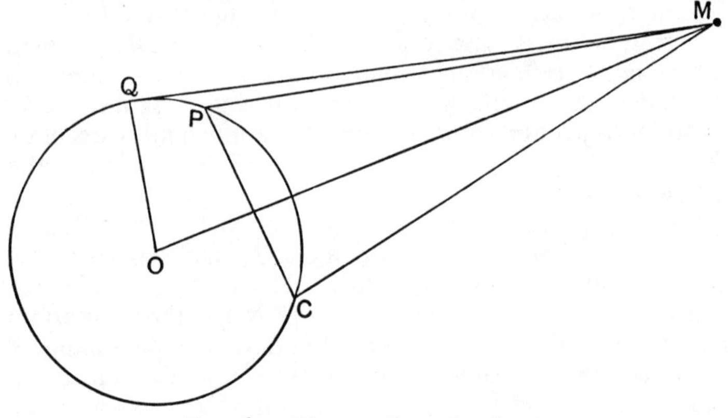

FIG. 69.—The parallax of a planet.

the length of one of its sides being thus known, the lengths of the other sides are readily calculated.

The result of an investigation of this sort is often most conveniently expressed by means of a certain angle, from which the distance in terms of the radius of the earth, and hence in miles, can readily be deduced when desired.

The parallax of a heavenly body such as the moon, the sun, or a planet, being in the first instance defined generally (chapter II., § 43) as the angle (O M P) between the lines joining the heavenly body to the observer and to the centre of the earth, varies in general with the position of the observer. It is evidently greatest when the observer is in such a position, as at Q, that the line M Q touches the earth; in this position M is on the observer's horizon. Moreover the angle O Q M being a right angle, the shape of the triangle and the ratio of its sides are completely known when the angle O M Q is known. Since this angle is the parallax of M, when on the observer's horizon, it is called the **horizontal parallax** of M, but the word horizontal is frequently omitted. It is easily seen by a figure that the more distant a body is the smaller is its horizontal parallax; and with the small parallaxes with which we are concerned in astronomy, the distance and the horizontal parallax can be treated as inversely proportional to one another; so that if, for example, one body is twice as distant as another, its parallax is half as great, and so on.

It may be convenient to point out here that the word "parallax" is used in a different though analogous sense when a fixed star is in question. The apparent displacement of a fixed star due to the earth's motion (chapter IV., § 92), which was not actually detected till long afterwards (chapter XIII., § 278), is called **annual** or **stellar parallax** (the adjective being frequently omitted); and the name is applied in particular to the greatest angle between the direction of the star as seen from the sun and as seen from the earth in the course of the year. If in fig. 69 we regard M as representing a star, O the sun, and the circle as being the earth's path round the sun, then the angle O M Q is the annual parallax of M.

In this particular case Cassini deduced from Richer's observations, by some rather doubtful processes, that the sun's parallax was about 9″·5, corresponding to a distance from the earth of about 87,000,000 miles, or about 360 times the distance of the moon, the most probable value, according to modern estimates (chapter XIII., § 284), being a little less than 93,000,000. Though not really an accurate result, this was an enormous improvement on anything that had gone before, as Ptolemy's estimate of the sun's distance, corresponding to a parallax of 3′, had survived up to the earlier part of the 17th century, and although it was generally believed to be seriously wrong, most corrections of it had been purely conjectural (chapter VII., §§ 145).

162. Another famous discovery associated with the early days of the Paris Observatory was that of the velocity of light. In 1671 Picard paid a visit to Denmark to examine what was left of Tycho Brahe's observatory at Hveen, and brought back a young Danish astronomer, *Olaus Roemer* (1644–1710), to help him at Paris. Roemer, in studying the motion of Jupiter's moons, observed (1675) that the intervals between successive eclipses of a moon (the eclipse being caused by the passage of the moon into Jupiter's shadow) were regularly less when Jupiter and the earth were approaching one another than when they were receding. This he saw to be readily explained by the supposition that light travels through space at a definite though very great speed. Thus if Jupiter is approaching the earth, the time which the light from one of his moons takes to reach the earth is gradually decreasing, and consequently the interval between successive eclipses as seen by us is apparently diminished. From the difference of the intervals thus observed and the known rates of motion of Jupiter and of the earth, it was thus possible to form a rough estimate of the rate at which light travels. Roemer also made a number of instrumental improvements of importance, but they are of too technical a character to be discussed here.

163. One great name belonging to the period dealt with in this chapter remains to be mentioned, that of *René Descartes* * (1596–1650). Although he ranks as a great philosopher, and also made some extremely important advances in pure mathematics, his astronomical writings were of little value and in many respects positively harmful. In his *Principles of Philosophy* (1644) he gave, among some wholly erroneous propositions, a fuller and more

* Also frequently referred to by the Latin name *Cartesius*.

general statement of the first law of motion discovered by Galilei (chapter VI., §§ 130, 133), but did not support it by any evidence of value. The same book contained an exposition of his famous theory of vortices, which was an attempt to explain the motions of the bodies of the solar system by means of a certain combination of vortices or eddies. The theory was unsupported by any experimental evidence, and it was not formulated accurately enough to be capable of being readily tested by comparison with actual observation; and, unlike many erroneous theories (such as the Greek epicycles), it in no way led up to or suggested the truer theories which followed it. But "Cartesianism," both in philosophy and in natural science, became extremely popular, especially in France, and its vogue contributed notably to the overthrow of the authority of Aristotle, already shaken by thinkers like Galilei and Bacon, and thus rendered men's minds a little more ready to receive new ideas: in this indirect way, as well as by his mathematical discoveries, Descartes probably contributed something to astronomical progress.

CHAPTER IX.

UNIVERSAL GRAVITATION.

"Nature and Nature's laws lay hid in night;
God said 'Let Newton be!' and all was light."
POPE.

164. NEWTON's life may be conveniently divided into three portions. First came 22 years (1643–1665) of boyhood and undergraduate life; then followed his great productive period, of almost exactly the same length, culminating in the publication of the *Principia* in 1687; while the rest of his life (1687–1727), which lasted nearly as long as the other two periods put together, was largely occupied with official work and studies of a non-scientific character, and was marked by no discoveries ranking with those made in his middle period, though some of his earlier work received important developments and several new results of decided interest were obtained.

165. Isaac Newton was born at Woolsthorpe, near Grantham, in Lincolnshire, on January 4th, 1643;* this was very nearly a year after the death of Galilei, and a few months after the beginning of our Civil Wars. His taste for study does not appear to have developed very early in life, but ultimately became so marked that, after

* According to the unreformed calendar (O.S.) then in use in England, the date was Christmas Day, 1642. To facilitate comparison with events occurring out of England, I have used throughout this and the following chapters the Gregorian Calendar (N.S.), which was at this time adopted in a large part of the Continent (cf. chapter II., § 22).

some unsuccessful attempts to turn him into a farmer, he was entered at Trinity College, Cambridge, in 1661.

Although probably at first rather more backward than most undergraduates, he made extremely rapid progress in mathematics and allied subjects, and evidently gave his teachers some trouble by the rapidity with which he absorbed what little they knew. He met with Euclid's *Elements of Geometry* for the first time while an undergraduate, but is reported to have soon abandoned it as being "a trifling book," in favour of more advanced reading. In January 1665 he graduated in the ordinary course as Bachelor of Arts.

166. The external events of Newton's life during the next 22 years may be very briefly dismissed. He was elected a Fellow in 1667, became M.A. in due course in the following year, and was appointed Lucasian Professor of Mathematics, in succession to his friend Isaac Barrow, in 1669. Three years later he was elected a Fellow of the recently founded Royal Society. With the exception of some visits to his Lincolnshire home, he appears to have spent almost the whole period in quiet study at Cambridge, and the history of his life is almost exclusively the history of his successive discoveries.

167. His scientific work falls into three main groups, astronomy (including dynamics), optics, and pure mathematics. He also spent a good deal of time on experimental work in chemistry, as well as on heat and other branches of physics, and in the latter half of his life devoted much attention to questions of chronology and theology; in none of these subjects, however, did he produce results of much importance.

168. In forming an estimate of Newton's genius it is of course important to bear in mind the range of subjects with which he dealt; from our present point of view, however, his mathematics only presents itself as a tool to be used in astronomical work; and only those of his optical discoveries which are of astronomical importance need be mentioned here. In 1668 he constructed a **reflecting telescope**, that is, a telescope in which the rays of light from the object viewed are concentrated by means of a curved mirror instead of by a lens, as in the **refracting telescopes**

of Galilei and Kepler. Telescopes on this principle, differing however in some important particulars from Newton's, had already been described in 1663 by *James Gregory* (1638–1675), with whose ideas Newton was acquainted, but it does not appear that Gregory had actually made an instrument. Owing to mechanical difficulties in construction, half a century elapsed before reflecting telescopes were made which could compete with the best refractors of the time, and no important astronomical discoveries were made with them before the time of William Herschel (chapter XII.), more than a century after the original invention.

Newton's discovery of the effect of a prism in resolving a beam of white light into different colours is in a sense the basis of the method of spectrum analysis (chapter XIII., § 299), to which so many astronomical discoveries of the last 40 years are due.

169. The ideas by which Newton is best known in each of his three great subjects—gravitation, his theory of colours, and fluxions—seem to have occurred to him and to have been partly thought out within less than two years after he took his degree, that is before he was 24. His own account—written many years afterwards—gives a vivid picture of his extraordinary mental activity at this time :—

"In the beginning of the year 1665 I found the method of approximating Series and the Rule for reducing any dignity of any Binomial into such a series. The same year in May I found the method of tangents of Gregory and Slusius, and in November had the direct method of Fluxions, and the next year in January had the Theory of Colours, and in May following I had entrance into the inverse method of Fluxions. And the same year I began to think of gravity extending to the orb of the Moon, and having found out how to estimate the force with which [a] globe revolving within a sphere presses the surface of the sphere, from Kepler's Rule of the periodical times of the Planets being in a sesquialterate proportion of their distances from the centers of their orbs I deduced that the forces which keep the Planets in their orbs must [be] reciprocally as the squares of their distances from the centers about which they revolve : and thereby compared the force requisite to keep the Moon in her orb with the force of gravity at the surface of the earth, and found them answer pretty nearly. All this was in the two plague years of 1665 and 1666, for in those days I was in the prime of my age for invention, and minded Mathematicks and Philosophy more than at any time since." *

170. He spent a considerable part of this time (1665–1666) at Woolsthorpe, on account of the prevalence of the plague.

The well-known story, that he was set meditating on gravity by the fall of an apple in the orchard, is based on good authority, and is perfectly credible in the sense that the apple may have reminded him at that particular time of certain problems connected with gravity. That the apple seriously suggested to him the existence of the problems or any key to their solution is wildly improbable.

Several astronomers had already speculated on the "cause" of the known motions of the planets and satellites; that is they had attempted to exhibit these motions as consequences of some more fundamental and more general laws. Kepler, as we have seen (chapter VII., § 150), had pointed out that the motions in question should not be considered as due to the influence of mere geometrical points, such as the centres of the old epicycles, but to that of other bodies; and in particular made some attempt to explain the motion of the planets as due to a special kind of influence emanating from the sun. He went, however, entirely wrong by looking for a force to keep up the motion of the planets and as it were push them along. Galilei's discovery that the motion of a body goes on indefinitely unless there is some cause at work to alter or stop it, at once put a new aspect on this as on other mechanical problems; but he himself did not develop his idea in this particular direction. *Giovanni Alfonso Borelli* (1608–1679), in a book on Jupiter's satellites published in 1666, and therefore about the time of Newton's first work on the subject, pointed out that a body revolving in a circle (or similar curve) had a tendency to recede from the centre, and that in the case of the planets this might be supposed to be counteracted by some kind of attraction towards the sun. We have then here the idea—

* From a MS. among the Portsmouth Papers, quoted in the Preface to the Catalogue of the Portsmouth Papers.

in a very indistinct form certainly—that the motion of a planet is to be explained, not by a force acting in the direction in which it is moving, but by a force directed towards the sun, that is about at right angles to the direction of the planet's motion. Huygens carried this idea much further—though without special reference to astronomy—and obtained (chapter VIII., § 158) a numerical measure for the tendency of a body moving in a circle to recede from the centre, a tendency which had in some way to be counteracted if the body was not to fly away. Huygens published his work in 1673, some years after Newton had obtained his corresponding result, but before he had published anything; and there can be no doubt that the two men worked quite independently.

171. Viewed as a purely general question, apart from its astronomical applications, the problem may be said to be to examine under what conditions a body can revolve with uniform speed in a circle.

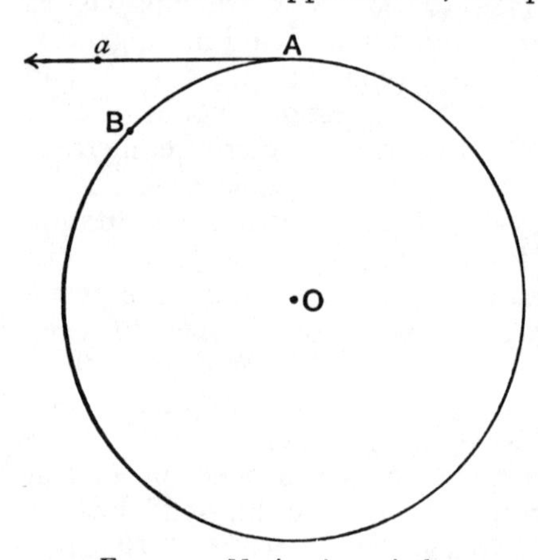

FIG. 70.—Motion in a circle.

Let A represent the position at a certain instant of a body which is revolving with uniform speed in a circle of centre O. Then at this instant the body is moving in the direction of the tangent A *a* to the circle. Consequently by Galilei's First Law (chapter VI., §§ 130, 133), if left to itself and uninfluenced by any other body, it would continue to move with the same speed and in the same direction, *i.e.* along the line A *a*, and consequently would be found after some time at such a point as *a*. But actually it is found to be at B on the circle. Hence some influence must have been at work to bring it to B instead of to *a*. But B is nearer to the centre of the circle than *a* is; hence some influence must be at work tending constantly to draw the body towards O, or counteracting the tendency which it has, in virtue of the First Law of Motion, to get farther and farther away from O. To express either of these tendencies numerically we want a more complex idea than that of velocity or rate of motion, namely **acceleration** or rate of change of velocity, an idea which Galilei added to science in his discussion of the law of falling bodies (chapter VI., §§ 116, 133). A falling body, for example, is moving after one second with the velocity of about 32 feet per second, after two seconds with the velocity of 64, after three seconds with the velocity of 96, and so on; thus in every second it gains a downward velocity of 32 feet per second; and this may be expressed otherwise by saying that the body has a downward acceleration of 32 feet per second per second. A further investigation of the motion in a circle shews that the motion is completely explained if the moving body has, in addition to its original velocity, an acceleration of a certain magnitude *directed towards the centre of the circle.* It can be shewn further that the acceleration may be numerically expressed by taking the square of the velocity of the moving body (expressed, say, in feet per second), and dividing this by the radius of the circle in feet. If, for example, the body is moving in a circle having a radius of four feet, at the rate of ten feet a second, then the acceleration towards the centre is $\left(\frac{10 \times 10}{4} =\right)$ 25 feet per second per second.

These results, with others of a similar character, were first published by Huygens—not of course precisely in this form—in his book on the *Pendulum Clock* (chapter VIII., § 158); and discovered independently by Newton in 1666.

If then a body is seen to move in a circle, its motion becomes intelligible if some other body can be discovered which produces this acceleration. In a common case, such as when a stone is tied to a string and whirled round, this acceleration is produced by the string which pulls the stone; in a spinning-top the acceleration of the outer parts is produced by the forces binding them on to the inner part, and so on.

172. In the most important cases of this kind which occur in astronomy, a planet is known to revolve round

the sun in a path which does not differ much from a circle. If we assume for the present that the path is actually a circle, the planet must have an acceleration towards the centre, and it is possible to attribute this to the influence of the central body, the sun. In this way arises the idea of attributing to the sun the power of influencing in some way a planet which revolves round it, so as to give it an acceleration towards the sun; and the question at once arises of how this "influence" differs at different distances. To answer this question Newton made use of Kepler's Third Law (chapter VII., § 144). We have seen that, according to this law, the squares of the times of revolution of any two planets are proportional to the cubes of their distances from the sun; but the velocity of the planet may be found by dividing the length of the path it travels in its revolution round the sun by the time of the revolution, and this length is again proportional to the distance of the planet from the sun. Hence the velocities of the two planets are proportional to their distances from the sun, divided by the times of revolution, and consequently the squares of the velocities are proportional to the squares of the distances from the sun divided by the squares of the times of revolution. Hence, by Kepler's law, the squares of the velocities are proportional to the squares of the distances divided by the cubes of the distances, that is the squares of the velocities are *inversely* proportional to the distances, the more distant planet having the less velocity and *vice versa*. Now by the formula of Huygens the acceleration is measured by the square of the velocity divided by the radius of the circle (which in this case is the distance of the planet from the sun). The accelerations of the two planets towards the sun are therefore inversely proportional to the distances each multiplied by itself, that is are inversely proportional to the squares of the distances. Newton's first result therefore is: that the motions of the planets—regarded as moving in circles, and in strict accordance with Kepler's Third Law—can be explained as due to the action of the sun, if the sun is supposed capable of producing on a planet an acceleration towards the sun itself which is proportional to the inverse square of its distance from the sun; *i.e.* at twice the distance it is $\frac{1}{4}$ as great, at three times the distance $\frac{1}{9}$ as great, at ten times the distance $\frac{1}{100}$ as great, and so on.

The argument may perhaps be made clearer by a numerical example. In round numbers Jupiter's distance from the sun is five times as great as that of the earth, and Jupiter takes 12 years to perform a revolution round the sun, whereas the earth takes one. Hence Jupiter goes in 12 years five times as far as the earth goes in one, and Jupiter's velocity is therefore about $\frac{5}{12}$ that of the earth's, or the two velocities are in the ratio of 5 to 12; the squares of the velocities are therefore as 5×5 to 12×12, or as 25 to 144. The accelerations of Jupiter and of the earth towards the sun are therefore as $25 \div 5$ to 144, or as 5 to 144; hence Jupiter's acceleration towards the sun is about $\frac{1}{28}$ that of the earth, and if we had taken more accurate figures this fraction would have come out more nearly $\frac{1}{25}$. Hence at five times the distance the acceleration is 25 times less.

This **law of the inverse square,** as it may be called, is also the law according to which the light emitted from the sun or any other bright body varies, and would on this account also be not unlikely to suggest itself in connection with any kind of influence emitted from the sun.

173. The next step in Newton's investigation was to see whether the motion of the moon round the earth could be explained in some similar way. By the same argument as before, the moon could be shewn to have an acceleration towards the earth. Now a stone if let drop falls downwards, that is in the direction of the centre of the earth, and, as Galilei had shewn (chapter VI., § 133), this motion is one of uniform acceleration; if, in accordance with the opinion generally held at that time, the motion is regarded as being due to the earth, we may say that the earth has the power of giving an acceleration towards its own centre to bodies near its surface. Newton noticed that this power extended at any rate to the tops of mountains, and it occurred to him that it might possibly extend as far as the moon and so give rise to the required acceleration. Although, however, the acceleration of falling bodies, as far as was known at the time, was the same for

terrestrial bodies wherever situated, it was probable that at such a distance as that of the moon the acceleration caused by the earth would be much less. Newton assumed as a working hypothesis that the acceleration diminished according to the same law which he had previously arrived at in the case of the sun's action on the planets, that is that the acceleration produced by the earth on any body is inversely proportional to the square of the distance of the body from the centre of the earth.

It may be noticed that a difficulty arises here which did not present itself in the corresponding case of the planets. The distances of the planets from the sun being large compared with the size of the sun, it makes little difference whether the planetary distances are measured from the centre of the sun or from any other point in it. The same is true of the moon and earth; but when we are comparing the action of the earth on the moon with that on a stone situated on or near the ground, it is clearly of the utmost importance to decide whether the distance of the stone is to be measured from the nearest point of the earth, a few feet off, from the centre of the earth, 4000 miles off, or from some other point. Provisionally at any rate Newton decided on measuring from the centre of the earth.

It remained to verify his conjecture in the case of the moon by a numerical calculation; this could easily be done if certain things were known, *viz.* the acceleration of a falling body on the earth, the distance of the surface of the earth from its centre, the distance of the moon, and the time taken by the moon to perform a revolution round the earth. The first of these was possibly known with fair accuracy; the last was well known; and it was also known that the moon's distance was about 60 times the radius of the earth. How accurately Newton at this time knew the size of the earth is uncertain. Taking moderately accurate figures, the calculation is easily performed. In a month of about 27 days the moon moves about 60 times as far as the distance round the earth; that is she moves about $60 \times 24{,}000$ miles in 27 days, which is equivalent to about 3,300 feet per second. The acceleration of the moon is therefore measured by the square of this, divided by the distance of the moon (which is 60 times the radius of the earth, or 20,000,000 feet); that is, it is $\frac{3,300 \times 3,300}{60 \times 20,000,000}$, which reduces to about $\frac{1}{110}$. Consequently, if the law of the inverse square holds, the acceleration of a falling body at the surface of the earth, which is 60 times nearer to the centre than the moon is, should be $\frac{60 \times 60}{110}$, or between 32 and 33; but the actual acceleration of falling bodies is rather more than 32. The argument is therefore satisfactory, and Newton's hypothesis is so far verified.

The analogy thus indicated between the motion of the moon round the earth and the motion of a falling stone may be illustrated by a comparison, due to Newton, of the moon to a bullet shot horizontally out of a gun from a high place on the earth. Let the bullet start from B in fig. 71, then moving at first horizontally it will describe a curved path and reach the ground at a point such as C, at some distance from the point A, vertically underneath its starting-point. If it were shot out with a greater velocity, its path at first would be flatter and it would reach the ground at a point C′ beyond C; if the velocity were greater still, it would reach the ground at C″ or at C‴; and it requires only a slight effort of the imagination to conceive that, with a still greater velocity to begin with, it would miss the earth altogether and describe a circuit round it, such as BDE. This is exactly what the moon does, the only difference being that the moon is at a much greater distance than we have supposed the bullet to be, and that her motion has not been produced by anything analogous to the gun; but the motion being once there it is immaterial how it was produced or whether it was ever produced in the past. We may in fact say of the moon "that she is a falling body, only she is going so fast and is so far off that she falls quite round to the other side of the earth, instead of hitting it; and so goes on for ever." *

In the memorandum already quoted (§ 169) Newton speaks of the hypothesis as fitting the facts "pretty nearly"; but in a letter of earlier date (June 20th, 1686)

* W. K. Clifford, *Aims and Instruments of Scientific Thought.*

he refers to the calculation as not having been made accurately enough. It is probable that he used a seriously inaccurate value of the size of the earth, having overlooked the measurements of Snell and Norwood (chapter VIII., § 159); it is known that even at a later stage he was unable

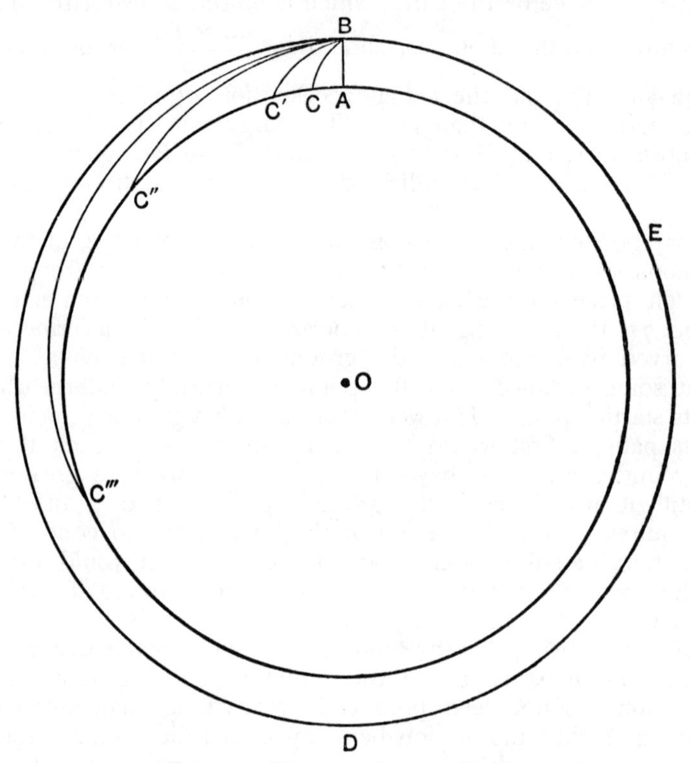

Fig. 71.—The moon as a projectile.

to deal satisfactorily with the difficulty above mentioned, as to whether the earth might for the purposes of the problem be identified with its centre; and he was of course aware that the moon's path differed considerably from a circle. The view, said to have been derived from Newton's conversation many years afterwards, that he was so dissatisfied with his results as to regard his hypothesis as substantially defective, is possible, but by no means certain; whatever the cause may have been, he laid the subject aside for some years without publishing anything on it, and devoted himself chiefly to optics and mathematics.

174. Meanwhile the problem of the planetary motions was one of the numerous subjects of discussion among the remarkable group of men who were the leading spirits of the Royal Society, founded in 1662. *Robert Hooke* (1635–1703), who claimed credit for most of the scientific discoveries of the time, suggested with some distinctness, not later than 1674, that the motions of the planets might be accounted for by attraction between them and the sun, and referred also to the possibility of the earth's attraction on bodies varying according to the law of the inverse square. *Christopher Wren* (1632–1723), better known as an architect than as a man of science, discussed some questions of this sort with Newton in 1677, and appears also to have thought of a law of attraction of this kind. A letter of Hooke's to Newton, written at the end of 1679, dealing amongst other things with the curve which a falling body would describe, the rotation of the earth being taken into account, stimulated Newton, who professed that at this time his "affection to philosophy" was "worn out," to go on with his study of the celestial motions. Picard's more accurate measurement of the earth (chapter VIII., § 159) was now well known, and Newton repeated his former calculation of the moon's motion, using Picard's improved measurement, and found the result more satisfactory than before.

175. At the same time (1679) Newton made a further discovery of the utmost importance by overcoming some of the difficulties connected with motion in a path other than a circle.

He shewed that if a body moved round a central body, in such a way that the line joining the two bodies sweeps out equal areas in equal times, as in Kepler's Second Law of planetary motion (chapter VII., § 141), then the moving body is acted on by an attraction directed exactly towards the central body; and further that if the path is an ellipse, with the central body in one focus, as in Kepler's First Law of planetary motion, then this attraction must vary in different parts of the path as the inverse square of the distance between the two bodies. Kepler's laws of planetary motion were in fact shewn to lead necessarily to the conclusions that the sun exerts on a planet an attraction inversely proportional to the square of the distance of the planet from the sun, and that such an attraction affords a sufficient explanation of the motion of the planet.

Once more, however, Newton published nothing and "threw his calculations by, being upon other studies."

176. Nearly five years later the matter was again brought to his notice, on this occasion by *Edmund Halley* (chapter x., §§ 199–205), whose friendship played henceforward an important part in Newton's life, and whose unselfish devotion to the great astronomer forms a pleasant contrast to the quarrels and jealousies prevalent at that time between so many scientific men. Halley, not knowing of Newton's work in 1666, rediscovered, early in 1684, the law of the inverse square, as a consequence of Kepler's Third Law, and shortly afterwards discussed with Wren and Hooke what was the curve in which a body would move if acted on by an attraction varying according to this law; but none of them could answer the question.* Later in the year Halley visited Newton at Cambridge and learnt from him the answer. Newton had, characteristically enough, lost his previous calculation, but was able to work it out again and sent it to Halley a few months afterwards. This time fortunately his attention was not diverted to other topics; he worked out at once a number of other problems of motion, and devoted his usual autumn course of University lectures to the subject. Perhaps the most interesting of the new results was that Kepler's Third Law, from which the law of the inverse square had been deduced in 1666, only on the supposition that the planets moved in circles, was equally consistent with Newton's law when the paths of the planets were taken to be ellipses.

177. At the end of the year 1684 Halley went to Cambridge again and urged Newton to publish his results. In accordance with this request Newton wrote out, and sent

* It is interesting to read that Wren offered a prize of 40s. to whichever of the other two should solve this the central problem of the solar system.

to the Royal Society, a tract called *Propositiones de Motu*, the 11 propositions of which contained the results already mentioned and some others relating to the motion of bodies under attraction to a centre. Although the propositions were given in an abstract form, it was pointed out that certain of them applied to the case of the planets. Further pressure from Halley persuaded Newton to give his results a more permanent form by embodying them in a larger book. As might have been expected, the subject grew under his hands, and the great treatise which resulted contained an immense quantity of material not contained in the *De Motu*. By the middle of 1686 the rough draft was finished, and some of it was ready for press. Halley not only undertook to pay the expenses, but superintended the printing and helped Newton to collect the astronomical data which were necessary. After some delay in the press, the book finally appeared early in July 1687, under the title *Philosophiae Naturalis Principia Mathematica*.

178. The *Principia*, as it is commonly called, consists of three books in addition to introductory matter: the first book deals generally with problems of the motion of bodies, solved for the most part in an abstract form without special reference to astronomy; the second book deals with the motion of bodies through media which resist their motion, such as ordinary fluids, and is of comparatively small astronomical importance, except that in it some glaring inconsistencies in the Cartesian theory of vortices are pointed out; the third book applies to the circumstances of the actual solar system the results already obtained, and is in fact an explanation of the motions of the celestial bodies on Newton's mechanical principles.

179. The introductory portion, consisting of "Definitions" and "Axioms, or Laws of Motion," forms a very notable contribution to dynamics, being in fact the first coherent statement of the fundamental laws according to which the motions of bodies are produced or changed. Newton himself does not appear to have regarded this part of his book as of very great importance, and the chief results embodied in it, being overshadowed as it were by the more striking discoveries in other parts of the book, attracted comparatively little attention. Much of it must be

passed over here, but certain results of special astronomical importance require to be mentioned.

Galilei, as we have seen (chapter VI., §§ 130, 133), was the first to enunciate the law that a body when once in motion continues to move in the same direction and at the same speed unless some cause is at work to make it change its motion. This law is given by Newton in the form already quoted in § 130, as the first of three fundamental laws, and is now commonly known as the First Law of Motion.

Galilei also discovered that a falling body moves with continually changing velocity, but with a uniform acceleration (chapter VI., § 133), and that this acceleration is the same for all bodies (chapter VI., § 116). The tendency of a body to fall having been generally recognised as due to the earth, Galilei's discovery involved the recognition that one effect of one body on another may be an acceleration produced in its motion. Newton extended this idea by shewing that the earth produced an acceleration in the motion of the moon, and the sun in the motion of the planets, and was led to the general idea of acceleration in a body's motion, which might be due in a variety of ways to the action of other bodies, and which could conveniently be taken as a measure of the effect produced by one body on another.

180. To these ideas Newton added the very important and difficult conception of **mass.**

If we are comparing two different bodies of the same material but of different sizes, we are accustomed to think of the larger one as heavier than the other. In the same way we readily think of a ball of lead as being heavier than a ball of wood of the same size. The most prominent idea connected with "heaviness" and "lightness" is that of the muscular effort required to support or to lift the body in question; a greater effort, for example, is required to hold the leaden ball than the wooden one. Again, the leaden ball if supported by an elastic string stretches it farther than does the wooden ball; or again, if they are placed in the scales of a balance, the lead sinks and the wood rises. All these effects we attribute to the "weight" of the two bodies, and the weight we are mostly accustomed to attribute in some way to the action of the earth on the bodies. The ordinary process of weighing a body in a balance shews, further, that we are accustomed to think of weight as a measurable quantity. On the other hand, we know from Galilei's result, which Newton tested very carefully by a series of pendulum experiments, that the leaden and the wooden ball, if allowed to drop, fall with the same acceleration. If therefore we measure the effect which the earth produces on the two balls by their acceleration, then the earth affects them equally; but if we measure it by the power which they have of stretching strings, or by the power which one has of supporting the other in a balance, then the effect which the earth produces on the leaden ball is greater than that produced on the wooden ball. Taken in this way, the action of the earth on either ball may be spoken of as weight, and the weight of a body can be measured by comparing it in a balance with standard bodies.

The difference between two such bodies as the leaden and wooden ball may, however, be recognised in quite a different way. We can easily see, for example, that a greater effort is needed to set the one in motion than the other; or that if each is tied to the end of a string of given kind and whirled round at a given rate, the one string is more tightly stretched than the other. In these cases the attraction of the earth is of no importance, and we recognise a distinction between the two bodies which is independent of the attraction of the earth. This distinction Newton regarded as due to a difference in the quantity of matter or material in the two bodies, and to this quantity he gave the name of mass. It may fairly be doubted whether anything is gained by this particular definition of mass, but the really important step was the distinct recognition of mass as a property of bodies, of fundamental importance in dynamical questions, and capable of measurement.

Newton, developing Galilei's idea, gave as one measurement of the action exerted by one body on another the product of the mass by the acceleration produced—a quantity for which he used different names, now replaced by **force**. The *weight* of a body was thus identified with the force exerted on it by the earth. Since the earth produces the same acceleration in all bodies at the same place, it follows that the masses of bodies at the same place are proportional to their weights; thus if two bodies are compared at the same place, and the weight of one (as shewn, for example, by a pair of scales) is found to be ten times that of the other, then its mass is also ten times as great. But such experiments as those of Richer at Cayenne (chapter VIII., § 161) shewed that the acceleration of falling bodies was less at the equator than in higher latitudes; so that if a body is carried from London or Paris to Cayenne, its weight is altered but its mass remains the same as before. Newton's conception of the earth's gravitation as extending as far as the moon gave further importance to the distinction between mass and weight; for if a body were removed from the earth to the moon, then its mass would be unchanged, but the acceleration due to the earth's attraction would be 60×60 times less, and its weight diminished in the same proportion.

Rules are also given for the effect produced on a body's motion by the simultaneous action of two or more forces.*

A further principle of great importance, of which only very indistinct traces are to be found before Newton's time, was given by him as the Third Law of Motion in the form: "To every action there is always an equal and contrary reaction; or, the mutual actions of any two bodies are always equal and oppositely directed." Here action and reaction are to be interpreted primarily in the sense of force. If a stone rests on the hand, the force with which the stone presses the hand downwards is equal to that with which the hand presses the stone upwards; if the earth attracts a stone downwards with a certain force, then the stone attracts the earth upwards with the same force, and so on. It is to be carefully noted that if, as in the last example, two bodies are acting on one another, the *accelerations* produced are not the same, but since force

* The familiar *parallelogram of forces*, of which earlier writers had had indistinct ideas, was clearly stated and proved in the introduction to the *Principia*, and was, by a curious coincidence, published also in the same year by *Varignon* and *Lami*.

is measured by the product of mass and acceleration, the body with the larger mass receives the lesser acceleration. In the case of a stone and the earth, the mass of the latter being enormously greater,* its acceleration is enormously less than that of the stone, and is therefore (in accordance with our experience) quite insensible.

181. When Newton began to write the *Principia* he had probably satisfied himself (§ 173) that the attracting power of the earth extended as far as the moon, and that the acceleration thereby produced in any body—whether the moon, or whether a body close to the earth—is inversely proportional to the square of the distance from the centre of the earth. With the ideas of force and mass this result may be stated in the form: *the earth attracts any body with a force inversely proportional to the square of the distance from the earth's centre, and also proportional to the mass of the body.*

In the same way Newton had established that the motions of the planets could be explained by an attraction towards the sun producing an acceleration inversely proportional to the square of the distance from the sun's centre, not only in the *same* planet in different parts of its path, but also in *different* planets. Again, it follows from this that the sun attracts any planet with a force inversely proportional to the square of the distance of the planet from the sun's centre, and also proportional to the mass of the planet.

But by the Third Law of Motion a body experiencing an attraction towards the earth must in turn exert an equal attraction on the earth; similarly a body experiencing an attraction towards the sun must exert an equal attraction on the sun. If, for example, the mass of Venus is seven times that of Mars, then the force with which the sun attracts Venus is seven times as great as that with which it would attract Mars if placed at the same distance; and therefore also the force with which Venus attracts the sun is seven times as great as that with which Mars would attract the sun if at an equal distance from it. Hence, in all the cases of attraction hitherto considered and in

* It is between 13 and 14 billion billion pounds. See chapter x., § 219.

which the comparison is possible, the force is proportional not only to the mass of the attracted body, but also to that of the attracting body, as well as being inversely proportional to the square of the distance. Gravitation thus appears no longer as a property peculiar to the central body of a revolving system, but as belonging to a planet in just the same way as to the sun, and to the moon or to a stone in just the same way as to the earth.

Moreover, the fact that separate bodies on the surface of the earth are attracted by the earth, and therefore in turn attract it, suggests that this power of attracting other bodies which the celestial bodies are shewn to possess does not belong to each celestial body as a whole, but to the separate particles making it up, so that, for example, the force with which Jupiter and the sun mutually attract one another is the result of compounding the forces with which the separate particles making up Jupiter attract the separate particles making up the sun. Thus is suggested finally the law of gravitation in its most general form: *every particle of matter attracts every other particle with a force proportional to the mass of each, and inversely proportional to the square of the distance between them.**

182. In all the astronomical cases already referred to the attractions between the various celestial bodies have been treated as if they were accurately directed towards their centres, and the distance between the bodies has been taken to be the distance between their centres. Newton's doubts on this point, in the case of the earth's attraction of bodies, have been already referred to (§ 173); but early in 1685 he succeeded in justifying this assumption. By a singularly beautiful and simple course of reasoning he shewed (*Principia*, Book I., propositions 70, 71) that, if a body is spherical in form and equally dense throughout, it attracts any particle external to it exactly as if its whole mass were concentrated at its centre. He shewed, further, that the same is true for a sphere of variable density, provided it can be regarded as made up of a series of spherical shells, having a common centre, each of uniform

* As far as I know Newton gives no short statement of the law in a perfectly complete and general form; separate parts of it are given in different passages of the *Principia*.

density throughout, different shells being, however, of different densities. For example, the result is true for a hollow indiarubber ball as well as for a solid one, but is not true for a sphere made up of a hemisphere of wood and a hemisphere of iron fastened together.

183. The law of gravitation being thus provisionally established, the great task which lay before Newton, and to which he devotes the greater part of the first and third books of the *Principia*, was that of deducing from it and the "laws of motion" the motions of the various members of the solar system, and of shewing, if possible, that the motions so calculated agreed with those observed. If this were successfully done, it would afford a verification of the most delicate and rigorous character of Newton's principles.

The conception of the solar system as a mechanism, each member of which influences the motion of every other member in accordance with one universal law of attraction, although extremely simple in itself, is easily seen to give rise to very serious difficulties when it is proposed actually to calculate the various motions. If in dealing with the motion of a planet such as Mars it were possible to regard Mars as acted on only by the attraction of the sun, and to ignore the effects of the other planets, then the problem would be completely solved by the propositions which Newton established in 1679 (§ 175), and by their means the position of Mars at any time could be calculated with any required degree of accuracy. But in the case which actually exists the motion of Mars is affected by the forces with which all the other planets (as well as the satellites) attract it, and these forces in turn depend on the position of Mars (as well as upon that of the other planets) and hence upon the motion of Mars. A problem of this kind in all its generality is quite beyond the powers of any existing mathematical methods. Fortunately, however, the mass of even the largest of the planets is so very much less than that of the sun, that the motion of any one planet is only slightly affected by the others; and it may be regarded as moving very nearly as it would move if the other planets did not exist, the effect of these being afterwards allowed for as producing disturbances or **perturbations** in its path. Although even in this simplified form the problem of the motion of the planets is one of extreme difficulty (cf. chapter XI., § 228), and Newton was unable to solve it with anything like completeness, yet he was able to point out certain general effects which must result from the mutual action of the planets, the most interesting being the slow forward motion of the apses of the earth's orbit, which had long ago been noticed by observing astronomers (chapter III., § 59). Newton also pointed out that Jupiter, on account of its great mass, must produce a considerable perturbation in the motion of its neighbour Saturn, and thus gave some explanation of an irregularity first noted by Horrocks (chapter VIII., § 156).

184. The motion of the moon presents special difficulties, but Newton, who was evidently much interested in the problems of lunar theory, succeeded in overcoming them much more completely than the corresponding ones connected with the planets.

The moon's motion round the earth is primarily due to the attraction of the earth; the perturbations due to the other planets are insignificant; but the sun, which though at a very great distance has an enormously greater mass than the earth, produces a very sensible disturbing effect on the moon's motion. Certain irregularities, as we have seen (chapter II., §§ 40, 48; chapter V., § 111), had already been discovered by observation. Newton was able to shew that the disturbing action of the sun would necessarily produce perturbations of the same general character as those thus recognised, and in the case of the motion of the moon's nodes and of her apogee he was able to get a very fairly accurate numerical result;* and he also discovered a number of other irregularities, for the most part very small, which had not hitherto been noticed. He indicated also the existence of certain irregularities in the motions of Jupiter's and Saturn's moons analogous to those which occur in the case of our moon.

* It is commonly stated that Newton's value of the motion of the moon's apses was only about half the true value. In a scholium of the *Principia* to prop. 35 of the third book, given in the first edition but afterwards omitted, he estimated the annual motion at 40°, the observed value being about 41°. In one of his unpublished papers, contained in the Portsmouth collection, he arrived at 39° by a process which he evidently regarded as not altogether satisfactory.

185. One group of results of an entirely novel character resulted from Newton's theory of gravitation. It became for the first time possible to estimate the *masses* of some of the celestial bodies, by comparing the attractions exerted by them on other bodies with that exerted by the earth.

The case of Jupiter may be given as an illustration. The time of revolution of Jupiter's outermost satellite is known to be about 16 days 16 hours, and its distance from Jupiter was estimated by Newton (not very correctly) at about four times the distance of the moon from the earth. A calculation exactly like that of § 172 or § 173 shews that the acceleration of the satellite due to Jupiter's attraction is about ten times as great as the acceleration of the moon towards the earth, and that therefore, the distance being four times as great, Jupiter attracts a body with a force $10 \times 4 \times 4$ times as great as that with which the earth attracts a body at the same distance; consequently Jupiter's mass is 160 times that of the earth. This process of reasoning applies also to Saturn, and in a very similar way a comparison of the motion of a planet, Venus for example, round the sun with the motion of the moon round the earth gives a relation between the masses of the sun and earth. In this way Newton found the mass of the sun to be 1067, 3021, and 169282 times greater than that of Jupiter, Saturn, and the earth, respectively. The corresponding figures now accepted are not far from 1047, 3530, 324439. The large error in the last number is due to the use of an erroneous value of the distance of the sun—then not at all accurately known—upon which depend the other distances in the solar system, except those connected with the earth-moon system. As it was necessary for the employment of this method to be able to observe the motion of some other body attracted by the planet in question, it could not be applied to the other three planets (Mars, Venus, and Mercury), of which no satellites were known.

186. From the equality of action and reaction it follows that, since the sun attracts the planets, they also attract the sun, and the sun consequently is in motion, though—owing to the comparative smallness of the planets—only to a very small extent. It follows that Kepler's Third Law is not strictly accurate, deviations from it becoming sensible in

the case of the large planets Jupiter and Saturn (cf. chapter VII., § 144). It was, however, proved by Newton that in any system of bodies, such as the solar system, moving about in any way under the influence of their mutual attractions, there is a particular point, called the **centre of gravity,** which can always be treated as at rest; the sun moves relatively to this point, but so little that the distance between the centre of the sun and the centre of gravity can never be much more than the diameter of the sun.

It is perhaps rather curious that this result was not seized upon by some of the supporters of the Church in the condemnation of Galilei, now rather more than half a century old; for if it was far from supporting the view that the earth is at the centre of the world, it at any rate negatived that part of the doctrine of Coppernicus and Galilei which asserted the sun to be *at rest* in the centre of the world. Probably no one who was capable of understanding Newton's book was a serious supporter of any anti-Coppernican system, though some still professed themselves obedient to the papal decrees on the subject.*

* Throughout the Coppernican controversy up to Newton's time it had been generally assumed, both by Coppernicans and by their opponents, that there was some meaning in speaking of a body simply as being "at rest" or "in motion," without any reference to any other body. But all that we can really observe is the motion of one body relative to one or more others. Astronomical observation tells us, for example, of a certain motion relative to one another of the earth and sun; and this motion was expressed in two quite different ways by Ptolemy and by Coppernicus. From a modern standpoint the question ultimately involved was whether the motions of the various bodies of the solar system relatively to the earth or relatively to the sun were the simpler to express. If it is found convenient to express them—as Coppernicus and Galilei did—in relation to the sun, some simplicity of statement is gained by speaking of the sun as "fixed" and omitting the qualification "relative to the sun" in speaking of any other body. The same motions might have been expressed relatively to any other body chosen at will: *e.g.* to one of the hands of a watch carried by a man walking up and down on the deck of a ship on a rough sea; in this case it is clear that the motions of the other bodies of the solar system relative to this body would be excessively complicated; and it would therefore be highly inconvenient though still possible to treat this particular body as "fixed."

A new aspect of the problem presents itself, however, when an attempt—like Newton's—is made to explain the motions of bodies of the solar system as the result of forces exerted on one another by

187. The variation of the time of oscillation of a pendulum in different parts of the earth, discovered by Richer in 1672 (chapter VIII., § 161), indicated that the earth was probably not a sphere. Newton pointed out that this departure from the spherical form was a consequence of the mutual gravitation of the particles making up the earth and of the earth's rotation. He supposed a canal of water to pass from the pole to the centre of the earth, and then from the centre to a point on the equator (B O *a* A in fig. 72), and then found the condition that these two columns of water O B, O A, each being attracted towards the centre of the earth, should balance. This method involved certain assumptions as to the inside of the earth, of which little can be said to be known even now, and consequently, though Newton's general result, that the earth is flattened at the poles and bulges out at the equator, was right, the actual numerical expression which he found was not very accurate. If, in the figure, the dotted line is a circle the radius of which is equal to the distance of the

those bodies. If, for example, we look at Newton's First Law of Motion (chapter VI., § 130), we see that it has no meaning, unless we know what are the body or bodies relative to which the motion is being expressed; a body at rest relatively to the earth is moving relatively to the sun or to the fixed stars, and the applicability of the First Law to it depends therefore on whether we are dealing with its motion relatively to the earth or not. For most terrestrial motions it is sufficient to regard the Laws of Motion as referring to motion relative to the earth; or, in other words, we may for this purpose treat the earth as "fixed." But if we examine certain terrestrial motions more exactly, we find that the Laws of Motion thus interpreted are not quite true; but that we get a more accurate explanation of the observed phenomena if we regard the Laws of Motion as referring to motion relative to the centre of the sun and to lines drawn from it to the stars; or, in other words, we treat the centre of the sun as a "fixed" point and these lines as "fixed" directions. But again when we are dealing with the solar system generally this interpretation is slightly inaccurate, and we have to treat the centre of gravity of the solar system instead of the sun as "fixed."

From this point of view we may say that Newton's object in the *Principia* was to shew that it was possible to choose a certain point (the centre of gravity of the solar system) and certain directions (lines joining this point to the fixed stars), as a base of reference, such that all motions being treated as relative to this base, the Laws of Motion and the law of gravitation afford a consistent explanation of the observed motions of the bodies of the solar system.

pole B from the centre of the earth O, then the actual surface of the earth extends at the equator beyond this circle as far as A, where, according to Newton, *a* A is about $\frac{1}{230}$ of O B or O A, and according to modern estimates, based on actual measurement of the earth as well as upon theory (chapter X., § 221), it is about $\frac{1}{293}$ of O A. Both Newton's fraction and the modern one are so small that the resulting flattening cannot be made sensible in a figure; in fig. 72

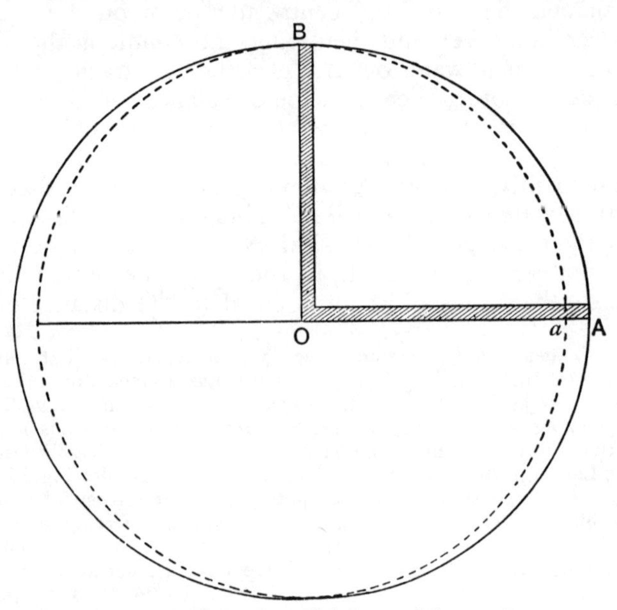

FIG. 72.—The spheroidal form of the earth.

the length *a* A is made, for the sake of distinctness, nearly 30 times as great as it should be.

Newton discovered also in a similar way the flattening of Jupiter, which, owing to its more rapid rotation, is considerably more flattened than the earth; this was also detected telescopically by Domenico Cassini four years after the publication of the *Principia*.

188. The discovery of the form of the earth led to an explanation of the precession of the equinoxes, a phenomenon which had been discovered 1,800 years before

(chapter II., § 42), but had remained a complete mystery ever since.

If the earth is a perfect sphere, then its attraction on any other body is exactly the same as if its mass were all concentrated at its centre (§ 182), and so also the attraction on it of any other body such as the sun or moon is equivalent to a single force passing through the centre O of the earth; but this is no longer true if the earth is not spherical. In fact the action of the sun or moon on the spherical part of the earth, inside the dotted circle in fig. 72, is equivalent to a force through O, and has no tendency to turn the earth in any way about its centre; but the attraction on the remaining portion is of a different character, and Newton shewed that from it resulted a motion of the axis of the earth of the same general character as precession. The amount of the precession as calculated by Newton did as a matter of fact agree pretty closely with the observed amount, but this was due to the accidental compensation of two errors, arising from his imperfect knowledge of the form and construction of the earth, as well as from erroneous estimates of the distance of the sun and of the mass of the moon, neither of which quantities Newton was able to measure with any accuracy.* It was further pointed out that the motion in question was necessarily not quite uniform, but that, owing to the unequal effects of the sun in different positions, the earth's axis would oscillate to and fro every six months, though to a very minute extent.

189. Newton also gave a general explanation of the tides as due to the disturbing action of the moon and sun, the former being the more important. If the earth be regarded as made of a solid spherical nucleus, covered by the ocean, then the moon attracts different parts unequally, and in particular the attraction, measured by the acceleration produced, on the water nearest to the moon is greater than

* He estimated the annual precession due to the sun to be about 9″, and that due to the moon to be about four and a half times as great, so that the total amount due to the two bodies came out about 50″, which agrees within a fraction of a second with the amount shewn by observation; but we know now that the moon's share is not much more than twice that of the sun.

that on the solid earth, and that on the water farthest from the moon is less. Consequently the water moves on the surface of the earth, the general character of the motion being the same as if the portion of the ocean on the side towards the moon were attracted and that on the opposite side repelled. Owing to the rotation of the earth and the moon's motion, the moon returns to nearly the same position with respect to any place on the earth in a period which exceeds a day by (on the average) about 50 minutes, and consequently Newton's argument shewed that low tides (or high tides) due to the moon would follow one another at any given place at intervals equal to about half this period; or, in other words, that two tides would in general occur daily, but that on each day any particular phase of the tides would occur on the average about 50 minutes later than on the preceding day, a result agreeing with observation. Similar but smaller tides were shewn by the same argument to arise from the action of the sun, and the actual tide to be due to the combination of the two. It was shewn that at new and full moon the lunar and solar tides would be added together, whereas at the half moon they would tend to counteract one another, so that the observed fact of greater tides every fortnight received an explanation. A number of other peculiarities of the tides were also shewn to result from the same principles.

Newton ingeniously used observations of the height of the tide when the sun and moon acted together and when they acted in opposite ways to compare the tide-raising powers of the sun and moon, and hence to estimate the mass of the moon in terms of that of the sun, and consequently in terms of that of the earth (§ 185). The resulting mass of the moon was about twice what it ought to be according to modern knowledge, but as before Newton's time no one knew of any method of measuring the moon's mass even in the roughest way, and this result had to be disentangled from the innumerable complications connected with both the theory and with observation of the tides, it cannot but be regarded as a remarkable achievement. Newton's theory of the tides was based on certain hypotheses which had to be made in order to render the problem at all manageable, but which were certainly not true, and consequently, as he was well aware, important modifications would necessarily have to be made, in order to bring his results into agreement with actual facts. The mere presence of land not covered by water is, for example, sufficient by itself to produce important alterations in tidal effects at different places. Thus Newton's theory was by no means equal to such a task as that of predicting the times of high tide at any required place, or the height of any required tide, though it gave a satisfactory explanation of many of the general characteristics of tides.

190. As we have seen (chapter v., § 103; chapter vii., § 146), comets until quite recently had been commonly regarded as terrestrial objects produced in the higher regions of our atmosphere, and even the more enlightened astronomers who, like Tycho, Kepler, and Galilei, recognised them as belonging to the celestial bodies, were unable to give an explanation of their motions and of their apparently quite irregular appearances and disappearances. Newton was led to consider whether a comet's motion could not be explained, like that of a planet, by gravitation towards the sun. If so then, as he had proved near the beginning of the *Principia*, its path must be either an ellipse or one of two other allied curves, the **parabola** and **hyperbola**. If a comet moved in an ellipse which only differed slightly from a circle, then it would never recede to any very great distance from the centre of the solar system, and would therefore be regularly visible, a result which was contrary to observation. If, however, the ellipse was very elongated, as shewn in fig. 73, then the period of revolution might easily be very great, and, during the greater part of it, the comet would be so far from the sun and consequently also from the earth as to be invisible. If so the comet would be seen for a short time and become invisible, only to reappear after a very long time, when it would naturally be regarded as a new comet. If again the path of the comet were a parabola (which may be regarded as an ellipse indefinitely elongated), the comet would not return at all, but would merely be seen once when in that part of its path which is near the sun. But if a comet moved in a parabola, with the sun in a focus,

then its positions when not very far from the sun would be almost the same as if it moved in an elongated ellipse (see fig. 73), and consequently it would hardly be possible to distinguish the two cases. Newton accordingly worked out the case of motion in a parabola, which is mathematically the simpler, and found that, in the case of a comet which had attracted much attention in the winter 1680–1, a parabolic path could be found, the calculated places of the comet in which agreed closely with those observed. In the later editions of the *Principia* the motions of a number of other comets were investigated with a similar

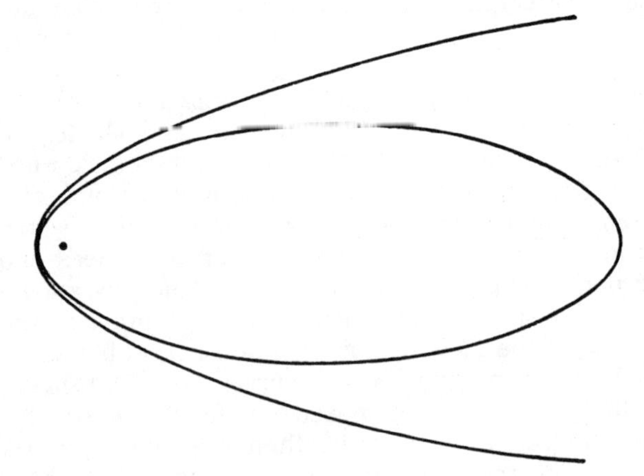

FIG. 73.—An elongated ellipse and a parabola.

result. It was thus established that in many cases a comet's path is either a parabola or an elongated ellipse, and that a similar result was to be expected in other cases. This reduction to rule of the apparently arbitrary motions of comets, and their inclusion with the planets in the same class of bodies moving round the sun under the action of gravitation, may fairly be regarded as one of the most striking of the innumerable discoveries contained in the *Principia*.

In the same section Newton discussed also at some length the nature of comets and in particular the structure of their tails, arriving at the conclusion, which is in general agreement with modern theories (chapter XIII., § 304), that the tail is formed by a stream of finely divided matter of the nature of smoke, rising up from the body of the comet, and so illuminated by the light of the sun when tolerably near it as to become visible.

191. The *Principia* was published, as we have seen, in 1687. Only a small edition seems to have been printed, and this was exhausted in three or four years. Newton's earlier discoveries, and the presentation to the Royal Society of the tract *De Motu* (§ 177), had prepared the scientific world to look for important new results in the *Principia*, and the book appears to have been read by the leading Continental mathematicians and astronomers, and to have been very warmly received in England. The Cartesian philosophy had, however, too firm a hold to be easily shaken; and Newton's fundamental principle, involving as it did the idea of an action between two bodies separated by an interval of empty space, seemed impossible of acceptance to thinkers who had not yet fully grasped the notion of judging a scientific theory by the extent to which its consequences agree with observed facts. Hence even so able a man as Huygens (chapter VIII., §§ 154, 157, 158), regarded the idea of gravitation as "absurd," and expressed his surprise that Newton should have taken the trouble to make such a number of laborious calculations with no foundation but this principle, a remark which shewed Huygens to have had no conception that the agreement of the results of these calculations with actual facts was proof of the soundness of the principle. Personal reasons also contributed to the Continental neglect of Newton's work, as the famous quarrel between Newton and Leibniz as to their respective claims to the invention of what Newton called fluxions and Leibniz the differential method (out of which the differential and integral calculus have developed) grew in intensity and fresh combatants were drawn into it on both sides. Half a century in fact elapsed before Newton's views made any substantial progress on the Continent (cf. chapter XI., § 229). In our country the case was different; not only was the *Principia* read with admiration by the few who were capable of understanding it, but scholars like Bentley, philosophers like Locke, and courtiers like Halifax all made attempts

to grasp Newton's general ideas, even though the details of his mathematics were out of their range. It was moreover soon discovered that his scientific ideas could be used with advantage as theological arguments.

192. One unfortunate result of the great success of the *Principia* was that Newton was changed from a quiet Cambridge professor, with abundant leisure and a slender income, into a public character, with a continually increasing portion of his time devoted to public business of one sort or another.

Just before the publication of the *Principia* he had been appointed one of the representatives of his University to defend its rights against the encroachments of James II., and two years later he sat as member for the University in the Convention Parliament, though he retired after its dissolution.

Notwithstanding these and many other distractions, he continued to work at the theory of gravitation, paying particular attention to the lunar theory, a difficult subject with his treatment of which he was never quite satisfied.* He was fortunately able to obtain from time to time first-rate observations of the moon (as well as of other bodies) from the Astronomer Royal Flamsteed (chapter x., §§ 197–8), though Newton's continual requests and Flamsteed's occasional refusals led to strained relations at intervals. It is possible that about this time Newton contemplated writing a new treatise, with more detailed treatment of various points discussed in the *Principia*; and in 1691 there was already some talk of a new edition of the *Principia*, possibly to be edited by some younger mathematician. In any case nothing serious in this direction was done for some years, perhaps owing to a serious illness, apparently some nervous disorder, which attacked Newton in 1692 and lasted about two years. During this illness, as he himself said, "he had not his usual consistency of mind," and it is by no means certain that he ever recovered his full mental activity and power.

Soon after recovering from this illness he made some

* He once told Halley in despair that the lunar theory "made his head ache and kept him awake so often that he would think of it no more."

NEWTON. [*To face p.* 240.

preparations for a new edition of the *Principia*, besides going on with the lunar theory, but the work was again interrupted in 1695, when he received the valuable appointment of Warden to the Mint, from which he was promoted to the Mastership four years later. He had, in consequence, to move to London (1696), and much of his time was henceforward occupied by official duties. In 1701 he resigned his professorship at Cambridge, and in the same year was for the second time elected the Parliamentary representative of the University. In 1703 he was chosen President of the Royal Society, an office which he held till his death, and in 1705 he was knighted on the occasion of a royal visit to Cambridge.

During this time he published (1704) his treatise on *Optics*, the bulk of which was probably written long before, and in 1709 he finally abandoned the idea of editing the *Principia* himself, and arranged for the work to be done by *Roger Cotes* (1682–1716), the brilliant young mathematician whose untimely death a few years later called from Newton the famous eulogy, "If Mr. Cotes had lived we might have known something." The alterations to be made were discussed in a long and active correspondence between the editor and author, the most important changes being improvements and additions to the lunar theory, and to the discussions of precession and of comets, though there were also a very large number of minor changes; and the new edition appeared in 1713. A third edition, edited by Pemberton, was published in 1726, but this time Newton, who was over 80, took much less part, and the alterations were of no great importance. This was Newton's last piece of scientific work, and his death occurred in the following year (March 3rd, 1727).

193. It is impossible to give an adequate idea of the immense magnitude of Newton's scientific discoveries except by a free use of the mathematical technicalities in which the bulk of them were expressed. The criticism passed on him by his personal enemy Leibniz that, "Taking mathematics from the beginning of the world to the time when Newton lived, what he had done was much the better half," and the remark of his great successor Lagrange (chapter XI., § 237), "Newton was the

greatest genius that ever existed, and the most fortunate, for we cannot find more than once a system of the world to establish," shew the immense respect for his work felt by those who were most competent to judge it.

With these magnificent eulogies it is pleasant to compare Newton's own grateful recognition of his predecessors, "If I have seen further than other men, it is because I have stood upon the shoulders of the giants," and his modest estimate of his own performances:—

"I do not know what I may appear to the world; but to myself I seem to have been only like a boy playing on the sea-shore, and diverting myself in now and then finding a smoother pebble or a prettier shell than ordinary, whilst the great ocean of truth lay all undiscovered before me."

194. It is sometimes said, in explanation of the difference between Newton's achievements and those of earlier astronomers, that whereas they discovered *how* the celestial bodies moved, he shewed *why* the motions were as they were, or, in other words, that they *described* motions while he *explained* them or ascertained their cause. It is, however, doubtful whether this distinction between How and Why, though undoubtedly to some extent convenient, has any real validity. Ptolemy, for example, represented the motion of a planet by a certain combination of epicycles; his scheme was equivalent to a particular method of describing the motion; but if any one had asked him why the planet would be in a particular position at a particular time, he might legitimately have answered that it was so because the planet was connected with this particular system of epicycles, and its place could be deduced from them by a rigorous process of calculation. But if any one had gone further and asked why the planet's epicycles were as they were, Ptolemy could have given no answer. Moreover, as the system of epicycles differed in some important respects from planet to planet, Ptolemy's system left unanswered a number of questions which obviously presented themselves. Then Coppernicus gave a partial answer to some of these questions. To the question why certain of the planetary motions, corresponding to certain epicycles, existed, he would have replied that it was because of certain motions of the earth, from which

these (apparent) planetary motions could be deduced as necessary consequences. But the same information could also have been given as a mere descriptive statement that the earth moves in certain ways and the planets move in certain other ways. But again, if Coppernicus had been asked why the earth rotated on its axis, or why the planets revolved round the sun, he could have given no answer; still less could he have said why the planets had certain irregularities in their motions, represented by his epicycles.

Kepler again described the same motions very much more simply and shortly by means of his three laws of planetary motion; but if any one had asked why a planet's motion varied in certain ways, he might have replied that it was because all planets moved in ellipses so as to sweep out equal areas in equal times. *Why* this was so Kepler was unable to say, though he spent much time in speculating on the subject. This question was, however, answered by Newton, who shewed that the planetary motions were necessary consequences of his law of gravitation and his laws of motion. Moreover from these same laws, which were extremely simple in statement and few in number, followed as necessary consequences the motion of the moon and many other astronomical phenomena, and also certain familiar terrestrial phenomena, such as the behaviour of falling bodies; so that a large number of groups of observed facts, which had hitherto been disconnected from one another, were here brought into connection as necessary consequences of certain fundamental laws. But again Newton's view of the solar system might equally well be put as a mere descriptive statement that the planets, etc., move with accelerations of certain magnitudes towards one another. As, however, the actual position or rate of motion of a planet at any time can only be deduced by an extremely elaborate calculation from Newton's laws, they are not at all obviously equivalent to the observed celestial motions, and we do not therefore at all easily think of them as being merely a description.

Again Newton's laws at once suggest the question why bodies attract one another in this particular way; and this question, which Newton fully recognised as legitimate, he was unable to answer. Or again we might ask why the planets are of certain sizes, at certain distances from the sun, etc., and to these questions again Newton could give no answer.

But whereas the questions left unanswered by Ptolemy, Coppernicus, and Kepler were in whole or in part answered by their successors, that is, their unexplained facts or laws were shewn to be necessary consequences of other simpler and more general laws, it happens that up to the present day no one has been able to answer, in any satisfactory way, these questions which Newton left unanswered. In this particular direction, therefore, Newton's laws mark the boundary of our present knowledge. But if any one were to succeed this year or next in shewing gravitation to be a consequence of some still more general law, this new law would still bring with it a new Why.

If, however, Newton's laws cannot be regarded as an ultimate explanation of the phenomena of the solar system, except in the historic sense that they have not yet been shewn to depend on other more fundamental laws, their success in "explaining," with fair accuracy; such an immense mass of observed results in all parts of the solar system, and their universal character, gave a powerful impetus to the idea of accounting for observed facts in other departments of science, such as chemistry and physics, in some similar way as the consequence of forces acting between bodies, and hence to the conception of the material universe as made up of a certain number of bodies, each acting on one another with definite forces in such a way that all the changes which can be observed to go on are necessary consequences of these forces, and are capable of prediction by any one who has sufficient knowledge of the forces and sufficient mathematical skill to develop their consequences.

Whether this conception of the material universe is adequate or not, it has undoubtedly exercised a very important influence on scientific discovery as well as on philosophical thought, and although it was never formulated by Newton, and parts of it would probably have been repudiated by him, there are indications that some such ideas were in his head, and those who held the conception most firmly undoubtedly derived their ideas directly or indirectly from him.

195. Newton's scientific method did not differ essentially from that followed by Galilei (chapter VI., § 134), which has been variously described as **complete induction** or as the **inverse deductive method,** the difference in name corresponding to a difference in the stress laid upon different parts of the same general process. Facts are obtained by observation or experiment; a hypothesis or provisional theory is devised to account for them; from this theory are obtained, if possible by a rigorous process of deductive reasoning, certain consequences capable of being compared with actual facts, and the comparison is then made. In some cases the first process may appear as the more important, but in Newton's work the really convincing part of the proof of his results lay in the verification involved in the two last processes. This has perhaps been somewhat obscured by his famous remark, *Hypotheses non fingo* (I do not invent hypotheses), dissociated from its context. The words occur in the conclusion of the *Principia*, after he has been speaking of universal gravitation :—

"I have not yet been able to deduce (*deducere*) from phenomena the reason of these properties of gravitation, and I do not invent hypotheses. For any thing which cannot be deduced from phenomena should be called a hypothesis."

Newton probably had in his mind such speculations as the Cartesian vortices, which could not be deduced directly from observations, and the consequences of which either could not be worked out and compared with actual facts or were inconsistent with them. Newton in fact rejected hypotheses which were unverifiable, but he constantly made hypotheses, suggested by observed facts, and verified by the agreement of their consequences with fresh observed facts. The extension of gravity to the moon (§ 173) is a good example: he was acquainted with certain facts as to the motion of falling bodies and the motion of the moon; it occurred to him that the earth's attraction might extend as far as the moon, and certain other facts connected with Kepler's Third Law suggested the law of the inverse square. If this were right, the moon's acceleration towards the earth ought to have a certain value, which could be obtained by calculation. The calculation was made and found to agree roughly with the actual motion of the moon.

Moreover it may be fairly urged, in illustration of the great importance of the process of verification, that Newton's fundamental laws were not rigorously established by him, but that the deficiencies in his proofs have been to a great extent filled up by the elaborate process of verification that has gone on since. For the motions of the solar system, as deduced by Newton from gravitation and the laws of motion, only agreed roughly with observation; many outstanding discrepancies were left; and though there was a strong presumption that these were due to the necessary imperfections of Newton's processes of calculation, an immense expenditure of labour and ingenuity on the part of a series of mathematicians has been required to remove these discrepancies one by one, and as a matter of fact there remain even to-day a few small ones which are unexplained (chapter XIII., § 290).

CHAPTER X.

OBSERVATIONAL ASTRONOMY IN THE 18TH CENTURY.

"Through Newton theory had made a great advance and was ahead of observation; the latter now made efforts to come once more level with theory."—BESSEL.

196. NEWTON virtually created a new department of astronomy, **gravitational astronomy**, as it is often called, and bequeathed to his successors the problem of deducing more fully than he had succeeded in doing the motions of the celestial bodies from their mutual gravitation.

To the solution of this problem Newton's own countrymen contributed next to nothing throughout the 18th century, and his true successors were a group of Continental mathematicians whose work began soon after his death, though not till nearly half a century after the publication of the *Principia*.

This failure of the British mathematicians to develop Newton's discoveries may be explained as due in part to the absence or scarcity of men of real ability, but in part also to the peculiarity of the mathematical form in which Newton presented his discoveries. The *Principia* is written almost entirely in the language of geometry, modified in a special way to meet the requirements of the case; nearly all subsequent progress in gravitational astronomy has been made by mathematical methods known as **analysis.** Although the distinction between the two methods cannot be fully appreciated except by those who have used them both, it may perhaps convey some impression of the differences between them to say that in the geometrical treatment of an astronomical problem each step of the reasoning is expressed in such a way as to be capable of being interpreted in terms of the original problem, whereas in the analytical treatment the problem is first expressed by means of algebraical symbols; these symbols are manipulated according to certain purely formal rules, no regard being paid to the interpretation of the intermediate steps, and the final algebraical result, if it can be obtained, yields on interpretation the solution of the original problem. The geometrical solution of a problem, if it can be obtained, is frequently shorter, clearer, and more elegant; but, on the other hand, each special problem has to be considered separately, whereas the analytical solution can be conducted to a great extent according to fixed rules applicable in a larger number of cases. In Newton's time modern analysis was only just coming into being, some of the most important parts of it being in fact the creation of Leibniz and himself, and although he sometimes used analysis to solve an astronomical problem, it was his practice to translate the result into geometrical language before publication; in doing so he was probably influenced to a large extent by a personal preference for the elegance of geometrical proofs, partly also by an unwillingness to increase the numerous difficulties contained in the *Principia*, by using mathematical methods which were comparatively unfamiliar. But though in the hands of a master like Newton geometrical methods were capable of producing astonishing results, the lesser men who followed him were scarcely ever capable of using his methods to obtain results beyond those which he himself had reached. Excessive reverence for Newton and all his ways, combined with the estrangement which long subsisted between British and foreign mathematicians, as the result of the fluxional controversy (chapter IX., § 191), prevented the former from using the analytical methods which were being rapidly perfected by Leibniz's pupils and other Continental mathematicians. Our mathematicians remained, therefore, almost isolated during the whole of the 18th century, and with the exception of some admirable work by *Colin Maclaurin* (1698–1746), which carried Newton's theory of the figure of the earth a stage further, nothing of importance was done in our country for nearly a century after Newton's death to develop the theory of gravitation beyond the point at which it was left in the *Principia*.

In other departments of astronomy, however, important progress was made both during and after Newton's lifetime, and by a curious inversion, while Newton's ideas were developed chiefly by French mathematicians, the Observatory of Paris, at which Picard and others had done such admirable work (chapter VIII., §§ 160–2), produced little of real importance for nearly a century afterwards, and a large part of the best observing work of the 18th century was done by Newton's countrymen. It will be convenient to separate these two departments of astronomical work, and to deal in the next chapter with the development of the theory of gravitation.

197. The first of the great English observers was Newton's contemporary *John Flamsteed*, who was born near Derby in 1646 and died at Greenwich in 1720.* Unfortunately the character of his work was such that, marked as it was by no brilliant discoveries, it is difficult to present it in an attractive form or to give any adequate idea of its real extent and importance. He was one of those laborious and careful investigators, the results of whose work are invaluable as material for subsequent research, but are not striking in themselves.

He made some astronomical observations while quite a boy, and wrote several papers, of a technical character, on astronomical subjects, which attracted some attention. In 1675 he was appointed a member of a Committee to report on a method for finding the longitude at sea which had been offered to the Government by a certain Frenchman of the name of *St. Pierre*. The Committee, acting largely on Flamsteed's advice, reported unfavourably on the method in question, and memorialised Charles II. in favour of founding a national observatory, in order that better knowledge of the celestial bodies might lead to a satisfactory method of finding the longitude, a problem which the rapid increase of English shipping rendered of great practical importance. The King having agreed, Flamsteed was in the same year appointed to the new

* December 31st, 1719, according to the unreformed calendar (O.S.) then in use in England.

office of Astronomer Royal, with a salary of £100 a year, and the warrant for building an Observatory at Greenwich was signed on June 12th, 1675. About a year was occupied in building it, and Flamsteed took up his residence there and began work in July 1676, five years after Cassini entered upon his duties at the Observatory of Paris (chapter VIII., § 160). The Greenwich Observatory was, however, on a very different scale from the magnificent sister institution. The King had, it is true, provided Flamsteed with a building and a very small salary, but furnished him neither with instruments nor with an assistant. A few instruments he possessed already, a few more were given to him by rich friends, and he gradually made at his own expense some further instrumental additions of importance. Some years after his appointment the Government provided him with "a silly, surly labourer" to help him with some of the rough work, but he was compelled to provide more skilled assistance out of his own pocket, and this necessity in turn compelled him to devote some part of his valuable time to taking pupils.

198. Flamsteed's great work was the construction of a more accurate and more extensive star catalogue than any that existed; he also made a number of observations of the moon, of the sun, and to a less extent of other bodies. Like Tycho, the author of the last great star catalogue (chapter V., § 107), he found problems continually presenting themselves in the course of his work which had to be solved before his main object could be accomplished, and we accordingly owe to him the invention of several improvements in practical astronomy, the best known being his method of finding the position of the first point of Aries (chapter II., § 42), one of the fundamental points with reference to which all positions on the celestial sphere are defined. He was the first astronomer to use a clock systematically for the determination of one of the two fundamental quantities (the right ascension) necessary to fix the position of a star, a method which was first suggested and to some extent used by Picard (chapter VIII., § 157), and, as soon as he could get the necessary instruments, he regularly used the telescopic sights of Gascoigne and Auzout (chapter VIII., § 155), instead of making naked-eye

observations. Thus while Hevel (chapter VIII., § 153) was the last and most accurate observer of the old school, employing methods not differing essentially from those which had been in use for centuries, Flamsteed belongs to the new school, and his methods differ rather in detail than in principle from those now in vogue for similar work at Greenwich, Paris, or Washington. This adoption of new methods, together with the most scrupulous care in details, rendered Flamsteed's observations considerably more accurate than any made in his time or earlier, the first definite advance afterwards being made by Bradley (§ 218).

Flamsteed compared favourably with many observers by not merely taking and recording observations, but by performing also the tedious process known as reduction (§ 218), whereby the results of the observation are put into a form suitable for use by other astronomers; this process is usually performed in modern observatories by assistants, but in Flamsteed's case had to be done almost exclusively by the astronomer himself. From this and other causes he was extremely slow in publishing observations; we have already alluded (chapter IX., § 192) to the difficulty which Newton had in extracting lunar observations from him, and after a time a feeling that the object for which the Observatory had been founded was not being fulfilled became pretty general among astronomers. Flamsteed always suffered from bad health as well as from the pecuniary and other difficulties which have been referred to; moreover he was much more anxious that his observations should be kept back till they were as accurate as possible, than that they should be published in a less perfect form and used for the researches which he once called "Mr. Newton's crotchets"; consequently he took remonstrances about the delay in the publication of his observations in bad part. Some painful quarrels occurred between Flamsteed on the one hand and Newton and Halley on the other. The last straw was the unauthorised publication in 1712, under the editorship of Halley, of a volume of Flamsteed's observations, a proceeding to which Flamsteed not unnaturally replied by calling Halley a "malicious thief." Three years later he succeeded in getting hold of all the unsold copies and in destroying them, but fortunately he was also stimulated to prepare for publication an authentic edition. The *Historia Coelestis Britannica*, as he called the book, contained an immense series of observations made both before and during his career at Greenwich, but the most important and permanently valuable part was a catalogue of the places of nearly 3,000 stars.*

Flamsteed himself only lived just long enough to finish the second of the three volumes; the third was edited by his assistants *Abraham Sharp* (1651–1742) and *Joseph Crosthwait;* and the whole was published in 1725. Four years later still appeared his valuable Star-Atlas, which long remained in common use.

The catalogue was not only three times as extensive as Tycho's, which it virtually succeeded, but was also very much more accurate. It has been estimated † that, whereas Tycho's determinations of the positions of the stars were on the average about 1′ in error, the corresponding errors in Flamsteed's case were about 10″. This quantity is the apparent diameter of a shilling seen from a distance of about 500 yards; so that if two marks were made at opposite points on the edge of the coin, and it were placed at a distance of 500 yards, the two marks might be taken to represent the true direction of an average star and its direction as given in Flamsteed's catalogue. In some cases of course the error might be much greater and in others considerably less.

Flamsteed contributed to astronomy no ideas of first-rate importance; he had not the ingenuity of Picard and of Roemer in devising instrumental improvements, and he took little interest in the theoretical work of Newton; ‡ but by unflagging industry and scrupulous care he succeeded in bequeathing to his successors an immense treasure of

* The apparent number is 2,935, but 12 of these are duplicates.

† By Bessel (chapter XIII., § 277).

‡ The relation between the work of Flamsteed and that of Newton was expressed with more correctness than good taste by the two astronomers themselves, in the course of some quarrel about the lunar theory: "Sir Isaac worked with the ore I had dug." "If he dug the ore, I made the gold ring."

observations, executed with all the accuracy that his instrumental means permitted.

199. Flamsteed was succeeded as Astronomer Royal by Edmund Halley, whom we have already met with (chapter IX., § 176) as Newton's friend and helper.

Born in 1656, ten years after Flamsteed, he studied astronomy in his schooldays, and published a paper on the orbits of the planets as early as 1676. In the same year he set off for St. Helena (in latitude 16° S.) in order to make observations of stars which were too near the south pole to be visible in Europe. The climate turned out to be disappointing, and he was only able after his return to publish (1678) a catalogue of the places of 341 southern stars, which constituted, however, an important addition to precise knowledge of the stars. The catalogue was also remarkable as being the first based on telescopic observation, though the observations do not seem to have been taken with all the accuracy which his instruments rendered attainable. During his stay at St. Helena he also took a number of pendulum observations which confirmed the results obtained a few years before by Richer at Cayenne (chapter VIII., § 161), and also observed a transit of Mercury across the sun, which occurred in November 1677.

After his return to England he took an active part in current scientific questions, particularly in those connected with astronomy, and made several small contributions to the subject. In 1684, as we have seen, he first came effectively into contact with Newton, and spent a good part of the next few years in helping him with the *Principia*.

200. Of his numerous contributions to astronomy, which touched almost every branch of the subject, his work on comets is the best known and probably the most important. He observed the comets of 1680 and 1682; he worked out the paths both of these and of a number of other recorded comets in accordance with Newton's principles, and contributed a good deal of the material contained in the sections of the *Principia* dealing with comets, particularly in the later editions. In 1705 he published a *Synopsis of Cometary Astronomy* in which no less than 24 cometary orbits were calculated. Struck by

the resemblance between the paths described by the comets of 1531, 1607, and 1682, and by the approximate equality in the intervals between their respective appearances and that of a fourth comet seen in 1456, he was shrewd enough to conjecture that the three later comets, if not all four, were really different appearances of the same comet, which revolved round the sun in an elongated ellipse in a period of about 75 or 76 years. He explained the difference between the 76 years which separate the appearances of the comet in 1531 and 1607, and the slightly shorter period which elapsed between 1607 and 1682, as probably due to the perturbations caused by planets near which the comet had passed; and finally predicted the probable reappearance of the same comet (which now deservedly bears his name) about 76 years after its last appearance, *i.e.* about 1758, though he was again aware that planetary perturbation might alter the time of its appearance; and the actual appearance of the comet about the predicted time (chapter XI., § 231) marked an important era in the progress of our knowledge of these extremely troublesome and erratic bodies.

201. In 1693 Halley read before the Royal Society a paper in which he called attention to the difficulty of reconciling certain ancient eclipses with the known motion of the moon, and referred to the possibility of some slight increase in the moon's average rate of motion round the earth.

This irregularity, now known as the **secular acceleration of the moon's mean motion,** was subsequently more definitely established as a fact of observation; and the difficulties met with in explaining it as a result of gravitation have rendered it one of the most interesting of the moon's numerous irregularities (cf. chapter XI., § 240, and chapter XIII., § 287).

202. Halley also rendered good service to astronomy by calling attention to the importance of the expected transits of Venus across the sun in 1761 and 1769 as a means of ascertaining the distance of the sun. The method had been suggested rather vaguely by Kepler, and more definitely by James Gregory in his *Optics* published in 1663. The idea was first suggested to Halley by

his observation of the transit of Mercury in 1677. In three papers published by the Royal Society he spoke warmly of the advantages of the method, and discussed in some detail the places and means most suitable for observing the transit of 1761. He pointed out that the desired result could be deduced from a comparison of the durations of the transit of Venus, as seen from different stations on the earth, *i.e.* of the intervals between the first appearance of Venus on the sun's disc and the final disappearance, as seen at two or more different stations. He estimated, moreover, that this interval of time, which would be several hours in length, could be measured with an error of only about two seconds, and that in consequence the method might be relied upon to give the distance of the sun to within about $\frac{1}{500}$ part of its true value. As the current estimates of the sun's distance differed among one another by 20 or 30 per cent., the new method, expounded with Halley's customary lucidity and enthusiasm, not unnaturally stimulated astronomers to take great trouble to carry out Halley's recommendations. The results, as we shall see (§ 227), were, however, by no means equal to Halley's expectations.

203. In 1718 Halley called attention to the fact that three well-known stars, Sirius, Procyon, and Arcturus, had changed their angular distances from the ecliptic since Greek times, and that Sirius had even changed its position perceptibly since the time of Tycho Brahe. Moreover comparison of the places of other stars shewed that the changes could not satisfactorily be attributed to any motion of the ecliptic, and although he was well aware that the possible errors of observation were such as to introduce a considerable uncertainty into the amounts involved, he felt sure that such errors could not wholly account for the discrepancies noticed, but that the stars in question must have really shifted their positions in relation to the rest; and he naturally inferred that it would be possible to detect similar **proper motions** (as they are now called) in other so-called "fixed" stars.

204. He also devoted a good deal of time to the standing astronomical problem of improving the tables of the moon and planets, particularly the former. He made

observations of the moon as early as 1683, and by means of them effected some improvement in the tables. In 1676 he had already noted defects in the existing tables of Jupiter and Saturn, and ultimately satisfied himself of the existence of certain irregularities in the motion of these two planets, suspected long ago by Horrocks (chapter viii., § 156); these irregularities he attributed correctly to the perturbations of the two planets by one another, though he was not mathematician enough to work out the theory; from observation, however, he was able to estimate the irregularities in question with fair accuracy and to improve the planetary tables by making allowance for them. But neither the lunar nor the planetary tables were ever completed in a form which Halley thought satisfactory. By 1719 they were printed, but kept back from publication, in hopes that subsequent improvements might be effected. After his appointment as Astronomer Royal in succession to Flamsteed (1720) he devoted special attention to getting fresh observations for this purpose, but he found the Observatory almost bare of instruments, those used by Flamsteed having been his private property, and having been removed as such by his heirs or creditors. Although Halley procured some instruments, and made with them a number of observations, chiefly of the moon, the age (63) at which he entered upon his office prevented him from initiating much, or from carrying out his duties with great energy, and the observations taken were in consequence only of secondary importance, while the tables for the improvement of which they were specially designed were only finally published in 1752, ten years after the death of their author. Although they thus appeared many years after the time at which they were virtually prepared and owed little to the progress of science during the interval, they at once became and for some time remained the standard tables for both the lunar and planetary motions (cf. § 226, and chapter xi., § 247).

205. Halley's remarkable versatility in scientific work is further illustrated by the labour which he expended in editing the writings of the great Greek geometer Apollonius (chapter ii., § 38) and the star catalogue of Ptolemy (chapter ii., § 50). He was also one of the first of modern

astronomers to pay careful attention to the effects to be observed during a total eclipse of the sun, and in the vivid description which he wrote of the eclipse of 1715, besides referring to the mysterious corona, which Kepler and others had noticed before (chapter VII., § 145), he called attention also to "a very narrow streak of a dusky but strong Red Light," which was evidently a portion of that remarkable envelope of the sun which has been so extensively studied in modern times (chapter XIII., § 301) under the name of the chromosphere.

It is worth while to notice, as an illustration of Halley's unselfish enthusiasm for science and of his power of looking to the future, that two of his most important pieces of work, by which certainly he is now best known, necessarily appeared during his lifetime as of little value, and only bore their fruit after his death (1742), for his comet only returned in 1759, when he had been dead 17 years, and the first of the pair of transits of Venus, from which he had shewn how to deduce the distance of the sun, took place two years later still (§ 227).

206. The third Astronomer Royal, *James Bradley*, is popularly known as the author of two memorable discoveries, *viz.* the aberration of light and the nutation of the earth's axis. Remarkable as these are both in themselves and on account of the ingenious and subtle reasoning and minutely accurate observations by means of which they were made, they were in fact incidents in a long and active astronomical career, which resulted in the execution of a vast mass of work of great value.

The external events of Bradley's life may be dealt with very briefly. Born in 1693, he proceeded in due course to Oxford (B.A. 1714, M.A. 1717), but acquired his first knowledge of astronomy and his marked taste for the subject from his uncle *James Pound*, for many years rector of Wansted in Essex, who was one of the best observers of the time. Bradley lived with his uncle for some years after leaving Oxford, and carried out a number of observations in concert with him. The first recorded observation of Bradley's is dated 1715, and by 1718 he was sufficiently well thought of in the scientific world to receive the honour of election as a Fellow of the Royal Society. But, as his biographer * remarks, "it could not be foreseen that his astronomical labours would lead to any establishment in life, and it became necessary for him to embrace a profession." He accordingly took orders, and was fortunate enough to be presented almost at once to two livings, the duties attached to which do not seem to have interfered appreciably with the prosecution of his astronomical studies at Wansted.

In 1721 he was appointed Savilian Professor of Astronomy at Oxford, and resigned his livings. The work of the professorship appears to have been very light, and for more than ten years he continued to reside chiefly at Wansted, even after his uncle's death in 1724. In 1732 he took a house in Oxford and set up there most of his instruments, leaving, however, at Wansted the most important of all, the "zenith-sector," with which his two famous discoveries were made. Ten years afterwards Halley's death rendered the post of Astronomer Royal vacant, and Bradley received the appointment.

The work of the Observatory had been a good deal neglected by Halley during the last few years of his life, and Bradley's first care was to effect necessary repairs in the instruments. Although the equipment of the Observatory with instruments worthy of its position and of the state of science at the time was a work of years, Bradley had some of the most important instruments in good working order within a few months of his appointment, and observations were henceforward made systematically. Although the 20 remaining years of his life (1742–1762) were chiefly spent at Greenwich in the discharge of the duties of his office and in researches connected with them, he retained his professorship at Oxford, and continued to make observations at Wansted at least up till 1747.

207. The discovery of aberration resulted from an attempt to detect the parallactic displacement of stars which should result from the annual motion of the earth. Ever since the Coppernican controversy had called attention to the importance of the problem (cf. chapter IV., § 92, and chapter VI., § 129), it had naturally exerted a fascination

* Rigaud, in the memoirs prefixed to Bradley's *Miscellaneous Works*.

BRADLEY. [*To face p.* 258.

on the minds of observing astronomers, many of whom had tried to detect the motion in question, and some of whom (including the "universal claimant" Hooke) professed to have succeeded. Actually, however, all previous attempts had been failures, and Bradley was no more successful than his predecessors in this particular undertaking, but was able to deduce from his observations two results of great interest and of an entirely unexpected character.

The problem which Bradley set himself was to examine whether any star could be seen to have in the course of the year a slight motion relative to others or relative to fixed points on the celestial sphere such as the pole. It was known that such a motion, if it existed, must be very small, and it was therefore evident that extreme delicacy in instrumental adjustments and the greatest care in observation would have to be employed. Bradley worked at first in conjunction with his friend *Samuel Molyneux* (1689–1728), who had erected a telescope at Kew. In accordance with the method adopted in a similar investigation by Hooke, whose results it was desired to test, the telescope was fixed in a nearly vertical position, so chosen that a particular star in the Dragon (γ Draconis) would be visible through it when it crossed the meridian, and the telescope was mounted with great care so as to maintain an invariable position throughout the year. If then the star in question were to undergo any motion which altered its distance from the pole, there would be a corresponding alteration in the position in which it would be seen in the field of view of the telescope. The first observations were taken on December 14th, 1725 (N.S.), and by December 28th Bradley believed that he had already noticed a slight displacement of the star towards the south. This motion was clearly verified on January 1st, and was then observed to continue; in the following March the star reached its extreme southern position, and then began to move northwards again. In September it once more altered its direction of motion, and by the end of the year had completed the cycle of its changes and returned to its original position, the greatest change in position amounting to nearly 40′.

The star was thus observed to go through *some* annual

motion. It was, however, at once evident to Bradley that this motion was not the parallactic motion of which he was in search, for the position of the star was such that parallax would have made it appear farthest south in December and farthest north in June, or in each case three months earlier than was the case in the actual observations. Another explanation which suggested itself was that the earth's axis might have a to-and-fro oscillatory motion or nutation which would alter the position of the celestial pole and hence produce a corresponding alteration in the position of the star. Such a motion of the celestial pole would evidently produce opposite effects on two stars situated on opposite sides of it, as any motion which brought the pole nearer to one star of such a pair would necessarily move it away from the other. Within a fortnight of the decisive observation made on January 1st a star * had already been selected for the application of this test, with the result which can best be given in Bradley's own words:—

"A nutation of the earth's axis was one of the first things that offered itself upon this occasion, but it was soon found to be insufficient; for though it might have accounted for the change of declination in γ Draconis, yet it would not at the same time agree with the phaenomena in other stars; particularly in a small one almost opposite in right ascension to γ Draconis, at about the same distance from the north pole of the equator: for though this star seemed to move the same way as a nutation of the earth's axis would have made it, yet, it changing its declination but about half as much as γ Draconis in the same time, (as appeared upon comparing the observations of both made upon the same days, at different seasons of the year,) this plainly proved that the apparent motion of the stars was not occasioned by a real nutation, since, if that had been the cause, the alteration in both stars would have been near equal."

One or two other explanations were tested and found insufficient, and as the result of a series of observations extending over about two years, the phenomenon in question, although amply established, still remained quite unexplained.

By this time Bradley had mounted an instrument of his

* A telescopic star named 37 Camelopardi in Flamsteed's catalogue.

own at Wansted, so arranged that it was possible to observe through it the motions of stars other than γ Draconis.

Several stars were watched carefully throughout a year, and the observations thus obtained gave Bradley a fairly complete knowledge of the geometrical laws according to which the motions varied both from star to star and in the course of the year.

208. The true explanation of **aberration**, as the phenomenon in question was afterwards called, appears to have occurred to him about September, 1728, and was published to the Royal Society, after some further verification, early in the following year. According to a well-known story,* he noticed, while sailing on the Thames, that a vane on the masthead appeared to change its direction every time that the boat altered its course, and was informed by the sailors that this change was not due to any alteration in the wind's direction, but to that of the boat's course. In fact the apparent direction of the wind, as shewn by the vane, was not the true direction of the wind, but resulted from a combination of the motions of the wind and of the boat, being more precisely that of the motion of the wind *relative* to the boat. Replacing in imagination the wind by light coming from a star, and the boat shifting its course by the earth moving round the sun and continually changing its direction of motion, Bradley arrived at an explanation which, when worked out in detail, was found to account most satisfactorily for the apparent changes in the direction of a star which he had been studying. His own account of the matter is as follows:—

"At last I conjectured that all the phaenomena hitherto mentioned proceeded from the progressive motion of light and the earth's annual motion in its orbit. For I perceived that, if light was propagated in time, the apparent place of a fixed object would not be the same when the eye is at rest, as when it is moving in any other direction than that of the line passing through the eye and object; and that when the eye is moving

* The story is given in T. Thomson's *History of the Royal Society*, published more than 80 years afterwards (1812), but I have not been able to find any earlier authority for it. Bradley's own account of his discovery gives a number of details, but has no allusion to this incident.

in different directions, the apparent place of the object would be different.

"I considered this matter in the following manner. I imagined C A to be a ray of light, falling perpendicularly upon the line B D; then if the eye is at rest at A, the object must appear in the direction A C, whether light be propagated in time or in an instant. But if the eye is moving from B towards A, and light is propagated in time, with a velocity that is to the velocity of the eye, as C A to B A; then light moving from C to A, whilst the eye moves from B to A, that particle of it by which the object will be discerned when the eye in its motion comes to A, is at C when the eye is at B. Joining the points B, C, I supposed the line C B to be a tube (inclined to the line B D in the angle D B C) of such a diameter as to admit of but one particle of light; then it was easy to conceive that the particle of light at C (by which the object must be seen when the eye, as it moves along, arrives at A) would pass through the tube B C, if it is inclined to B D in the angle D B C, and accompanies the eye in its motion from B to A; and that it could not come to the eye, placed behind such a tube, if it had any other inclination to the line B D. . . .

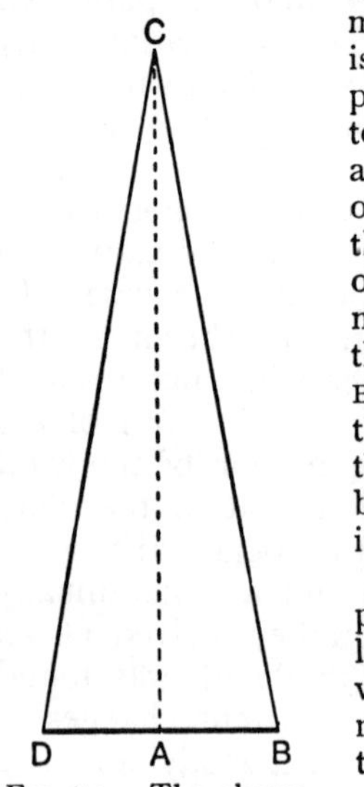

FIG. 74.—The aberration of light. From Bradley's paper in the *Phil. Trans.*

"Although therefore the true or real place of an object is perpendicular to the line in which the eye is moving, yet the visible place will not be so, since that, no doubt, must be in the direction of the tube; but the difference between the true and apparent place will be (caeteris paribus) greater or less, according to the different proportion between the velocity of light and that of the eye. So that if we could suppose that light was propagated in an instant, then there would be no difference between the real and visible place of an object, although the eye were in motion; for in that case, A C being infinite with respect to A B, the angle A C B (the difference between the true and visible place) vanishes. But if light be propagated in time, (which I presume will readily be allowed by most of the philosophers of this age,) then it is evident from the foregoing considerations, that there will be always a difference between the real and visible place of an object, unless the eye is moving either directly towards or from the object."

Bradley's explanation shews that the apparent position of a star is determined by the motion of the star's light *relative* to the earth, so that the star appears slightly nearer to the point on the celestial sphere towards which the earth is moving than would otherwise be the case. A familiar illustration of a precisely analogous effect may perhaps be of service. Any one walking on a rainy but windless day protects himself most effectually by holding his umbrella, not immediately over his head, but a little in front, exactly as he would do if he were at rest and there were a slight wind blowing in his face. In fact, if he were to ignore his own motion and pay attention only to the direction in which he found it advisable to point his umbrella, he would believe that there was a slight head-wind blowing the rain towards him.

209. The passage quoted from Bradley's paper deals only with the simple case in which the star is at right angles to the direction of the earth's motion. He shews elsewhere that if the star is in any other direction the effect is of the same kind but less in amount. In Bradley's figure (fig. 74) the amount of the star's displacement from its true position is represented by the angle B C A, which depends on the proportion between the lines A C and A B; but if (as in fig. 75) the earth is moving (without change of speed) in the direction A B′ instead of A B, so that the direction of the star is oblique to it, it is evident from the figure that the star's displacement, represented by the angle A C B′, is less than before; and the amount varies according to a simple mathematical law* with the angle between the two directions. It follows therefore that the displacement in question is different for different stars, as Bradley's observations had already shewn, and is, moreover, different for the same star in the course of the year, so that a star appears to describe a curve which is very nearly an ellipse (fig. 76), the centre (s)

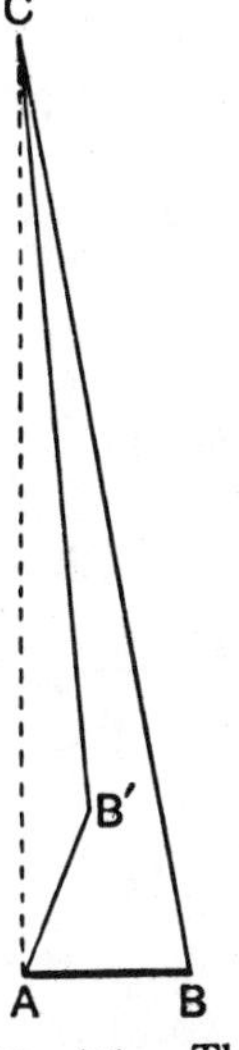

FIG. 75.—The aberration of light.

* It is $k \sin$ C A B, where k is the constant of aberration.

corresponding to the position which the star would occupy if aberration did not exist. It is not difficult to see that, wherever a star is situated, the earth's motion is twice a year, at intervals of six months, at right angles to the direction of the star, and that at these times the star receives the greatest possible displacement from its mean position, and is consequently at the ends of the greatest axis of the ellipse which it describes, as at A and A′, whereas at intermediate times it undergoes its least displacement, as at B and B′. The greatest displacement S A, or half of A A′, which is the same for all stars, is known as the **constant of aberration,** and was fixed by Bradley at between 20″ and 20½″, the value at present accepted being 20″·47. The least displacement, on the other hand, S B, or half of B B′, was shewn to depend in a simple way upon the star's distance from the ecliptic, being greatest for stars farthest from the ecliptic.

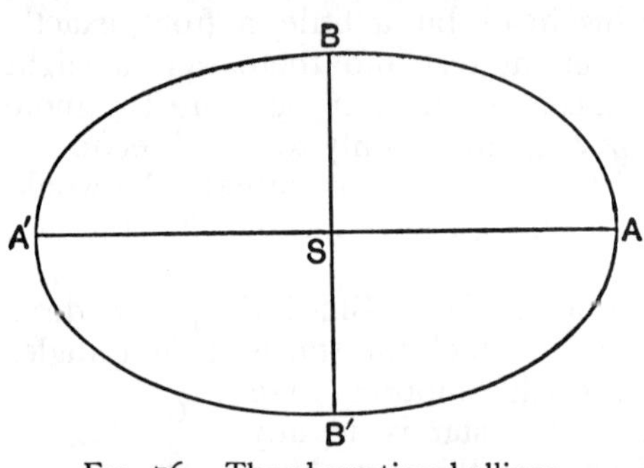

FIG. 76.—The aberrational ellipse.

210. The constant of aberration, which is represented by the angle A C B in fig. 74, depends only on the ratio between A C and A B, which are in turn proportional to the velocities of light and of the earth. Observations of aberration give then the ratio of these two velocities. From Bradley's value of the constant of aberration it follows by an easy calculation that the velocity of light is about 10,000 times that of the earth; Bradley also put this result into the form that light travels from the sun to the earth in 8 minutes 13 seconds. From observations of the eclipses of Jupiter's moons, Roemer and others had estimated the same interval at from 8 to 11 minutes (chapter VIII., § 162); and Bradley was thus able to get a satisfactory confirmation of the truth of his discovery. Aberration being once established, the same calculation could be used to give the most accurate

measure of the velocity of light in terms of the dimensions of the earth's orbit, the determination of aberration being susceptible of considerably greater accuracy than the corresponding measurements required for Roemer's method.

211. One difficulty in the theory of aberration deserves mention. Bradley's own explanation, quoted above, refers to light as a material substance shot out from the star or other luminous body. This was in accordance with the corpuscular theory of light, which was supported by the great weight of Newton's authority and was commonly accepted in the 18th century. Modern physicists, however, have entirely abandoned the corpuscular theory, and regard light as a particular form of wave-motion transmitted through ether. From this point of view Bradley's explanation and the physical illustrations given are far less convincing; the question becomes in fact one of considerable difficulty, and the most careful and elaborate of modern investigations cannot be said to be altogether satisfactory. The curious inference may be drawn that, if the more correct modern notions of the nature of light had prevailed in Bradley's time, it must have been very much more difficult, if not impracticable, for him to have thought of his explanation of the stellar motions which he was studying; and thus an erroneous theory led to a most important discovery.

212. Bradley had of course not forgotten the original object of his investigation. He satisfied himself, however, that the agreement between the observed positions of γ Draconis and those which resulted from aberration was so close that any displacement of a star due to parallax which might exist must certainly be less than $2''$, and probably not more than $\frac{1}{2}''$, so that the large parallax amounting to nearly $30''$, which Hooke claimed to have detected, must certainly be rejected as erroneous.

From the point of view of the Coppernican controversy, however, Bradley's discovery was almost as good as the discovery of a parallax; since if the earth were at rest no explanation of the least plausibility could be given of aberration.

213. The close agreement thus obtained between theory and observation would have satisfied an astronomer less

accurate and careful than Bradley. But in his paper on aberration (1729) we find him writing:—

"I have likewise met with some small varieties in the declination of other stars in different years which do not seem to proceed from the same cause. . . . But whether these small alterations proceed from a regular cause, or are occasioned by any change in the materials, etc., of my instrument, I am not yet able fully to determine."

The slender clue thus obtained was carefully followed up and led to a second striking discovery, which affords one of the most beautiful illustrations of the important results which can be deduced from the study of "residual phenomena." Aberration causes a star to go through a cyclical series of changes in the course of a year; if therefore at the end of a year a star is found not to have returned to its original place, some other explanation of the motion has to be sought. Precession was one known cause of such an alteration; but Bradley found, at the end of his first year's set of observations at Wansted, that the alterations in the positions of various stars differed by a minute amount (not exceeding 2″) from those which would have resulted from the usual estimate of precession; and that, although an alteration in the value of precession would account for the observed motions of some of these stars, it would have increased the discrepancy in the case of others. A nutation or nodding of the earth's axis had, as we have seen (§ 207), already presented itself to him as a possibility; and although it had been shewn to be incapable of accounting for the main phenomenon—due to aberration—it might prove to be a satisfactory explanation of the much smaller residual motions. It soon occurred to Bradley that such a nutation might be due to the action of the moon, as both observation and the Newtonian explanation of precession indicated:—

"I suspected that the moon's action upon the equatorial parts of the earth might produce these effects: for if the precession of the equinox be, according to Sir Isaac Newton's principles, caused by the actions of the sun and moon upon those parts, the plane of the moon's orbit being at one time above ten degrees more inclined to the plane of the equator than at another, it was reasonable to conclude, that the part of the whole annual precession, which arises from her action, would in different years be varied in its quantity; whereas the plane of the ecliptic, wherein the sun appears, keeping always nearly the same inclination to the equator, that part of the precession which is owing to the sun's action may be the same every year; and from hence it would follow, that although the mean annual precession, proceeding from the joint actions of the sun and moon, were 50″, yet the apparent annual precession might sometimes exceed and sometimes fall short of that mean quantity, according to the various situations of the nodes of the moon's orbit."

Newton in his discussion of precession (chapter IX., § 188; *Principia*, Book III., proposition 21) had pointed out the existence of a small irregularity with a period of six months. But it is evident, on looking at this discussion of the effect of the solar and lunar attractions on the protuberant parts of the earth, that the various alterations in the positions of the sun and moon relative to the earth might be expected to produce irregularities, and that the uniform precessional motion known from observation and deduced from gravitation by Newton was, as it were, only a smoothing out of a motion of a much more complicated character. Except for the allusion referred to, Newton made no attempt to discuss these irregularities, and none of them had as yet been detected by observation.

Of the numerous irregularities of this class which are now known, and which may be referred to generally as **nutation**, that indicated by Bradley in the passage just quoted is by far the most important. As soon as the idea of an irregularity depending on the position of the moon's nodes occurred to him, he saw that it would be desirable to watch the motions of several stars during the whole period (about 19 years) occupied by the moon's nodes in performing the circuit of the ecliptic and returning to the same position. This inquiry was successfully carried out between 1727 and 1747 with the telescope mounted at Wansted. When the moon's nodes had performed half their revolution, *i.e.* after about nine years, the correspondence between the displacements of the stars and the changes in the moon's orbit was so close that Bradley was satisfied with the general correctness of his theory, and in 1737 he communicated the result privately to Maupertuis (§ 221), with whom he had

had some scientific correspondence. Maupertuis appears to have told others, but Bradley himself waited patiently for the completion of the period which he regarded as necessary for the satisfactory verification of his theory, and only published his results definitely at the beginning of 1748.

214. Bradley's observations established the existence of certain alterations in the positions of various stars, which

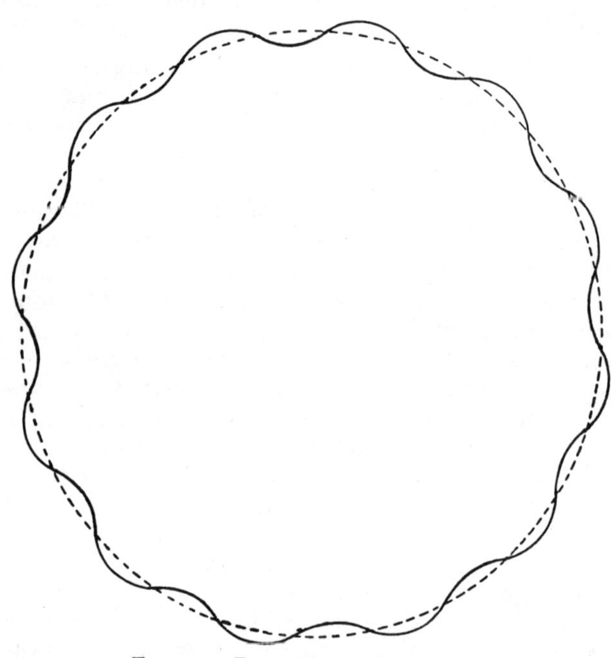

FIG. 77.—Precession and nutation.

could be accounted for by supposing that, on the one hand, the distance of the pole from the ecliptic fluctuated, and that, on the other, the precessional motion of the pole was not uniform, but varied slightly in speed. *John Machin* (? –1751), one of the best English mathematicians of the time, pointed out that these effects would be produced if the pole were supposed to describe on the celestial sphere a minute circle in a period of rather less than 19 years—being that of the revolution of the nodes of the moon's orbit—round the position which it would occupy if there were no nutation, but a uniform precession. Bradley found that this hypothesis fitted his observations, but that it would be better to replace the circle by a slightly flattened ellipse, the greatest and least axes of which he estimated at about 18″ and 16″ respectively.* This ellipse would be about as large as a shilling placed in a slightly oblique position at a distance of 300 yards from the eye. The motion of the pole was thus shewn to be a double one; as the result of precession and nutation combined it describes round the pole of the ecliptic "a gently undulated ring," as represented in the figure, in which, however, the undulations due to nutation are enormously exaggerated.

215. Although Bradley was aware that nutation must be produced by the action of the moon, he left the theoretical investigation of its cause to more skilled mathematicians than himself.

In the following year (1749) the French mathematician D'Alembert (chapter XI., § 232) published a treatise † in which not only precession, but also a motion of nutation agreeing closely with that observed by Bradley, were shewn by a rigorous process of analysis to be due to the attraction of the moon on the protuberant parts of the earth round the equator (cf. chapter IX., § 187), while Newton's explanation of precession was confirmed by the same piece of work. Euler (chapter XI., § 236) published soon afterwards another investigation of the same subject; and it has been studied afresh by many mathematical astronomers since that time, with the result that Bradley's nutation is found to be only the most important of a long series of minute irregularities in the motion of the earth's axis.

216. Although aberration and nutation have been discussed first, as being the most important of Bradley's

* His observations as a matter of fact point to a value rather greater than 18″, but he preferred to use round numbers. The figures at present accepted are 18″·42 and 13″·75, so that his ellipse was decidedly less flat than it should have been.

† *Recherches sur la précession des équinoxes et sur la nutation de l'axe de la terre.*

discoveries, other investigations were carried out by him before or during the same time.

The earliest important piece of work which he accomplished was in connection with Jupiter's satellites. His uncle had devoted a good deal of attention to this subject, and had drawn up some tables dealing with the motion of the first satellite, which were based on those of Domenico Cassini, but contained a good many improvements. Bradley seems for some years to have made a practice of frequently observing the eclipses of Jupiter's satellites, and of noting discrepancies between the observations and the tables; and he was thus able to detect several hitherto unnoticed peculiarities in the motions, and thereby to form improved tables. The most interesting discovery was that of a period of 437 days, after which the motions of the three inner satellites recurred with the same irregularities. Bradley, like Pound, made use of Roemer's suggestion (chapter VIII., § 162) that light occupied a finite time in travelling from Jupiter to the earth, a theory which Cassini and his school long rejected. Bradley's tables of Jupiter's satellites were embodied in Halley's planetary and lunar tables, printed in 1719, but not published till more than 30 years afterwards (§ 204). Before that date the Swedish astronomer *Pehr Vilhelm Wargentin* (1717–1783) had independently discovered the period of 437 days, which he utilised for the construction of an extremely accurate set of tables for the satellites published in 1746.

In this case as in that of nutation Bradley knew that his mathematical powers were unequal to giving an explanation on gravitational principles of the inequalities which observation had revealed to him, though he was well aware of the importance of such an undertaking, and definitely expressed the hope "that some geometer,* in imitation of the great Newton, would apply himself to the investigation of these irregularities, from the certain and demonstrative principles of gravity."

On the other hand, he made in 1726 an interesting practical application of his superior knowledge of Jupiter's

* The word "geometer" was formerly used, as "geomètre" still is in French, in the wider sense in which "mathematician" is now customary.

satellites by determining, in accordance with Galilei's method (chapter VI., § 127), but with remarkable accuracy, the longitudes of Lisbon and of New York.

217. Among Bradley's minor pieces of work may be mentioned his observations of several comets and his calculation of their respective orbits according to Newton's method; the construction of improved tables of refraction, which remained in use for nearly a century; a share in pendulum experiments carried out in England and Jamaica with the object of verifying the variation of gravity in different latitudes; a careful testing of Mayer's lunar tables (§ 226), together with improvements of them; and lastly, some work in connection with the reform of the calendar made in 1752 (cf. chapter II., § 22).

218. It remains to give some account of the magnificent series of observations carried out during Bradley's administration of the Greenwich Observatory.

These observations fall into two chief divisions of unequal merit, those after 1749 having been made with some more accurate instruments which a grant from the government enabled him at that time to procure.

The main work of the Observatory under Bradley consisted in taking observations of fixed stars, and to a lesser extent of other bodies, as they passed the meridian, the instruments used (the "mural quadrant" and the "transit instrument") being capable of motion only in the meridian, and being therefore steadier and susceptible of greater accuracy than those with more freedom of movement. The most important observations taken during the years 1750–1762, amounting to about 60,000, were published long after Bradley's death in two large volumes which appeared in 1798 and 1805. A selection of them had been used earlier as the basis of a small star catalogue, published in the *Nautical Almanac* for 1773; but it was not till 1818 that the publication of Bessel's *Fundamenta Astronomiae* (chapter XIII., § 277), a catalogue of more than 3000 stars based on Bradley's observations, rendered these observations thoroughly available for astronomical work. One reason for this apparently excessive delay is to be found in Bradley's way of working. Allusion has already been made to a variety of causes which prevent the apparent place of a star, as seen in the telescope and noted at the time, from being a satisfactory permanent record of its position. There are various instrumental errors, and errors due to refraction; again, if a star's places at two different times are to be compared, precession must be taken into account; and Bradley himself unravelled in aberration and nutation two fresh sources of error. In order therefore to put into a form satisfactory for permanent reference a number of star observations, it is necessary to make corrections which have the effect of allowing for these various sources of error. This process of **reduction**, as it is technically called, involves a certain amount of rather tedious calculation, and though in modern observatories the process has been so far systematised that it can be carried out almost according to fixed rules by comparatively unskilled assistants, in Bradley's time it required more judgment, and it is doubtful if his assistants could have performed the work satisfactorily, even if their time had not been fully occupied with other duties. Bradley himself probably found the necessary calculations tedious, and preferred devoting his energies to work of a higher order. It is true that Delambre, the famous French historian of astronomy, assures his readers that he had never found the reduction of an observation tedious if performed the same day, but a glance at any of his books is enough to shew his extraordinary fondness for long calculations of a fairly elementary character, and assuredly Bradley is not the only astronomer whose tastes have in this respect differed fundamentally from Delambre's. Moreover reducing an observation is generally found to be a duty that, like answering letters, grows harder to perform the longer it is neglected; and it is not only less interesting but also much more difficult for an astronomer to deal satisfactorily with some one else's observations than with his own. It is not therefore surprising that after Bradley's death a long interval should have elapsed before an astronomer appeared with both the skill and the patience necessary for the complete reduction of Bradley's 60,000 observations.

A variety of circumstances combined to make Bradley's observations decidedly superior to those of his predecessors. He evidently possessed in a marked degree the personal characteristics—of eye and judgment—which make a first-rate observer; his instruments were mounted in the best known way for securing accuracy, and were constructed by the most skilful makers; he made a point of studying very carefully the defects of his instruments, and of allowing for them; his discoveries of aberration and nutation enabled him to avoid sources of error, amounting to a considerable number of seconds, which his predecessors could only have escaped imperfectly by taking the average of a number of observations; and his improved tables of refraction still further added to the correctness of his results.

Bessel estimates that the errors in Bradley's observations of the declination of stars were usually less than 4″, while the corresponding errors in right ascension, a quantity which depends ultimately on a time-observation, were less than 15″, or one second of time. His observations thus shewed a considerable advance in accuracy compared with those of Flamsteed (§ 198), which represented the best that had hitherto been done.

219. The next Astronomer Royal was *Nathaniel Bliss* (1700–1764), who died after two years. He was in turn succeeded by *Nevil Maskelyne* (1732–1811), who carried on for nearly half a century the tradition of accurate observation which Bradley had established at Greenwich, and made some improvements in methods.

To him is also due the first serious attempt to measure the density and hence the mass of the earth. By comparing the attraction exerted by the earth with that of the sun and other bodies, Newton, as we have seen (chapter IX., § 185), had been able to connect the masses of several of the celestial bodies with that of the earth. To connect the mass of the whole earth with that of a given terrestrial body, and so express it in pounds or tons, was a problem of quite a different kind. It is of course possible to examine portions of the earth's surface and compare their density with that of, say, water; then to make some conjecture, based on rough observations in mines, etc., as to the rate at which density increases as we go from the surface towards the centre of the earth, and hence to infer the average density of the earth. Thus the mass of the whole earth is compared with that of a globe of water of the same size, and, the size being known, is expressible in pounds or tons.

By a process of this sort Newton had in fact, with extraordinary insight, estimated that the density of the earth was between five and six times as great as that of water.*

It was, however, clearly desirable to solve the problem in a less conjectural manner, by a direct comparison of the gravitational attraction exerted by the earth with that exerted by a known mass—a method that would at the same time afford a valuable test of Newton's theory of the gravitating properties of portions of the earth, as distinguished from the whole earth. In their Peruvian expedition (§ 221), Bouguer and La Condamine had noticed certain small deflections of the plumb-line, which indicated an attraction by Chimborazo, near which they were working; but the observations were too uncertain to be depended on. Maskelyne selected for his purpose Schehallien in Perthshire, a narrow ridge running east and west. The direction of the plumb-line was observed (1774) on each side of the ridge, and a change in direction amounting to about 12″ was found to be caused by the attraction of the mountain. As the direction of the plumb-line depends on the attraction of the earth as a whole and on that of the mountain, this deflection at once led to a comparison of the two attractions. Hence an intricate calculation performed by *Charles Hutton* (1737–1823) led to a comparison of the average densities of the earth and mountain, and hence to the final conclusion (published in 1778) that the earth's density was about $4\frac{1}{2}$ times that of water. As Hutton's estimate of the density of the mountain was avowedly almost conjectural, this result was of course correspondingly uncertain.

A few years later *John Michell* (1724–1793) suggested, and the famous chemist and electrician *Henry Cavendish* (1731–1810) carried out (1798), an experiment in which the mountain was replaced by a pair of heavy balls, and their attraction on another body was compared with that of the earth, the result being that the density of the earth was found to be about $5\frac{1}{2}$ times that of water.

* *Principia*, Book III., proposition 10.

The **Cavendish experiment**, as it is often called, has since been repeated by various other experimenters in modified forms, and one or two other methods, too technical to be described here, have also been devised. All the best modern experiments give for the density numbers converging closely on $5\frac{1}{2}$, thus verifying in a most striking way both Newton's conjecture and Cavendish's original experiment.

With this value of the density the mass of the earth is a little more than 13 billion billion pounds, or more precisely 13,136,000,000,000,000,000,000,000 lbs.

220. While Greenwich was furnishing the astronomical world with a most valuable series of observations, the Paris Observatory had not fulfilled its early promise. It was in fact suffering, like English mathematics, from the evil effects of undue adherence to the methods and opinions of a distinguished man. Domenico Cassini happened to hold several erroneous opinions in important astronomical matters; he was too good a Catholic to be a genuine Coppernican, he had no belief in gravitation, he was firmly persuaded that the earth was flattened at the equator instead of at the poles, and he rejected Roemer's discovery of the velocity of light. After his death in 1712 the directorship of the Observatory passed in turn to three of his descendants, the last of whom resigned office in 1793; and several members of the Maraldi family, into which his sister had married, worked in co-operation with their cousins. Unfortunately a good deal of their energy was expended, first in defending, and afterwards in gradually withdrawing from, the errors of their distinguished head. *Jacques Cassini*, for example, the second of the family (1677–1756), although a Coppernican, was still a timid one, and rejected Kepler's law of areas; his son again, commonly known as *Cassini de Thury* (1714–1784), still defended the ancestral errors as to the form of the earth; while the fourth member of the family, *Count Cassini* (1748–1845), was the first of the family to accept the Newtonian idea of gravitation.

Some planetary and other observations of value were made by the Cassini-Maraldi school, but little of this work was of first-rate importance.

221. A series of important measurements of the earth,

in which the Cassinis had a considerable share, were made during the 18th century, almost entirely by Frenchmen, and resulted in tolerably exact knowledge of the earth's size and shape.

The variation of the length of the seconds pendulum observed by Richer in his Cayenne expedition (chapter VIII., § 161) had been the first indication of a deviation of the earth from a spherical form. Newton inferred, both from these pendulum experiments and from an independent theoretical investigation (chapter IX., § 187), that the earth was spheroidal, being flattened towards the poles; and this view was strengthened by the satisfactory explanation of precession to which it led (chapter IX., § 188).

On the other hand, a comparison of various measurements of arcs of the meridian in different latitudes gave some support to the view that the earth was elongated towards the poles and flattened towards the equator, a view championed with great ardour by the Cassini school. It was clearly important that the question should be settled by more extensive and careful earth-measurements.

The essential part of an ordinary measurement of the earth consists in ascertaining the distance in miles between two places on the same meridian, the latitudes of which differ by a known amount. From these two data the length of an arc of a meridian corresponding to a difference of latitude of 1° at once follows. The **latitude** of a place is the angle which the vertical at the place makes with the equator, or, expressed in a slightly different form, is the angular distance of the zenith from the celestial equator. The **vertical** at any place may be defined as a direction perpendicular to the surface of still water at the place in question, and may be regarded as perpendicular to the true surface of the earth, accidental irregularities in its form such as hills and valleys being ignored.*

The difference of latitude between two places, north and south of one another, is consequently the angle between the verticals there. Fig. 78 shews the verticals, marked by the arrowheads, at places on the same meridian in

* It is important for the purposes of this discussion to notice that the vertical is *not* the line drawn from the centre of the earth to the place of observation.

latitudes differing by 10° ; so that two consecutive verticals are inclined in every case at an angle of 10°.

If, as in fig. 78, the shape of the earth is drawn in accordance with Newton's views, the figure shews at once that the arcs $A A_1$, $A_1 A_2$, etc., each of which corresponds to 10° of latitude, steadily *increase* as we pass from a point A on the equator to the pole B. If the opposite hypothesis be

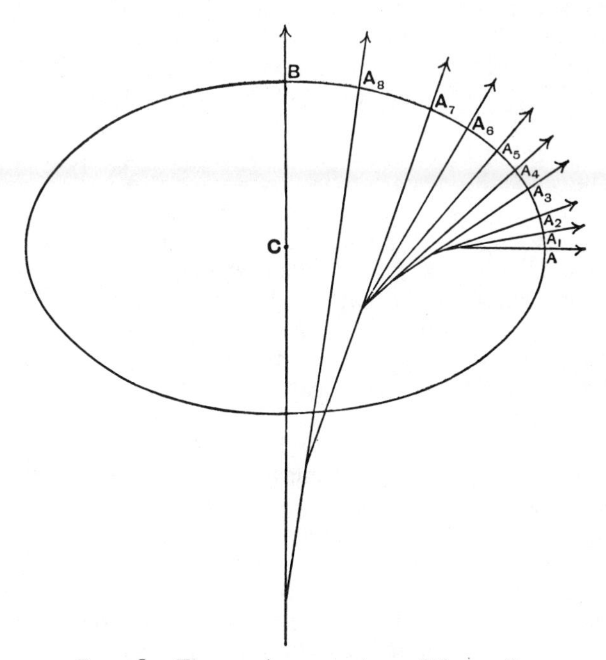

FIG. 78.—The varying curvature of the earth.

adopted, which will be illustrated by the same figure if we now regard A as the pole and B as a point on the equator, then the successive arcs *decrease* as we pass from equator to pole. A comparison of the measurements made by Eratosthenes in Egypt (chapter II., § 36) with some made in Europe (chapter VIII., § 159) seemed to indicate that a degree of the meridian near the equator was longer than one in higher latitudes; and a similar conclusion was indicated by a comparison of different portions of an extensive

French arc, about 9° in length, extending from Dunkirk to the Pyrenees, which was measured under the superintendence of the Cassinis in continuation of Picard's arc, the result being published by J. Cassini in 1720. In neither case, however, were the data sufficiently accurate to justify the conclusion; and the first decisive evidence was obtained by measurement of arcs in places differing far more widely in latitude than any that had hitherto been available. The French Academy organised an expedition to Peru, under the management of three Academicians, *Pierre Bouguer* (1698–1758), *Charles Marie de La Condamine* (1701–1774), and *Louis Godin* (1704–1760), with whom two Spanish naval officers also co-operated.

The expedition started in 1735, and, owing to various difficulties, the work was spread out over nearly ten years. The most important result was the measurement, with very fair accuracy, of an arc of about 3° in length, close to the equator; but a number of pendulum experiments of value were also performed, and a good many miscellaneous additions to knowledge were made.

But while the Peruvian party were still at their work a similar expedition to Lapland, under the Academician *Pierre Louis Moreau de Maupertuis* (1698–1759), had much more rapidly (1736–7), if somewhat carelessly, effected the measurement of an arc of nearly 1° close to the arctic circle.

From these measurements it resulted that the lengths of a degree of a meridian about latitude 2° S. (Peru), about latitude 47° N. (France) and about latitude 66° N. (Lapland) were respectively 362,800 feet, 364,900 feet, and 367,100 feet.* There was therefore clear evidence, from a comparison of any two of these arcs, of an increase of the length of a degree of a meridian as the latitude increases; and the general correctness of Newton's views as against Cassini's was thus definitely established.

The extent to which the earth deviates from a sphere is usually expressed by a fraction known as the **ellipticity**, which is the difference between the lines C A, C B of fig. 78 divided by the greater of them. From comparison of the three arcs just mentioned several very different values of the

* 69 miles is 364,320 feet, so that the two northern degrees were a little more and the Peruvian are a little less than 69 miles.

ellipticity were deduced, the discrepancies being partly due to different theoretical methods of interpreting the results and partly to errors in the arcs.

A measurement, made by *Jöns Svanberg* (1771–1851) in 1801–3, of an arc near that of Maupertuis has in fact shewn that his estimate of the length of a degree was about 1,000 feet too large.

A large number of other arcs have been measured in different parts of the earth at various times during the 18th and 19th centuries. The details of the measurements need not be given, but to prevent recurrence to the subject it is convenient to give here the results, obtained by a comparison of these different measurements, that the ellipticity is very nearly $\frac{1}{292}$, and the greatest radius of the earth (C A in fig. 78) a little less than 21,000,000 feet or 4,000 miles. It follows from these figures that the length of a degree in the latitude of London contains, to use Sir John Herschel's ingenious mnemonic, almost exactly as many thousand feet as the year contains days.

222. Reference has already been made to the supremacy of Greenwich during the 18th century in the domain of exact observation. France, however, produced during this period one great observing astronomer who actually accomplished much, and under more favourable external conditions might almost have rivalled Bradley.

Nicholas Louis de Lacaille was born in 1713. After he had devoted a good deal of time to theological studies with a view to an ecclesiastical career, his interests were diverted to astronomy and mathematics. He was introduced to Jacques Cassini, and appointed one of the assistants at the Paris Observatory.

In 1738 and the two following years he took an active part in the measurement of the French arc, then in process of verification. While engaged in this work he was appointed (1739) to a poorly paid professorship at the Mazarin College, at which a small observatory was erected. Here it was his regular practice to spend the whole night, if fine, in observation, while "to fill up usefully the hours of leisure which bad weather gives to observers only too often" he undertook a variety of extensive calculations and wrote innumerable scientific memoirs. It is therefore not surprising that he died comparatively early (1762) and that his death was generally attributed to overwork.

223. The monotony of Lacaille's outward life was broken by the scientific expedition to the Cape of Good Hope (1750–1754) organised by the Academy of Sciences and placed under his direction.

The most striking piece of work undertaken during this expedition was a systematic survey of the southern skies, in the course of which more than 10,000 stars were observed.

These observations, together with a carefully executed catalogue of nearly 2,000 of the stars * and a star-map, were published posthumously in 1763 under the title *Coelum Australe Stelliferum*, and entirely superseded Halley's much smaller and less accurate catalogue (§ 199). Lacaille found it necessary to make 14 new constellations (some of which have since been generally abandoned), and to restore to their original places the stars which the loyal Halley had made into King Charles's Oak. Incidentally Lacaille observed and described 42 nebulae, nebulous stars, and star-clusters, objects the systematic study of which was one of Herschel's great achievements (chapter XII., §§ 259–261).

He made a large number of pendulum experiments, at Mauritius as well as at the Cape, with the usual object of determining in a new part of the world the acceleration due to gravity, and measured an arc of the meridian extending over rather more than a degree. He made also careful observations of the positions of Mars and Venus, in order that from comparison of them with simultaneous observations in northern latitudes he might get the parallax of the sun (chapter VIII., § 161). These observations of Mars compared with some made in Europe by Bradley and others, and a similar treatment of Venus, both pointed to a solar parallax slightly in excess of 10″, a result less accurate than Cassini's (chapter VIII., § 161), though obtained by more reliable processes.

A large number of observations of the moon, of which

* The remaining 8,000 stars were not "reduced" by Lacaille. The whole number were first published in the "reduced" form by the British Association in 1845.

those made by him at the Cape formed an important part, led, after an elaborate discussion in which the spheroidal form of the earth was taken into account, to an improved value of the moon's distance, first published in 1761.

Lacaille also used his observations of fixed stars to improve our knowledge of refraction, and obtained a number of observations of the sun in that part of its orbit which it traverses in our winter months (the summer of the southern hemisphere), and in which it is therefore too near the horizon to be observed satisfactorily in Europe.

The results of this—one of the most fruitful scientific expeditions ever undertaken—were published in separate memoirs or embodied in various books published after his return to Paris.

224. In 1757, under the title *Astronomiae Fundamenta*, appeared a catalogue of 400 of the brightest stars, observed and reduced with the most scrupulous care, so that, notwithstanding the poverty of Lacaille's instrumental outfit, the catalogue was far superior to any of its predecessors, and was only surpassed by Bradley's observations as they were gradually published. It is characteristic of Lacaille's unselfish nature that he did not have the *Fundamenta* sold in the ordinary way, but distributed copies gratuitously to those interested in the subject, and earned the money necessary to pay the expenses of publication by calculating some astronomical almanacks.

Another catalogue, of rather more than 500 stars situated in the zodiac, was published posthumously.

In the following year (1758) he published an excellent set of Solar Tables, based on an immense series of observations and calculations. These were remarkable as the first in which planetary perturbations were taken into account.

Among Lacaille's minor contributions to astronomy may be mentioned: improved methods of calculating cometary orbits and the actual calculation of the orbits of a large number of recorded comets, the calculation of all eclipses visible in Europe since the year 1, a warning that the transit of Venus would be capable of far less accurate observation than Halley had expected (§ 202), observations of the actual transit of 1761 (§ 227), and a number of improvements in methods of calculation and of utilising observations.

In estimating the immense mass of work which Lacaille accomplished during an astronomical career of about 22 years, it has also to be borne in mind that he had only moderately good instruments at his observatory, and *no assistant*, and that a considerable part of his time had to be spent in earning the means of living and of working.

225. During the period under consideration Germany also produced one astronomer, primarily an observer, of great merit, *Tobias Mayer* (1723–1762). He was appointed professor of mathematics and political economy at Göttingen in 1751, apparently on the understanding that he need not lecture on the latter subject, of which indeed he seems to have professed no knowledge; three years later he was put in charge of the observatory, which had been erected 20 years before. He had at least one fine instrument,* and following the example of Tycho, Flamsteed, and Bradley, he made a careful study of its defects, and carried further than any of his predecessors the theory of correcting observations for instrumental errors.†

He improved Lacaille's tables of the sun, and made a catalogue of 998 zodiacal stars, published posthumously in 1775; by a comparison of star places recorded by Roemer (1706) with his own and Lacaille's observations he obtained evidence of a considerable number of proper motions (§ 203); and he made a number of other less interesting additions to astronomical knowledge.

226. But Mayer's most important work was on the moon. At the beginning of his career he made a careful study of the position of the craters and other markings, and was thereby able to get a complete geometrical explanation of the various librations of the moon (chapter VI., § 133), and to fix with accuracy the position of the axis about which the moon rotates. A map of the moon based on his observations was published with other posthumous works in 1775.

* A mural quadrant.

† The ordinary approximate theory of the *collimation error*, *level error*, and *deviation error* of a transit, as given in text-books of spherical and practical astronomy, is substantially his.

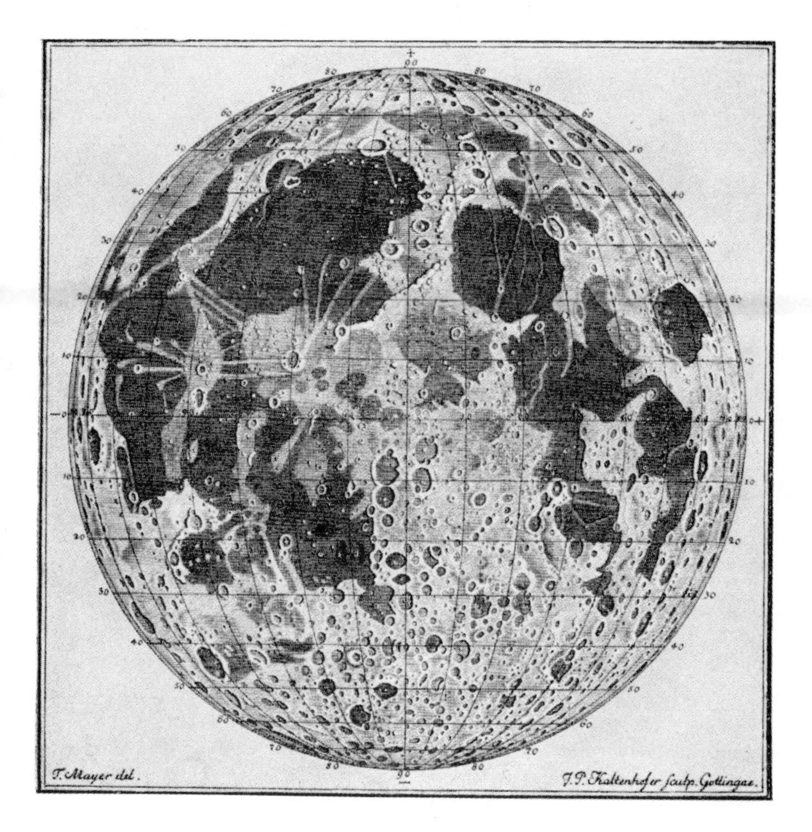

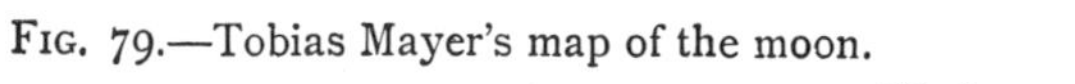
FIG. 79.—Tobias Mayer's map of the moon.

[*To face p.* 282.

Much more important, however, were his lunar theory and the tables based on it. The intrinsic mathematical interest of the problem of the motion of the moon, and its practical importance for the determination of longitude, had caused a great deal of attention to be given to the subject by the astronomers of the 18th century. A further stimulus was also furnished by the prizes offered by the British Government in 1713 for a method of finding the longitude at sea, *viz.* £20,000 for a method reliable to within half a degree, and smaller amounts for methods of less accuracy.

All the great mathematicians of the period made attempts at deducing the moon's motions from gravitational principles. Mayer worked out a theory in accordance with methods used by Euler (chapter XI., § 233), but made a much more liberal and also more skilful use of observations to determine various numerical quantities, which pure theory gave either not at all or with considerable uncertainty. He accordingly succeeded in calculating tables of the moon (published with those of the sun in 1753) which were a notable improvement on those of any earlier writer. After making further improvements, he sent them in 1755 to England. Bradley, to whom the Admiralty submitted them for criticism, reported favourably of their accuracy; and a few years later, after making some alterations in the tables on the basis of his own observations, he recommended to the Admiralty a longitude method based on their use which he estimated to be in general capable of giving the longitude within about half a degree.

Before anything definite was done, Mayer died at the early age of 39, leaving behind him a new set of tables, which were also sent to England. Ultimately £3,000 was paid to his widow in 1765; and both his *Theory of the Moon** and his improved Solar and Lunar Tables were published in 1770 at the expense of the Board of Longitude. A later edition, improved by Bradley's former assistant *Charles Mason* (1730–1787), appeared in 1787.

A prize was also given to Euler for his theoretical work; while £3,000 and subsequently £10,000 more were awarded to *John Harrison* for improvements in the chronometer,

* The title-page is dated 1767; but it is known not to have been actually published till three years later.

which rendered practicable an entirely different method of finding the longitude (chapter vi., § 127).

227. The astronomers of the 18th century had two opportunities of utilising a transit of Venus for the determination of the distance of the sun, as recommended by Halley (§ 202).

A passage or transit of Venus across the sun's disc is a phenomenon of the same nature as an eclipse of the sun by the moon, with the important difference that the apparent magnitude of the planet is too small to cause any serious diminution in the sun's light, and it merely appears as a small black dot on the bright surface of the sun.

If the path of Venus lay in the ecliptic, then at every inferior conjunction, occurring once in 584 days, she would necessarily pass between the sun and earth and would appear to transit. As, however, the paths of Venus and the earth are inclined to one another, at inferior conjunction Venus is usually far enough "above" or "below" the ecliptic for no transit to occur. With the present position of the two paths—which planetary perturbations are only very gradually changing—transits of Venus occur in pairs eight years apart, while between the latter of one pair and the earlier of the next pair elapse alternately intervals of 105½ and of 121½ years. Thus transits have taken place in December 1631 and 1639, June 1761 and 1769, December 1874 and 1882, and will occur again in 2004 and 2012, 2117 and 2125, and so on.

The method of getting the distance of the sun from a transit of Venus may be said not to differ essentially from that based on observations of Mars (chapter viii., § 161).

The observer's object in both cases is to obtain the difference in direction of the planet as seen from different places on the earth. Venus, however, when at all near the earth, is usually too near the sun in the sky to be capable of minutely exact observation, but when a transit occurs the sun's disc serves as it were as a dial-plate on which the position of the planet can be noted. Moreover the measurement of minute angles, an art not yet carried to very great perfection in the 18th century, can be avoided by time-observations, as the difference in the times at which Venus enters (or leaves) the sun's disc as seen at different stations, or the difference in the durations of the transit, can be without difficulty translated into difference of direction, and the distances of Venus and the sun can be deduced.*

Immense trouble was taken by Governments, Academies, and private persons in arranging for the observation of the transits of 1761 and 1769. For the former observing parties were sent as far as to Tobolsk, St. Helena, the Cape of Good Hope, and India, while observations were also made by astronomers at Greenwich, Paris, Vienna, Upsala, and elsewhere in Europe. The next transit was observed on an even larger scale, the stations selected ranging from Siberia to California, from the Varanger Fjord to Otaheiti (where no less famous a person than Captain Cook was placed), and from Hudson's Bay to Madras.

The expeditions organised on this occasion by the American Philosophical Society may be regarded as the first of the contributions made by America to the science which has since owed so much to her; while the Empress Catherine bore witness to the newly acquired civilisation of her country by arranging a number of observing stations on Russian soil.

The results were far more in accordance with Lacaille's anticipations than with Halley's. A variety of causes prevented the moments of contact between the discs of Venus and the sun from being observed with the precision that had been hoped. By selecting different sets of observations, and by making different allowances for the various probable sources of error, a number of discordant results were obtained by various calculators. The values of the parallax (chapter VIII., § 161) of the sun deduced from the earlier of the two transits ranged between about 8″ and 10″; while those obtained in 1769, though much more consistent, still varied between about 8″ and 9″, corresponding to a variation of about 10,000,000 miles in the distance of the sun.

The whole set of observations were subsequently very elaborately discussed in 1822–4 and again in 1835 by *Johann Franz Encke* (1791–1865), who deduced a parallax of 8″·571, corresponding to a distance of 95,370,000 miles,

* For a more detailed discussion of the transit of Venus, see Airy's *Popular Astronomy* and Newcomb's *Popular Astronomy*.

a number which long remained classical. The uncertainty of the data is, however, shewn by the fact that other equally competent astronomers have deduced from the observations of 1769 parallaxes of 8″·8 and 8″·9.

No account has yet been given of William Herschel, perhaps the most famous of all observers, whose career falls mainly into the last quarter of the 18th century and the earlier part of the 19th century. As, however, his work was essentially different from that of almost all the astronomers of the 18th century, and gave a powerful impulse to a department of astronomy hitherto almost ignored, it is convenient to postpone to a later chapter (XII.) the discussion of his work.

CHAPTER XI.

GRAVITATIONAL ASTRONOMY IN THE 18TH CENTURY.

"Astronomy, considered in the most general way, is a great problem of mechanics, the arbitrary data of which are the elements of the celestial movements; its solution depends both on the accuracy of observations and on the perfection of analysis."

LAPLACE, Preface to the *Mécanique Céleste*.

228. THE solar system, as it was known at the beginning of the 18th century, contained 18 recognised members: the sun, six planets, ten satellites (one belonging to the earth, four to Jupiter, and five to Saturn), and Saturn's ring.

Comets were known to have come on many occasions into the region of space occupied by the solar system, and there were reasons to believe that one of them at least (chapter x., § 200) was a regular visitor; they were, however, scarcely regarded as belonging to the solar system, and their action (if any) on its members was ignored, a neglect which subsequent investigation has completely justified. Many thousands of fixed stars had also been observed, and their places on the celestial sphere determined; they were known to be at very great though unknown distances from the solar system, and their influence on it was regarded as insensible.

The motions of the 18 members of the solar system were tolerably well known; their actual distances from one another had been roughly estimated, while the *proportions* between most of the distances were known with considerable accuracy. Apart from the entirely anomalous ring of Saturn, which may for the present be left out of consideration, most of the bodies of the system were known from observation to be nearly spherical in form, and the rest were generally supposed to be so also.

Newton had shewn, with a considerable degree of probability, that these bodies attracted one another according to the law of gravitation; and there was no reason to suppose that they exerted any other important influence on one another's motions.*

The problem which presented itself, and which may conveniently be called Newton's problem, was therefore:—

Given these 18 *bodies, and their positions and motions at any time, to deduce from their mutual gravitation by a process of mathematical calculation their positions and motions at any other time; and to shew that these agree with those actually observed.*

Such a calculation would necessarily involve, among other quantities, the masses of the several bodies; it was evidently legitimate to assume these at will in such a way as to make the results of calculation agree with those of observation. If this were done successfully the masses would thereby be determined. In the same way the commonly accepted estimates of the dimensions of the solar system and of the shapes of its members might be modified in any way not actually inconsistent with direct observation.

The general problem thus formulated can fortunately be reduced to somewhat simpler ones.

Newton had shewn (chapter IX., § 182) that an ordinary sphere attracted other bodies and was attracted by them, as if its mass were concentrated at its centre; and that the effects of deviation from a spherical form became very small at a considerable distance from the body. Hence, except in special cases, the bodies of the solar system could be treated as spheres, which could again be regarded as concentrated at their respective centres. It will be convenient for the sake of brevity to assume for the future that all "bodies" referred to are of this sort, unless the contrary is stated or implied. The effects of deviations from spherical form could then be treated separately

* *Some* other influences are known—*e.g.* the sun's heat causes various motions of our air and water, and has a certain minute effect on the earth's rate of rotation, and presumably produces similar effects on other bodies.

when required, as in the cases of precession and of other motions of a planet or satellite about its centre, and of the corresponding action of a non-spherical planet on its satellites; to this group of problems belongs also that of the tides and other cases of the motion of parts of a body of any form relative to the rest.

Again, the solar system happens to be so constituted that each body's motion can be treated as determined primarily by one other body only. A planet, for example, moves nearly as if no other body but the sun existed, and the moon's motion relative to the earth is roughly the same as if the other bodies of the solar system were non-existent.

The problem of the motion of two mutually gravitating spheres was completely solved by Newton, and was shewn to lead to Kepler's first two laws. Hence each body of the solar system could be regarded as moving nearly in an ellipse round some one body, but as slightly disturbed by the action of others. Moreover, by a general mathematical principle applicable in problems of motion, the effect of a number of small disturbing causes acting conjointly is nearly the same as that which results from adding together their separate effects. Hence each body could, without great error, be regarded as disturbed by one body at a time; the several disturbing effects could then be added together, and a fresh calculation could be made to further diminish the error. The kernel of Newton's problem is thus seen to be a special case of the so-called **problem of three bodies,** viz.:—

Given at any time the positions and motions of three mutually gravitating bodies, to determine their positions and motions at any other time.

Even this apparently simple problem in its general form entirely transcends the powers, not only of the mathematical methods of the early 18th century, but also of those that have been devised since. Certain special cases have been solved, so that it has been shewn to be possible to suppose three bodies initially moving in such a way that their future motion can be completely determined. But these cases do not occur in nature.

In the case of the solar system the problem is simplified, not only by the consideration already mentioned that one of the three bodies can always be regarded as exercising only a small influence on the relative motion of the other two, but also by the facts that the orbits of the planets and satellites do not differ much from circles, and that the planes of their orbits are in no case inclined at large angles to any one of them, such as the ecliptic; in other words, that the eccentricities and inclinations are small quantities.

Thus simplified, the problem has been found to admit of solutions of considerable accuracy by methods of approximation.*

In the case of the system formed by the sun, earth, and moon, the characteristic feature is the great distance of the sun, which is the disturbing body, from the other two bodies; in the case of the sun and two planets, the enormous mass of the sun as compared with the disturbing planet is the important factor. Hence the methods of treatment suitable for the two cases differ, and two substantially distinct branches of the subject, **lunar theory** and **planetary theory,** have developed. The problems presented by the motions of the satellites of Jupiter and Saturn, though allied to those of the lunar theory, differ in some important respects, and are usually treated separately.

229. As we have seen, Newton made a number of important steps towards the solution of his problem, but little was done by his successors in his own country. On the Continent also progress was at first very slow. The *Principia* was read and admired by most of the leading mathematicians of the time, but its principles were not accepted, and Cartesianism remained the prevailing philosophy. A forward step is marked by the publication by the Paris Academy of Sciences in 1720 of a memoir written by the *Chevalier de Louville* (1671–1732) on the basis of Newton's principles; ten years later the Academy awarded a prize to an essay on the planetary motions written by *John Bernouilli* (1667–1748) on Cartesian principles, a Newtonian essay being put second. In 1732 Maupertuis (chapter x., § 221) published a treatise on the figure of the

* The arithmetical processes of working out, figure by figure, a non-terminating decimal or a square root are simple cases of successive approximation.

earth on Newtonian lines, and the appearance six years later of Voltaire's extremely readable *Éléments de la Philosophie de Newton* had a great effect in popularising the new ideas. The last official recognition of Cartesianism in France seems to have been in 1740, when the prize offered by the Academy for an essay on the tides was shared between a Cartesian and three eminent Newtonians (§ 230).

The rapid development of gravitational astronomy that ensued between this time and the beginning of the 19th century was almost entirely the work of five great Continental mathematicians, Euler, Clairaut, D'Alembert, Lagrange, and Laplace, of whom the eldest was born in 1707 and the youngest died in 1827, within a month of the centenary of Newton's death. Euler was a Swiss, Lagrange was of Italian birth but French by extraction and to a great extent by adoption, and the other three were entirely French. France therefore during nearly the whole of the 18th century reigned supreme in gravitational astronomy, and has not lost her supremacy even to-day, though during the present century America, England, Germany, Italy, and other countries have all made substantial contributions to the subject.

It is convenient to consider first the work of the three first-named astronomers, and to treat later Lagrange and Laplace, who carried gravitational astronomy to a decidedly higher stage of development than their predecessors.

230. *Leonhard Euler* was born at Basle in 1707, 14 years later than Bradley and six years earlier than Lacaille. He was the son of a Protestant minister who had studied mathematics under *James Bernouilli* (1654–1705), the first of a famous family of mathematicians. Leonhard Euler himself was a favourite pupil of John Bernouilli (the younger brother of James), and was an intimate friend of his two sons, one of whom, *Daniel* (1700–1782), was not only a distinguished mathematician like his father and uncle, but was also the first important Newtonian outside Great Britain. Like so many other astronomers, Euler began by studying theology, but was induced both by his natural tastes and by the influence of the Bernouillis to turn his attention to mathematics. Through the influence of Daniel Bernouilli, who had recently been appointed to a professorship at

St. Petersburg, Euler received and accepted an invitation to join the newly created Academy of Sciences there (1727). This first appointment carried with it a stipend, and the duties were the general promotion of science; subsequently Euler undertook more definite professorial work, but most of his energy during the whole of his career was devoted to writing mathematical papers, the majority of which were published by the St. Petersburg Academy. Though he took no part in politics, Russian autocracy appears to have been oppressive to him, reared as he had been among Swiss and Protestant surroundings; and in 1741 he accepted an invitation from Frederick the Great, a despot of a less pronounced type, to come to Berlin, and assist in reorganising the Academy of Sciences there. On being reproached one day by the Queen for his taciturn and melancholy demeanour, he justified his silence on the ground that he had just come from a country where speech was liable to lead to hanging;* but notwithstanding this frank criticism he remained on good terms with the Russian court, and continued to draw his stipend as a member of the St. Petersburg Academy and to contribute to its Transactions. Moreover, after 25 years spent at Berlin, he accepted a pressing invitation from the Empress Catherine II. and returned to Russia (1766).

He had lost the use of one eye in 1735, a disaster which called from him the remark that he would henceforward have less to distract him from his mathematics; the second eye went soon after his return to Russia, and with the exception of a short time during which an operation restored the partial use of one eye he remained blind till the end of his life. But this disability made little difference to his astounding scientific activity; and it was only after nearly 17 years of blindness that as a result of a fit of apoplexy "he ceased to live and to calculate" (1783).

Euler was probably the most versatile as well as the most prolific of mathematicians of all time. There is scarcely any branch of modern analysis to which he was not a large contributor, and his extraordinary powers of devising and applying methods of calculation were employed by him with great success in each of the existing branches of applied

* "C'est que je viens d'un pays où, quand on parle, on est pendu."

mathematics; problems of abstract dynamics, of optics, of the motion of fluids, and of astronomy were all in turn subjected to his analysis and solved. The extent of his writings is shewn by the fact that, in addition to several books, he wrote about 800 papers on mathematical and physical subjects; it is estimated that a complete edition of his works would occupy 25 quarto volumes of about 600 pages each.

Euler's first contribution to astronomy was an essay on the tides which obtained a share of the Academy prize for 1740 already referred to, Daniel Bernouilli and Maclaurin (chapter x., § 196) being the other two Newtonians. The problem of the tides was, however, by no means solved by any of the three writers.

He gave two distinct solutions of the problem of three bodies in a form suitable for the lunar theory, and made a number of extremely important and suggestive though incomplete contributions to planetary theory. In both subjects his work was so closely connected with that of Clairaut and D'Alembert that it is more convenient to discuss it in connection with theirs.

231. *Alexis Claude Clairaut*, born at Paris in 1713, belongs to the class of precocious geniuses. He read the Infinitesimal Calculus and Conic Sections at the age of ten, presented a scientific memoir to the Academy of Sciences before he was 13, and published a book containing some important contributions to geometry when he was 18, thereby winning his admission to the Academy.

Shortly afterwards he took part in Maupertuis' expedition to Lapland (chapter x., § 221), and after publishing several papers of minor importance produced in 1743 his classical work on the figure of the earth. In this he discussed in a far more complete form than either Newton or Maclaurin the form which a rotating body like the earth assumes under the influence of the mutual gravitation of its parts, certain hypotheses of a very general nature being made as to the variations of density in the interior; and deduced formulae for the changes in different latitudes of the acceleration due to gravity, which are in satisfactory agreement with the results of pendulum experiments.

Although the subject has since been more elaborately

and more generally treated by later writers, and a good many additions have been made, few if any results of fundamental importance have been added to those contained in Clairaut's book.

He next turned his attention to the problem of three bodies, obtained a solution suitable for the moon, and made some progress in planetary theory.

Halley's comet (chapter x., § 200) was "due" about

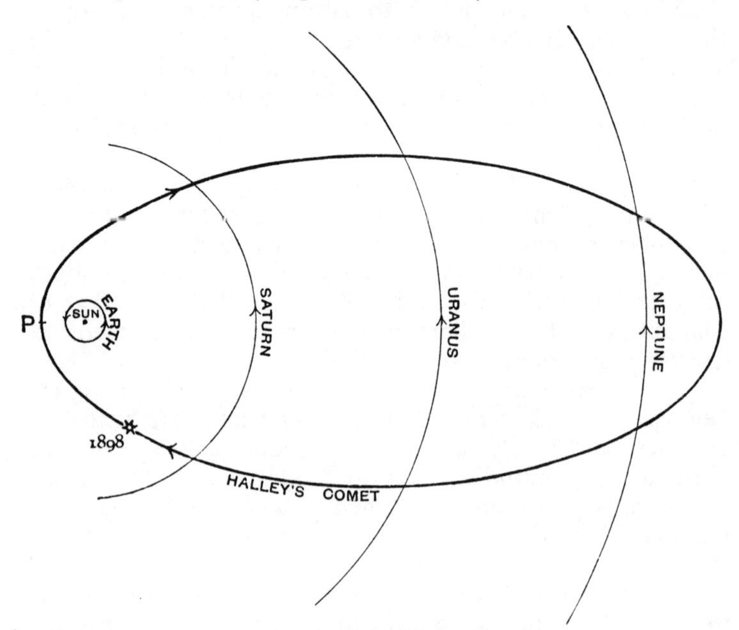

Fig. 80.—The path of Halley's comet.

1758; as the time approached Clairaut took up the task of computing the perturbations which it would probably have experienced since its last appearance, owing to the influence of the two great planets, Jupiter and Saturn, close to both of which it would have passed. An extremely laborious calculation shewed that the comet would have been retarded about 100 days by Saturn and about 518 days by Jupiter, and he accordingly announced to the Academy towards the end of 1758 that the comet might be expected to pass its perihelion (the point of its orbit nearest the sun, P in fig. 80) about April 13th of the following year, though owing to various defects in his calculation there might be an error of a month either way. The comet was anxiously watched for by the astronomical world, and was actually discovered by an amateur, *George Palitzsch* (1723–1788) of Saxony, on Christmas Day, 1758; it passed its perihelion just a month and a day before the time assigned by Clairaut.

Halley's brilliant conjecture was thus justified; a new member was added to the solar system, and hopes were raised—to be afterwards amply fulfilled—that in other cases also the motions of comets might be reduced to rule, and calculated according to the same principles as those of less erratic bodies. The superstitions attached to comets were of course at the same time still further shaken.

Clairaut appears to have had great personal charm and to have been a conspicuous figure in Paris society. Unfortunately his strength was not equal to the combined claims of social and scientific labours, and he died in 1765 at an age when much might still have been hoped from his extraordinary abilities.*

232. *Jean-le-Rond D'Alembert* was found in 1717 as an infant on the steps of the church of St. Jean-le-Rond in Paris, but was afterwards recognised, and to some extent provided for, by his father, though his home was with his foster-parents. After receiving a fair school education, he studied law and medicine, but then turned his attention to mathematics. He first attracted notice in mathematical circles by a paper written in 1738, and was admitted to the Academy of Sciences two years afterwards. His earliest important work was the *Traité de Dynamique* (1743), which contained, among other contributions to the subject, the first statement of a dynamical principle which bears his name, and which, though in one sense only a corollary from Newton's Third Law of Motion, has proved to be of immense service in nearly all general dynamical problems,

* Longevity has been a remarkable characteristic of the great mathematical astronomers: Newton died in his 85th year; Euler, Lagrange, and Laplace lived to be more than 75, and D'Alembert was almost 66 at his death.

astronomical or otherwise. During the next few years he made a number of contributions to mathematical physics, as well as to the problem of three bodies; and published in 1749 his work on precession and nutation, already referred to (chapter x., § 215). From this time onwards he began to give an increasing part of his energies to work outside mathematics. For some years he collaborated with Diderot in producing the famous French Encyclopaedia, which began to appear in 1751, and exercised so great an influence on contemporary political and philosophic thought. D'Alembert wrote the introduction, which was read to the *Académie Française* * in 1754 on the occasion of his admission to that distinguished body, as well as a variety of scientific and other articles. In the later part of his life, which ended in 1783, he wrote little on mathematics, but published a number of books on philosophical, literary, and political subjects; † as secretary of the Academy he also wrote obituary notices (*éloges*) of some 70 of its members. He was thus, in Carlyle's words, "of great faculty, especially of great clearness and method; famous in Mathematics; no less so, to the wonder of some, in the intellectual provinces of Literature."

D'Alembert and Clairaut were great rivals, and almost every work of the latter was severely criticised by the former, while Clairaut retaliated though with much less zeal and vehemence. The great popular reputation acquired by Clairaut through his work on Halley's comet appears to have particularly excited D'Alembert's jealousy. The rivalry, though not a pleasant spectacle, was, however, useful in leading to the detection and subsequent improvement of various weak points in the work of each. In other respects D'Alembert's personal characteristics appear to have been extremely pleasant. He was always a poor man, but nevertheless declined magnificent offers made to him by both Catherine II. of Russia and Frederick the

* This body, which is primarily literary, has to be distinguished from the much less famous Paris Academy of Sciences, constantly referred to (often simply as the Academy) in this chapter and the preceding.

† E.g. *Mélanges de Philosophie, de l'Histoire, et de Littérature; Élements de Philosophie; Sur la Destruction des Jésuites.*

Great of Prussia, and preferred to keep his independence, though he retained the friendship of both sovereigns and accepted a small pension from the latter. He lived extremely simply, and notwithstanding his poverty was very generous to his foster-mother, to various young students, and to many others with whom he came into contact.

233. Euler, Clairaut, and D'Alembert all succeeded in obtaining independently and nearly simultaneously solutions of the problem of three bodies in a form suitable for lunar theory. Euler published in 1746 some rather imperfect Tables of the Moon, which shewed that he must have already obtained his solution. Both Clairaut and D'Alembert presented to the Academy in 1747 memoirs containing their respective solutions, with applications to the moon as well as to some planetary problems. In each of these memoirs occurred the same difficulty which Newton had met with: the calculated motion of the moon's apogee was only about half the observed result. Clairaut at first met this difficulty by assuming an alteration in the law of gravitation, and got a result which seemed to him satisfactory by assuming gravitation to vary partly as the inverse square and partly as the inverse cube of the distance.* Euler also had doubts as to the correctness of the inverse square. Two years later, however (1749), on going through his original calculation again, Clairaut discovered that certain terms, which had appeared unimportant at the beginning of the calculation and had therefore been omitted, became important later on. When these were taken into account, the motion of the apogee as deduced from theory agreed very nearly with that observed. This was the first of several cases in which a serious discrepancy between theory and observation has at first discredited the law of gravitation, but has subsequently been explained away, and has thereby given a new verification of its accuracy. When Clairaut had announced his discovery, Euler arrived by a fresh calculation at substantially the same result, while D'Alembert by carrying the approximation further obtained one that was slightly more accurate. A fresh calculation of the motion of the moon by Clairaut won the prize on the subject offered by the St. Petersburg Academy, and was

* *I.e.* he assumed a law of attraction represented by $\mu/r^2 + \nu/r^3$.

published in 1752, with the title *Théorie de la Lune.* Two years later he published a set of lunar tables, and just before his death (1765) he brought out a revised edition of the *Théorie de la Lune* in which he embodied a new set of tables.

D'Alembert followed his paper of 1747 by a complete lunar theory (with a moderately good set of tables), which, though substantially finished in 1751, was only published in 1754 as the first volume of his *Recherches sur différens points importans du système du Monde.* In 1756 he published an improved set of tables, and a few months afterward a third volume of *Recherches* with some fresh developments of the theory. The second volume of his *Opuscules Mathématiques* (1762) contained another memoir on the subject with a third set of tables, which were a slight improvement on the earlier ones.

Euler's first lunar theory (*Theoria Motuum Lunae*) was published in 1753, though it had been sent to the St. Petersburg Academy a year or two earlier. In an appendix * he points out with characteristic frankness the defects from which his treatment seems to him to suffer, and suggests a new method of dealing with the subject. It was on this theory that Tobias Mayer based his tables, referred to in the preceding chapter (§ 226). Many years later Euler devised an entirely new way of attacking the subject, and after some preliminary papers dealing generally with the method and with special parts of the problem, he worked out the lunar theory in great detail, with the help of one of his sons and two other assistants, and published the whole, together with tables, in 1772. He attempted, but without success, to deal in this theory with the secular acceleration of the mean motion which Halley had detected (chapter X., § 201).

In any mathematical treatment of an astronomical problem some data have to be borrowed from observation, and of the three astronomers Clairaut seems to have been the most skilful in utilising observations, many of which he obtained from Lacaille. Hence his tables represented the actual

* This appendix is memorable as giving for the first time the method of *variation of parameters* which Lagrange afterwards developed and used with such success.

motions of the moon far more accurately than those of D'Alembert, and were even superior in some points to those based on Euler's very much more elaborate second theory; Clairaut's last tables were seldom in error more than $1\frac{1}{2}'$, and would hence serve to determine the longitude to within about $\frac{3}{4}^\circ$. Clairaut's tables were, however, never much used, since Tobias Mayer's as improved by Bradley were found in practice to be a good deal more accurate; but Mayer borrowed so extensively from observation that his formulae cannot be regarded as true deductions from gravitation in the same sense in which Clairaut's were. Mathematically Euler's second theory is the most interesting and was of the greatest importance as a basis for later developments. The most modern lunar theory * is in some sense a return to Euler's methods.

234. Newton's lunar theory may be said to have given a *qualitative* account of the lunar inequalities known by observation at the time when the *Principia* was published, and to have indicated others which had not yet been observed. But his attempts to explain these irregularities *quantitatively* were only partially successful.

Euler, Clairaut, and D'Alembert threw the lunar theory into an entirely new form by using analytical methods instead of geometrical; one advantage of this was that by the expenditure of the necessary labour calculations could in general be carried further when required and lead to a higher degree of accuracy. The result of their more elaborate development was that—with one exception—the inequalities known from observation were explained with a considerable degree of accuracy quantitatively as well as qualitatively; and thus tables, such as those of Clairaut, based on theory, represented the lunar motions very closely. The one exception was the secular acceleration: we have just seen that Euler failed to explain it; D'Alembert was equally unsuccessful, and Clairaut does not appear to have considered the question.

235. The chief inequalities in planetary motion which observation had revealed up to Newton's time were the forward motion of the apses of the earth's orbit and a very

* That of the distinguished American astronomer Dr. G. W. Hill (chapter XIII., § 286).

slow diminution in the obliquity of the ecliptic. To these may be added the alterations in the rates of motion of Jupiter and Saturn discovered by Halley (chapter x., § 204).

Newton had shewn generally that the perturbing effect of another planet would cause displacements in the apses of any planetary orbit, and an alteration in the relative positions of the planes in which the disturbing and disturbed planet moved; but he had made no detailed calculations. Some effects of this general nature, in addition to those already known, were, however, indicated with more or less distinctness as the result of observation in various planetary tables published between the date of the *Principia* and the middle of the 18th century.

The irregularities in the motion of the earth, shewing themselves as irregularities in the apparent motion of the sun, and those of Jupiter and Saturn, were the most interesting and important of the planetary inequalities, and prizes for essays on one or another subject were offered several times by the Paris Academy.

The perturbations of the moon necessarily involved—by the principle of action and reaction—corresponding though smaller perturbations of the earth; these were discussed on various occasions by Clairaut and Euler, and still more fully by D'Alembert.

In Clairaut's paper of 1747 (§ 233) he made some attempt to apply his solution of the problem of three bodies to the case of the sun, earth, and Saturn, which on account of Saturn's great distance from the sun (nearly ten times that of the earth) is the planetary case most like that of the earth, moon, and sun (cf. § 228).

Ten years later he discussed in some detail the perturbations of the earth due to Venus and to the moon. This paper was remarkable as containing the first attempt to estimate masses of celestial bodies by observation of perturbations due to them. Clairaut applied this method to the moon and to Venus, by calculating perturbations in the earth's motion due to their action (which necessarily depended on their masses), and then comparing the results with Lacaille's observations of the sun. The mass of the moon was thus found to be about $\frac{1}{67}$ and that of Venus $\frac{2}{3}$ that of the earth; the first result was a considerable improvement on Newton's estimate from tides (chapter IX., § 189), and the second, which was entirely new, previous estimates having been merely conjectural, is in tolerable agreement with modern measurements.* It is worth noticing as a good illustration of the reciprocal influence of observation and mathematical theory that, while Clairaut used Lacaille's observations for his theory, Lacaille in turn used Clairaut's calculations of the perturbations of the earth to improve his tables of the sun published in 1758.

Clairaut's method of solving the problem of three bodies was also applied by *Joseph Jérôme Le François Lalande* (1732–1807), who is chiefly known as an admirable populariser of astronomy but was also an indefatigable calculator and observer, to the perturbations of Mars by Jupiter, of Venus by the earth, and of the earth by Mars, but with only moderate success.

D'Alembert made some progress with the general treatment of planetary perturbations in the second volume of his *Recherches*, and applied his methods to Jupiter and Saturn.

236. Euler carried the general theory a good deal further in a series of papers beginning in 1747. He made several attempts to explain the irregularities of Jupiter and Saturn, but never succeeded in representing the observations satisfactorily. He shewed, however, that the perturbations due to the other planets would cause the earth's apse line to advance about 13″ annually, and the obliquity of the ecliptic to diminish by about 48″ annually, both results being in fair accordance both with observations and with more elaborate calculations made subsequently. He indicated also the existence of various other planetary irregularities, which for the most part had not previously been observed.

In an essay to which the Academy awarded a prize in 1756, but which was first published in 1771, he developed with some completeness a method of dealing with perturbations which he had indicated in his lunar theory of 1753. As this method, known as that of the **variation of the elements** or **parameters**, played a very important part

* They give about ·78 for the mass of Venus compared to that of the earth.

in subsequent researches, it may be worth while to attempt to give a sketch of it.

If perturbations are ignored, a planet can be regarded as moving in an ellipse with the sun in one focus. The size and shape of the ellipse can be defined by the length of its axis and by the eccentricity; the plane in which the ellipse is situated is determined by the position of the line, called the line of **nodes,** in which it cuts a fixed plane, usually taken to be the ecliptic, and by the inclination of the two planes. When these four quantities are fixed, the ellipse may still turn about its focus in its own plane, but if the direction of the apse line is also fixed the ellipse is completely determined. If, further, the position of the planet in its ellipse at any one time is known, the motion is completely determined and its position at any other time can be calculated. There are thus six quantities known as **elements** which completely determine the motion of a planet not subject to perturbation.

When perturbations are taken into account, the path described by a planet in any one revolution is no longer an ellipse, though it differs very slightly from one; while in the case of the moon the deviations are a good deal greater. But if the motions of a planet at two widely different epochs are compared, though on each occasion the path described is very nearly an ellipse, the ellipses differ in some respects. For example, between the time of Ptolemy (A.D. 150) and that of Euler the direction of the apse line of the earth's orbit altered by about 5°, and some of the other elements also varied slightly. Hence in dealing with the motion of a planet through a long period of time it is convenient to introduce the idea of an elliptic path which is gradually changing its position and possibly also its size and shape. One consequence is that the actual path described in the course of a considerable number of revolutions is a curve no longer bearing much resemblance to an ellipse. If, for example, the apse line turns round uniformly while the other elements remain unchanged, the path described is like that shewn in the figure.

Euler extended this idea so as to represent any perturbation of a planet, whether experienced in the course of one revolution or in a longer time, by means of changes in an elliptic orbit. For wherever a planet may be and whatever (within certain limits *) be its speed or direction of motion some ellipse can be found, having the sun in one focus, such that the planet can be regarded as moving in it for a short time. Hence as the planet describes a perturbed orbit it can be regarded as moving at any instant

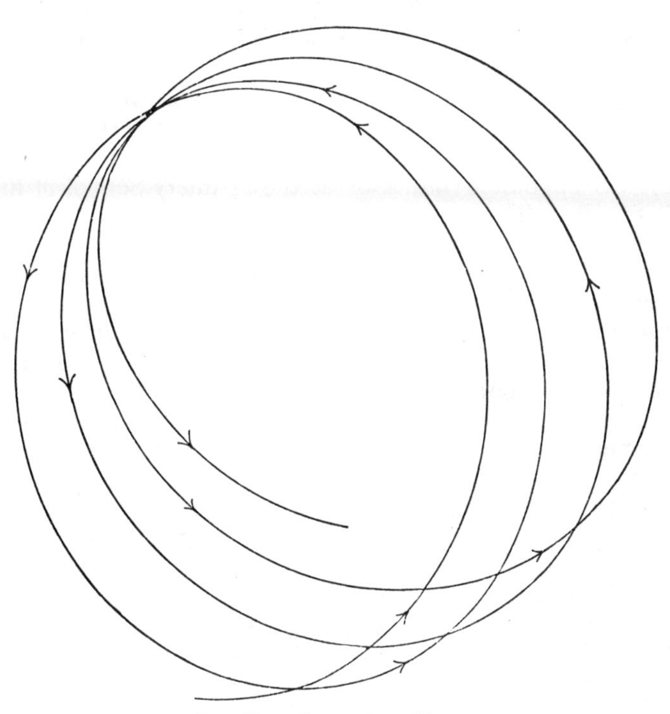

FIG. 81.—A varying ellipse.

in an ellipse, which, however, is continually altering its position or other characteristics. Thus the problem of discussing the planet's motion becomes that of determining the elements of the ellipse which represents its motion at any time. Euler shewed further how, when the position of the perturbing planet was known, the corresponding

* The orbit might be a parabola or hyperbola, though this does not occur in the case of any known planet.

rates of change of the elements of the varying ellipse could be calculated, and made some progress towards deducing from these data the actual elements; but he found the mathematical difficulties too great to be overcome except in some of the simpler cases, and it was reserved for the next generation of mathematicians, notably Lagrange, to shew the full power of the method.

237. *Joseph Louis Lagrange* was born at Turin in 1736, when Clairaut was just starting for Lapland and D'Alembert was still a child; he was descended from a French family three generations of which had lived in Italy. He shewed extraordinary mathematical talent, and when still a mere boy was appointed professor at the Artillery School of his native town, his pupils being older than himself. A few years afterwards he was the chief mover in the foundation of a scientific society, afterwards the Turin Academy of Sciences, which published in 1759 its first volume of Transactions, containing several mathematical articles by Lagrange, which had been written during the last few years. One of these* so impressed Euler, who had made a special study of the subject dealt with, that he at once obtained for Lagrange the honour of admission to the Berlin Academy.

In 1764 Lagrange won the prize offered by the Paris Academy for an essay on the libration of the moon. In this essay he not only gave the first satisfactory, though still incomplete, discussion of the librations (chapter VI., § 133) of the moon due to the non-spherical forms of both the earth and moon, but also introduced an extremely general method of treating dynamical problems,† which is the basis of nearly all the higher branches of dynamics which have been developed up to the present day.

Two years later (1766) Frederick II., at the suggestion of D'Alembert, asked Lagrange to succeed Euler (who had just returned to St. Petersburg) as the head of the mathematical section of the Berlin Academy, giving as a reason that the greatest king in Europe wished to have the greatest mathematician in Europe at his court.

* On the *Calculus of Variations*.

† The establishment of the general equations of motion by a combination of *virtual velocities* and *D'Alembert's principle*.

LAGRANGE. [*To face p.* 305.

Lagrange accepted this magnificently expressed invitation and spent the next 21 years at Berlin.

During this period he produced an extraordinary series of papers on astronomy, on general dynamics, and on a variety of subjects in pure mathematics. Several of the most important of the astronomical papers were sent to Paris and obtained prizes offered by the Academy; most of the other papers—about 60 in all—were published by the Berlin Academy. During this period he wrote also his great *Mécanique Analytique*, one of the most beautiful of all mathematical books, in which he developed fully the general dynamical ideas contained in the earlier paper on libration. Curiously enough he had great difficulty in finding a publisher for his masterpiece, and it only appeared in 1788 in Paris. A year earlier he had left Berlin in consequence of the death of Frederick, and accepted an invitation from Louis XVI. to join the Paris Academy. About this time he suffered from one of the fits of melancholy with which he was periodically seized and which are generally supposed to have been due to overwork during his career at Turin. It is said that he never looked at the *Mécanique Analytique* for two years after its publication, and spent most of the time over chemistry and other branches of natural science as well as in non-scientific pursuits. In 1790 he was made president of the Commission appointed to draw up a new system of weights and measures, which resulted in the establishment of the metric system; and the scientific work connected with this undertaking gradually restored his interest in mathematics and astronomy. He always avoided politics, and passed through the Revolution uninjured, unlike his friend Lavoisier the great chemist and Bailly the historian of astronomy, both of whom were guillotined during the Terror. He was in fact held in great honour by the various governments which ruled France up to the time of his death; in 1793 he was specially exempted from a decree of banishment directed against all foreigners; subsequently he was made professor of mathematics, first at the École Normale (1795), and then at the École Polytechnique (1797), the last appointment being retained till his death in 1813. During this period of his life he published, in addition to a large number of papers on astronomy and mathematics, three important books on pure mathematics,* and at the time of his death had not quite finished a second edition of the *Mécanique Analytique*, the second volume appearing posthumously.

238. *Pierre Simon Laplace*, the son of a small farmer, was born at Beaumont in Normandy in 1749, being thus 13 years younger than his great rival Lagrange. Thanks to the help of well-to-do neighbours, he was first a pupil and afterwards a teacher at the Military School of his native town. When he was 18 he went to Paris with a letter of introduction to D'Alembert, and, when no notice was taken of it, wrote him a letter on the principles of mechanics which impressed D'Alembert so much that he at once took interest in the young mathematician and procured him an appointment at the Military School at Paris. From this time onwards Laplace lived continuously at Paris, holding various official positions. His first paper (on pure mathematics) was published in the Transactions of the Turin Academy for the years 1766–69, and from this time to the end of his life he produced an uninterrupted series of papers and books on astronomy and allied departments of mathematics.

Laplace's work on astronomy was to a great extent incorporated in his *Mécanique Céleste*, the five volumes of which appeared at intervals between 1799 and 1825. In this great treatise he aimed at summing up all that had been done in developing gravitational astronomy since the time of Newton. The only other astronomical book which he published was the *Exposition du Système du Monde* (1796), one of the most perfect and charmingly written popular treatises on astronomy ever published, in which the great mathematician never uses either an algebraical formula or a geometrical diagram. He published also in 1812 an elaborate treatise on the theory of probability or chance,† on which nearly all later developments of the subject have been based, and in 1819 a more popular *Essai Philosophique* on the same subject.

* *Théorie des Fonctions Analytiques* (1797); *Résolution des Équations Numériques* (1798); *Leçons sur le Calcul des Fonctions* (1805).

† *Théorie Analytique des Probabilités.*

LAPLACE

[*To face p.* 307.

Laplace's personality seems to have been less attractive than that of Lagrange. He was vain of his reputation as a mathematician and not always generous to rival discoverers. To Lagrange, however, he was always friendly, and he was also kind in helping young mathematicians of promise. While he was perfectly honest and courageous in upholding his scientific and philosophical opinions, his politics bore an undoubted resemblance to those of the Vicar of Bray, and were professed by him with great success. He was appointed a member of the Commission for Weights and Measures, and afterwards of the Bureau des Longitudes, and was made professor at the École Normale when it was founded. When Napoleon became First Consul, Laplace asked for and obtained the post of Home Secretary, but—fortunately for science—was considered quite incompetent, and had to retire after six weeks (1799)*; as a compensation he was made a member of the newly created Senate. The third volume of the *Mécanique Céleste*, published in 1802, contained a dedication to the "Heroic Pacificator of Europe," at whose hand he subsequently received various other distinctions, and by whom he was created a Count when the Empire was formed. On the restoration of the Bourbons in 1814 he tendered his services to them, and was subsequently made a Marquis. In 1816 he also received a very unusual honour for a mathematician (shared, however, by D'Alembert) by being elected one of the Forty "Immortals" of the *Académie Française*; this distinction he seems to have owed in great part to the literary excellence of the *Système du Monde.*

Notwithstanding these distractions he worked steadily at mathematics and astronomy, and even after the completion of the *Mécanique Céleste* wrote a supplement to it which was published after his death (1827).

His last words, "*Ce que nous connaissons est peu de chose, ce que nous ignorons est immense*," coming as they did from one who had added so much to knowledge, shew his character in a pleasanter aspect than it sometimes presented during his career.

* The fact that the post was then given by Napoleon to his brother Lucien suggests some doubts as to the unprejudiced character of the verdict of incompetence pronounced by Napoleon against Laplace.

239. With the exception of Lagrange's paper on libration, nearly all his and Laplace's important contributions to astronomy were made when Clairaut's and D'Alembert's work was nearly finished, though Euler's activity continued for nearly 20 years more. Lagrange, however, survived him by 30 years and Laplace by more than 40; and together they carried astronomical science to a far higher stage of development than their three predecessors.

240. To the lunar theory Lagrange contributed comparatively little except general methods, applicable to this as to other problems of astronomy; but Laplace devoted great attention to it. Of his special discoveries in the subject the most notable was his explanation of the secular acceleration of the moon's mean motion (chapter x., § 201), which had puzzled so many astronomers. Lagrange had attempted to explain it (1774), and had failed so completely that he was inclined to discredit the early observations on which the existence of the phenomenon was based. Laplace, after trying ordinary methods without success, attempted to explain it by supposing that gravitation was an effect not transmitted instantaneously, but that, like light, it took time to travel from the attracting body to the attracted one; but this also failed. Finally he traced it (1787) to an indirect planetary effect. For, as it happens, certain perturbations which the moon experiences owing to the action of the sun depend among other things on the eccentricity of the earth's orbit; this is one of the elements (§ 236) which is being altered by the action of the planets, and has for many centuries been very slowly decreasing; the perturbation in question is therefore being very slightly altered, and the moon's average rate of motion is in consequence very slowly increasing, or the length of the month decreasing. The whole effect is excessively minute, and only becomes perceptible in the course of a long time. Laplace's calculation shewed that the moon would, in the course of a century, or in about 1,300 complete revolutions, gain about 10″ (more exactly 10″·2) owing to this cause, so that her place in the sky would differ by that amount from what it would be if this disturbing cause did not exist; in two centuries the angle gained would be 40″, in three centuries 90″, and so on.

This may be otherwise expressed by saying that the length of the month diminishes by about one-thirtieth of a second in the course of a century. Moreover, as Laplace shewed (§ 245), the eccentricity of the earth's orbit will not go on diminishing indefinitely, but after an immense period to be reckoned in thousands of years will begin to increase, and the moon's motion will again become slower in consequence.

Laplace's result agreed almost exactly with that indicated by observation; and thus the last known discrepancy of importance in the solar system between theory and observation appeared to be explained away; and by a curious coincidence this was effected just a hundred years after the publication of the *Principia*.

Many years afterwards, however, Laplace's explanation was shewn to be far less complete than it appeared at the time (chapter XIII., § 287).

The same investigation revealed to Laplace the existence of alterations of a similar character, and due to the same cause, of other elements in the moon's orbit, which, though not previously noticed, were found to be indicated by ancient eclipse observations.

241. The third volume of the *Mécanique Céleste* contains a general treatment of the lunar theory, based on a method entirely different from any that had been employed before, and worked out in great detail. "My object," says Laplace, "in this book is to exhibit in the one law of universal gravitation the source of all the inequalities of the motion of the moon, and then to employ this law as a means of discovery, to perfect the theory of this motion and to deduce from it several important elements in the system of the moon." Laplace himself calculated no lunar tables, but the Viennese astronomer *John Tobias Bürg* (1766–1834) made considerable use of his formulae, together with an immense number of Greenwich observations, for the construction of lunar tables, which were sent to the Institute of France in 1801 (before the publication of Laplace's complete lunar theory), and published in a slightly amended form in 1806. A few years later (1812) *John Charles Burckhardt* (1773–1825), a German who had settled in Paris and worked under Laplace and Lalande, produced a new set of tables based directly on the formulae

of the *Mécanique Céleste.* These were generally accepted in lieu of Bürg's, which had been in their turn an improvement on Mason's and Mayer's.

Later work on lunar theory may conveniently be regarded as belonging to a new period of astronomy (chapter XIII., § 286).

242. Observation had shewn the existence of inequalities in the planetary and lunar motions which seemed to belong to two different classes. On the one hand were inequalities, such as most of those of the moon, which went through their cycle of changes in a single revolution or a few revolutions of the disturbing body; and on the other such inequalities as the secular acceleration of the moon's mean motion or the motion of the earth's apses, in which a continuous disturbance was observed always acting in the same direction, and shewing no signs of going through a periodic cycle of changes.

The mathematical treatment of perturbations soon shewed the desirability of adopting different methods of treatment for two classes of inequalities, which corresponded roughly, though not exactly, to those just mentioned, and to which the names of **periodic** and **secular** gradually came to be attached. The distinction plays a considerable part in Euler's work (§ 236), but it was Lagrange who first recognised its full importance, particularly for planetary theory, and who made a special study of secular inequalities.

When the perturbations of one planet by another are being studied, it becomes necessary to obtain a mathematical expression for the disturbing force which the second planet exerts. This expression depends in general both on the elements of the two orbits, and on the positions of the planets at the time considered. It can, however, be divided up into two parts, one of which depends on the positions of the planets (as well as on the elements), while the other depends only on the elements of the two orbits, and is independent of the positions in their paths which the planets may happen to be occupying at the time. Since the positions of planets in their orbits change rapidly, the former part of the disturbing force changes rapidly, and produces in general, at short intervals of time, effects in opposite directions, first, for example, accelerating and then retarding the motion of the disturbed planet; and the corresponding inequalities of motion are the periodic inequalities, which for the most part go through a complete cycle of changes in the course of a few revolutions of the planets, or even more rapidly. The other part of the disturbing force remains nearly unchanged for a considerable period, and gives rise to changes in the elements which, though in general very small, remain for a long time without sensible alteration, and therefore continually accumulate, becoming considerable with the lapse of time: these are the secular inequalities.

Speaking generally, we may say that the periodical inequalities are temporary and the secular inequalities permanent in their effects, or as Sir John Herschel expresses it:—

> "The secular inequalities are, in fact, nothing but what remains after the mutual destruction of a much larger amount (as it very often is) of periodical. But these are in their nature transient and temporary; they disappear in short periods, and leave no trace. The planet is temporarily withdrawn from its orbit (its slowly varying orbit), but forthwith returns to it, to deviate presently as much the other way, while the varied orbit accommodates and adjusts itself to the average of these excursions on either side of it." *

"Temporary" and "short" are, however, relative terms. Some periodical inequalities, notably in the case of the moon, have periods of only a few days, and the majority which are of importance extend only over a few years; but some are known which last for centuries or even thousands of years, and can often be treated as secular when we only want to consider an interval of a few years. On the other hand, most of the known secular inequalities are not really permanent, but fluctuate like the periodical ones, though only in the course of immense periods of time to be reckoned usually by tens of thousands of years.

One distinction between the lunar and planetary theories is that in the former periodic inequalities are comparatively large and, especially for practical purposes such as computing the position of the moon a few months hence, of great

* *Outlines of Astronomy*, § 656.

importance; whereas the periodic inequalities of the planets are generally small and the secular inequalities are the most interesting.

The method of treating the elements of the elliptic orbits as variable is specially suitable for secular inequalities; but for periodic inequalities it is generally better to treat the body as being disturbed from an elliptic path, and to study these deviations.

"The simplest way of regarding these various perturbations consists in imagining a planet moving in accordance with the laws of elliptic motion, on an ellipse the elements of which vary by insensible degrees; and to conceive at the same time that the true planet oscillates round this fictitious planet in a very small orbit the nature of which depends on its periodic perturbations." *

The former method, due as we have seen in great measure to Euler, was perfected and very generally used by Lagrange, and often bears his name.

243. It was at first naturally supposed that the slow alteration in the rates of the motions of Jupiter and Saturn (§§ 235, 236, and chapter x., § 204) was a secular inequality; Lagrange in 1766 made an attempt to explain it on this basis which, though still unsuccessful, represented the observations better than Euler's work. Laplace in his first paper on secular inequalities (1773) found by the use of a more complete analysis that the secular alterations in the rates of motions of Jupiter and Saturn appeared to vanish entirely, and attempted to explain the motions by the hypothesis, so often used by astronomers when in difficulties, that a comet had been the cause.

In 1773 *John Henry Lambert* (1728–1777) discovered from a study of observations that, whereas Halley had found Saturn to be moving more slowly than in ancient times, it was now moving faster than in Halley's time—a conclusion which pointed to a fluctuating or periodic cause of some kind.

Finally in 1784 Laplace arrived at the true explanation. Lagrange had observed in 1776 that if the times of revolution of two planets are exactly proportional to two whole

* Laplace, *Système du Monde*.

numbers, then part of the periodic disturbing force produces a secular change in their motions, acting continually in the same direction; though he pointed out that such a case did not occur in the solar system. If moreover the times of revolution are *nearly* proportional to two whole numbers (neither of which is very large), then part of the periodic disturbing force produces an irregularity that is not strictly secular, but has a very long period; and a disturbing force so small as to be capable of being ordinarily overlooked may, if it is of this kind, be capable of producing a considerable effect.* Now Jupiter and Saturn revolve round the sun in about 4,333 days and 10,759 days respectively; five times the former number is 21,665, and twice the latter is 21,518, which is very little less. Consequently the exceptional case occurs; and on working it out Laplace found an appreciable inequality with a period of about 900 years, which explained the observations satisfactorily.

The inequalities of this class, of which several others have been discovered, are known as **long inequalities**, and may be regarded as connecting links between secular inequalities and periodical inequalities of the usual kind.

244. The discovery that the observed inequality of Jupiter and Saturn was not secular may be regarded as the first step in a remarkable series of investigations on secular inequalities carried out by Lagrange and Laplace, for the most part between 1773 and 1784, leading to some of the most interesting and general results in the whole of gravitational astronomy. The two astronomers, though living respectively in Berlin and Paris, were in constant

* If n, n' are the mean motions of the two planets, the expression for the disturbing force contains terms of the type $= {}^{sin}_{cos}\,(n\,p \pm n'\,p')\,t$, where p, p' are integers, and the coefficient is of the order $p \sim p'$ in the eccentricities and inclinations. If now p and p' are such that $np \sim n'\,p'$ is small, the corresponding inequality has a period $2\,\pi\,/\,(np \sim n'\,p')$, and though its coefficient is of order $p \sim p'$, it has the small factor $n\,p \sim n'\,p'$ (or its square) in the denominator and may therefore be considerable. In the case of Jupiter and Saturn, for example, $n = 109{,}257$ in seconds of arc per annum, $n' = 43{,}996$; $5\,n' - 2\,n = 1{,}466$; there is therefore an inequality of the *third* order, with a period (in years) $= \dfrac{360^\circ}{1{,}466''} = 900$.

communication, and scarcely any important advance was made by the one which was not at once utilised and developed by the other.

The central problem was that of the secular alterations in the elements of a planet's orbit regarded as a varying ellipse. Three of these elements, the axis of the ellipse, its eccentricity, and the inclination of its plane to a fixed plane (usually the ecliptic), are of much greater importance than the other three. The first two are the elements on which the size and shape of the orbit depend, and the first also determines (by Kepler's Third Law) the period of revolution and average rate of motion of the planet; * the third has an important influence on the mutual relations of the two planets. The other three elements are chiefly of importance for periodical inequalities.

It should be noted moreover that the eccentricities and inclinations were in all cases (except those specially mentioned) considered as small quantities; and thus all the investigations were approximate, these quantities and the disturbing forces themselves being treated as small.

245. The basis of the whole series of investigations was a long paper published by Lagrange in 1766, in which he explained the method of variation of elements, and gave formulae connecting their rates of change with the disturbing forces.

In his paper of 1773 Laplace found that what was true of Jupiter and Saturn had a more general application, and proved that in the case of any planet, disturbed by any other, the axis was not only undergoing no secular change at the present time, but could not have altered appreciably since "the time when astronomy began to be cultivated."

In the next year Lagrange obtained an expression for the secular change in the inclination, *valid for all time*. When this was applied to the case of Jupiter and Saturn, which on account of their superiority in size and great distance from the other planets could be reasonably treated as forming with the sun a separate system, it appeared that the changes in the inclinations would always be of a periodic nature, so

* This statement requires some qualification when perturbations are taken into account. But the point is not very important, and is too technical to be discussed.

that they could never pass beyond certain fixed limits, not differing much from the existing values. The like result held for the system formed by the sun, Venus, the earth, and Mars. Lagrange noticed moreover that there were cases, which, as he said, fortunately did not appear to exist in the system of the world, in which, on the contrary, the inclinations might increase indefinitely. The distinction depended on the masses of the bodies in question; and although all the planetary masses were somewhat uncertain, and those assumed by Lagrange for Venus and Mars almost wholly conjectural, it did not appear that any reasonable alteration in the estimated masses would affect the general conclusion arrived at.

Two years later (1775) Laplace, much struck by the method which Lagrange had used, applied it to the discussion of the secular variations of the eccentricity, and found that these were also of a periodic nature, so that the eccentricity also could not increase or decrease indefinitely.

In the next year Lagrange, in a remarkable paper of only 14 pages, proved that whether the eccentricities and inclinations were treated as small or not, and whatever the masses of the planets might be, the changes in the length of the axis of any planetary orbit were necessarily all periodic, so that for all time the length of the axis could only fluctuate between certain definite limits. This result was, however, still based on the assumption that the disturbing forces could be treated as small.

Next came a series of five papers published between 1781 and 1784 in which Lagrange summed up his earlier work, revised and improved his methods, and applied them to periodical inequalities and to various other problems.

Lastly in 1784 Laplace, in the same paper in which he explained the long inequality of Jupiter and Saturn, established by an extremely simple method two remarkable relations between the eccentricities and inclinations of the planets, or any similar set of bodies.

The first relation is:—

If the mass of each planet be multiplied by the square root of the axis of its orbit and by the square of the eccentricity, then the sum of these products for all the planets is invariable save for periodical inequalities.

The second is precisely similar, save that eccentricity is replaced by inclination.*

The first of these propositions establishes the existence of what may be called a stock or fund of eccentricity shared by the planets of the solar system. If the eccentricity of any one orbit increases, that of some other orbit must undergo a corresponding decrease. Also the fund can never be overdrawn. Moreover observation shews that the eccentricities of all the planetary orbits are small; consequently the whole fund is small, and the share owned at any time by any one planet must be small.† Consequently the eccentricity of the orbit of a planet of which the mass and distance from the sun are considerable can never increase much, and a similar conclusion holds for the inclinations of the various orbits.

One remarkable characteristic of the solar system is presupposed in these two propositions; namely, that all the planets revolve round the sun in the same direction, which to an observer supposed to be on the north side of the orbits appears to be contrary to that in which the hands of a clock move. If any planet moved in the opposite direction, the corresponding parts of the eccentricity and inclination funds would have to be subtracted instead of being added; and there would be nothing to prevent the fund from being overdrawn.

A somewhat similar restriction is involved in Laplace's earlier results as to the impossibility of permanent changes in the eccentricities, though a system might exist in which his result would still be true if one or more of its members revolved in a different direction from the rest, but in this case there would have to be certain restrictions on the proportions of the orbits not required in the other case.

* $\Sigma\ e^2 m \sqrt{a} = c$, $\Sigma\ tan^2 i m \sqrt{a} = c'$, where m is the mass of any planet, a, e, i are the semi-major axis, eccentricity, and inclination of the orbit. The equation is true as far as squares of small quantities, and therefore it is indifferent whether or not $tan\, i$ is replaced as in the text by i.

† Nearly the whole of the "eccentricity fund" and of the "inclination fund" of the solar system is shared between Jupiter and Saturn. If Jupiter were to absorb the whole of each fund, the eccentricity of its orbit would only be increased by about 25 per cent., and the inclination to the ecliptic would not be doubled.

Stated briefly, the results established by the two astronomers were that the changes in axis, eccentricity, and inclination of any planetary orbit are all permanently restricted within certain definite limits. The perturbations caused by the planets make all these quantities undergo fluctuations of limited extent, some of which, caused by the periodic disturbing forces, go through their changes in comparatively short periods, while others, due to secular forces, require vast intervals of time for their completion.

It may thus be said that the stability of the solar system was established, as far as regards the particular astronomical causes taken into account.

Moreover, if we take the case of the earth, as an inhabited planet, any large alteration in the axis, that is in the average distance from the sun, would produce a more than proportional change in the amount of heat and light received from the sun ; any great increase in the eccentricity would increase largely that part (at present very small) of our seasonal variations of heat and cold which are due to varying distance from the sun ; while any change in position of the ecliptic, which was unaccompanied by a corresponding change of the equator, and had the effect of increasing the angle between the two, would largely increase the variations of temperature in the course of the year. The stability shewn to exist is therefore a guarantee against certain kinds of great climatic alterations which might seriously affect the habitability of the earth.

It is perhaps just worth while to point out that the results established by Lagrange and Laplace were mathematical consequences, obtained by processes involving the neglect of certain small quantities and therefore not perfectly rigorous, of certain definite hypotheses to which the actual conditions of the solar system bear a tolerably close resemblance. Apart from causes at present unforeseen, it is therefore not unreasonable to expect that for a very considerable period of time the motions of the actual bodies forming the solar system may be very nearly in accordance with these results ; but there is no valid reason why certain disturbing causes, ignored or rejected by Laplace and Lagrange on account of their insignificance, should not sooner or later produce quite appreciable effects (cf. chapter XIII., § 293).

246. A few of Laplace's numerical results as to the secular variations of the elements may serve to give an idea of the magnitudes dealt with.

The line of apses of each planet moves in the same direction; the most rapid motion, occurring in the case of Saturn, amounted to about 15″ per annum, or rather less than half a degree in a century. If this motion were to continue uniformly, the line of apses would require no less than 80,000 years to perform a complete circuit and return to its original position. The motion of the line of nodes (or line in which the plane of the planet's orbit meets that of the ecliptic) was in general found to be rather more rapid. The annual alteration in the inclination of any orbit to the ecliptic in no case exceeded a fraction of a second; while the change of eccentricity of Saturn's orbit, which was considerably the largest, would, if continued for four centuries, have only amounted to $\frac{1}{1000}$.

247. The theory of the secular inequalities has been treated at some length on account of the general nature of the results obtained. For the purpose of predicting the places of the planets at moderate distances of time the periodical inequalities are, however, of greater importance. These were also discussed very fully both by Lagrange and Laplace, the detailed working out in a form suitable for numerical calculation being largely due to the latter. From the formulæ given by Laplace and collected in the *Mécanique Céleste* several sets of solar and planetary tables were calculated, which were in general found to represent closely the observed motions, and which superseded the earlier tables based on less developed theories.*

248. In addition to the lunar and planetary theories nearly all the minor problems of gravitational astronomy were rediscussed by Laplace, in many cases with the aid of methods due to Lagrange, and their solution was in all cases advanced.

The theory of Jupiter's satellites, which with Jupiter form

* Of tables based on Laplace's work and published up to the time of his death, the chief solar ones were those of *von Zach* (1804) and *Delambre* (1806); and the chief planetary ones were those of *Lalande* (1771), of *Lindenau* for Venus, Mars, and Mercury (1810–13), and of *Bouvard* for Jupiter, Saturn, and Uranus (1808 and 1821).

a sort of miniature solar system but with several characteristic peculiarities, was fully dealt with; the other satellites received a less complete discussion. Some progress was also made with the theory of Saturn's ring by shewing that it could not be a uniform solid body.

Precession and nutation were treated much more completely than by D'Alembert; and the allied problems of the irregularities in the rotation of the moon and of Saturn's ring were also dealt with.

The figure of the earth was considered in a much more general way than by Clairaut, without, however, upsetting the substantial accuracy of his conclusions; and the theory of the tides was entirely reconstructed and greatly improved, though a considerable gap between theory and observation still remained.

The theory of perturbations was also modified so as to be applicable to comets, and from observation of a comet (known as Lexell's) which had appeared in 1770 and was found to have passed close to Jupiter in 1767 it was inferred that its orbit had been completely changed by the attraction of Jupiter, but that, on the other hand, it was incapable of exercising any appreciable disturbing influence on Jupiter or its satellites.

As, on the one hand, the complete calculation of the perturbations of the various bodies of the solar system presupposes a knowledge of their masses, so reciprocally if the magnitudes of these disturbances can be obtained from observation they can be used to determine or to correct the values of the several masses. In this way the masses of Mars and of Jupiter's satellites, as well as of Venus (§ 235), were estimated, and those of the moon and the other planets revised. In the case of Mercury, however, no perturbation of any other planet by it could be satisfactorily observed, and—except that it was known to be small—its mass remained for a long time a matter of conjecture. It was only some years after Laplace's death that the effect produced by it on a comet enabled its mass to be estimated (1842), and the mass is even now very uncertain.

249. By the work of the great mathematical astronomers of the 18th century, the results of which were summarised in the *Mécanique Céleste*, it was shewn to be possible to

account for the observed motions of the bodies of the solar system with a tolerable degree of accuracy by means of the law of gravitation.

Newton's problem (§ 228) was therefore approximately solved, and the agreement between theory and observation was in most cases close enough for the practical purpose of predicting for a moderate time the places of the various celestial bodies. The outstanding discrepancies between theory and observation were for the most part so small as compared with those that had already been removed as to leave an almost universal conviction that they were capable of explanation as due to errors of observation, to want of exactness in calculation, or to some similar cause.

250. Outside the circle of professed astronomers and mathematicians Laplace is best known, not as the author of the *Mécanique Céleste*, but as the inventor of the **Nebular Hypothesis.**

This famous speculation was published (in 1796) in his popular book the *Système du Monde* already mentioned, and was almost certainly independent of a somewhat similar but less detailed theory which had been suggested by the philosopher *Immanuel Kant* in 1755.

Laplace was struck with certain remarkable characteristics of the solar system. The seven planets known to him when he wrote revolved round the sun in the same direction, the fourteen satellites revolved round their primaries still in the same direction,* and such motions of rotation of sun, planets, and satellites about their axes as were known followed the same law. There were thus some 30 or 40 motions all in the same direction. If these motions of the several bodies were regarded as the result of chance and were independent of one another, this uniformity would be a coincidence of a most extraordinary character, as unlikely as that a coin when tossed the like number of times should invariably come down with the same face uppermost.

These motions of rotation and revolution were moreover all in planes but slightly inclined to one another; and the

* The motion of the satellites of Uranus (chapter XII., §§ 253, 255) is in the opposite direction. When Laplace first published his theory their motion was doubtful, and he does not appear to have thought it worth while to notice the exception in later editions of his book.

eccentricities of all the orbits were quite small, so that they were nearly circular.

Comets, on the other hand, presented none of these peculiarities; their paths were very eccentric, they were inclined at all angles to the ecliptic, and were described in either direction.

Moreover there were no known bodies forming a connecting link in these respects between comets and planets or satellites.*

From these remarkable coincidences Laplace inferred that the various bodies of the solar system must have had some common origin. The hypothesis which he suggested was that they had condensed out of a body that might be regarded either as the sun with a vast atmosphere filling the space now occupied by the solar system, or as a fluid mass with a more or less condensed central part or nucleus; while at an earlier stage the central condensation might have been almost non-existent.

Observations of Herschel's (chapter XII., §§ 259–61) had recently revealed the existence of many hundreds of bodies known as nebulae, presenting very nearly such appearances as might have been expected from Laplace's primitive body. The differences in structure which they shewed, some being apparently almost structureless masses of some extremely diffused substance, while others shewed decided signs of central condensation, and others again looked like ordinary stars with a slight atmosphere round them, were also strongly suggestive of successive stages in some process of condensation.

Laplace's suggestion then was that the solar system had been formed by condensation out of a nebula; and a similar explanation would apply to the fixed stars, with the planets (if any) which surrounded them.

He then sketched, in a somewhat imaginative way, the process whereby a nebula, if once endowed with a rotatory motion, might, as it condensed, throw off a series of rings,

* This statement again has to be modified in consequence of the discoveries, beginning on January 1st, 1801, of the minor planets (chapter XIII., § 294), many of which have orbits that are far more eccentric than those of the other planets and are inclined to the ecliptic at considerable angles.

and each of these might in turn condense into a planet with or without satellites ; and gave on this hypothesis plausible reasons for many of the peculiarities of the solar system.

So little is, however, known of the behaviour of a body like Laplace's nebula when condensing and rotating that it is hardly worth while to consider the details of the scheme.

That Laplace himself, who has never been accused of underrating the importance of his own discoveries, did not take the details of his hypothesis nearly as seriously as many of its expounders, may be inferred both from the fact that he only published it in a popular book, and from his remarkable description of it as "these conjectures on the formation of the stars and of the solar system, conjectures which I present with all the distrust (*défiance*) which everything which is not a result of observation or of calculation ought to inspire." *

* *Systeme du Monde*, Book V., chapter VI.

CHAPTER XII.

HERSCHEL.

"Coelorum perrupit claustra."

HERSCHEL'S *Epitaph.*

251. *Frederick William Herschel* was born at Hanover on November 15th, 1738, two years after Lagrange and nine years before Laplace. His father was a musician in the Hanoverian army, and the son, who shewed a remarkable aptitude for music as well as a decided taste for knowledge of various sorts, entered his father's profession as a boy (1753). On the breaking out of the Seven Years' War he served during part of a campaign, but his health being delicate his parents "determined to remove him from the service—a step attended by no small difficulties," and he was accordingly sent to England (1757), to seek his fortune as a musician.

After some years spent in various parts of the country, he moved (1766) to Bath, then one of the great centres of fashion in England. At first oboist in Linley's orchestra, then organist of the Octagon Chapel, he rapidly rose to a position of great popularity and distinction, both as a musician and as a music-teacher. He played, conducted, and composed, and his private pupils increased so rapidly that the number of lessons which he gave was at one time 35 a week. But this activity by no means exhausted his extraordinary energy; he had never lost his taste for study, and, according to a contemporary biographer, "after a fatiguing day of 14 or 16 hours spent in his vocation, he would retire at night with the greatest avidity to *unbend the mind*, if it may be so called, with a few propositions in Maclaurin's Fluxions, or other books of that sort." His musical studies had long ago given him an interest in mathematics, and it seems likely that the study of Robert Smith's *Harmonics* led him to the *Compleat System of Optics* of the same author, and so to an interest in the construction and use of telescopes. The astronomy that he read soon gave him a desire to see for himself what the books described; first he hired a small reflecting telescope, then thought of buying a larger instrument, but found that the price was prohibitive. Thus he was gradually led to attempt the construction of his own telescopes (1773). His brother Alexander, for whom he had found musical work at Bath, and who seems to have had considerable mechanical talent but none of William's perseverance, helped him in this undertaking, while his devoted sister *Caroline* (1750–1848), who had been brought over to England by William in 1772, not only kept house, but rendered a multitude of minor services. The operation of grinding and polishing the mirror for a telescope was one of the greatest delicacy, and at a certain stage required continuous labour for several hours. On one occasion Herschel's hand never left the polishing tool for 16 hours, so that "by way of keeping him alive" Caroline was "obliged to feed him by putting the victuals by bits into his mouth," and in less extreme cases she helped to make the operation less tedious by reading aloud: it is with some feeling of relief that we hear that on these occasions the books read were not on mathematics, optics, or astronomy, but were such as *Don Quixote*, the *Arabian Nights*, and the novels of Sterne and Fielding.

252. After an immense number of failures Herschel succeeded in constructing a tolerable reflecting telescope—soon to be followed by others of greater size and perfection—and with this he made his first recorded observation, of the Orion nebula, in March 1774.

This observation, made when he was in his 36th year, may be conveniently regarded as the beginning of his astronomical career, though for several years more music remained his profession, and astronomy could only be cultivated in such leisure time as he could find or make for himself; his biographers give vivid pictures of his extraordinary activity during this period, and of his zeal in using odd fragments of time, such as intervals between the acts at a theatre, for his beloved telescopes.

A letter written by him in 1783 gives a good account of the spirit in which he was at this time carrying out his astronomical work:—

"I determined to accept nothing on faith, but to see with my own eyes what others had seen before me. . . . I finally succeeded in completing a so-called Newtonian instrument, 7 feet in length. From this I advanced to one of 10 feet, and at last to one of 20, for I had fully made up my mind to carry on the improvement of my telescopes as far as it could possibly be done. When I had carefully and thoroughly perfected the great instrument in all its parts, I made systematic use of it in my observations of the heavens, first forming a determination never to pass by any, the smallest, portion of them without due investigation."

In accordance with this last resolution he executed on four separate occasions, beginning in 1775, each time with an instrument of greater power than on the preceding, a review of the whole heavens, in which everything that appeared in any way remarkable was noticed and if necessary more carefully studied. He was thus applying to astronomy methods comparable with those of the naturalist who aims at drawing up a complete list of the flora or fauna of a country hitherto little known.

253. In the course of the second of these reviews, made with a telescope of the Newtonian type, 7 feet in length, he made the discovery (March 13th, 1781) which gave him a European reputation and enabled him to abandon music as a profession and to devote the whole of his energies to science.

"In examining the small stars in the neighbourhood of H *Geminorum* I perceived one that appeared visibly larger than the rest; being struck with its uncommon appearance I compared it to H *Geminorum* and the small star in the quartile between *Auriga* and *Gemini*, and finding it so much larger than either of them, I suspected it to be a comet."

If Herschel's suspicion had been correct the discovery would have been of far less interest than it actually was, for when the new body was further observed and attempts were made to calculate its path, it was found that no ordinary cometary orbit would in any way fit its motion, and within three or four months of its discovery it was recognised—first by *Anders Johann Lexell* (1740–1784)—as being no comet but a new planet, revolving round the sun in a nearly circular path, at a distance about 19 times that of the earth and nearly double that of Saturn.

No new planet had been discovered in historic times, and Herschel's achievement was therefore absolutely unique; even the discovery of satellites inaugurated by Galilei (chapter VI., § 121) had come to a stop nearly a century before (1684), when Cassini had detected his second pair of satellites of Saturn (chapter VIII., § 160). Herschel wished to exercise the discoverer's right of christening by calling the new planet after his royal patron *Georgium Sidus*, but though the name was used for some time in England, Continental astronomers never accepted it, and after an unsuccessful attempt to call the new body *Herschel*, it was generally agreed to give a name similar to those of the other planets, and *Uranus* was proposed and accepted.

Although by this time Herschel had published two or three scientific papers and was probably known to a slight extent in English scientific circles, the complete obscurity among Continental astronomers of the author of this memorable discovery is curiously illustrated by a discussion in the leading astronomical journal (Bode's *Astronomisches Jahrbuch*) as to the way to spell his name, *Hertschel* being perhaps the best and *Mersthel* the worst of several attempts.

254. This obscurity was naturally dissipated by the discovery of Uranus. Distinguished visitors to Bath, among them the Astronomer Royal Maskelyne (chapter X., § 219), sought his acquaintance; before the end of the year he was elected a Fellow of the Royal Society, in addition to receiving one of its medals, and in the following spring he was summoned to Court to exhibit himself, his telescopes, and his stars to George III. and to various members of the royal family. As the outcome of this visit he received from the King an appointment as royal astronomer, with a salary of £200 a year.

With this appointment his career as a musician came to an end, and in August 1782 the brother and sister left Bath for good, and settled first in a dilapidated house at

WILLIAM HERSCHEL.

[*To face p.* 327.

Datchet, then, after a few months (1785–6) spent at Clay Hall in Old Windsor, at Slough in a house now known as Observatory House and memorable in Arago's words as "le lieu du monde où il a été fait le plus de découvertes."

255. Herschel's modest salary, though it would have sufficed for his own and his sister's personal wants, was of course insufficient to meet the various expenses involved in making and mounting telescopes. The skill which he had now acquired in the art was, however, such that his telescopes were far superior to any others which were available, and, as his methods were his own, there was a considerable demand for instruments made by him. Even while at Bath he had made and sold a number, and for years after moving to the neighbourhood of Windsor he derived a considerable income from this source, the royal family and a number of distinguished British and foreign astronomers being among his customers.

The necessity for employing his valuable time in this way fortunately came to an end in 1788, when he married a lady with a considerable fortune; Caroline lived henceforward in lodgings close to her brother, but worked for him with unabated zeal.

By the end of 1783 Herschel had finished a telescope 20 feet in length with a great mirror 18 inches in diameter, and with this instrument most of his best work was done; but he was not yet satisfied that he had reached the limit of what was possible. During the last winter at Bath he and his brother had spent a great deal of labour in an unsuccessful attempt to construct a 30-foot telescope; the discovery of Uranus and its consequences prevented the renewal of the attempt for some time, but in 1785 he began a 40-foot telescope with a mirror four feet in diameter, the expenses of which were defrayed by a special grant from the King. While it was being made Herschel tried a new form of construction of reflecting telescopes, suggested by Lemaire in 1732 but never used, by which a considerable gain of brilliancy was effected, but at the cost of some loss of distinctness. This **Herschelian** or **front-view** construction, as it is called, was first tried with the 20-foot, and led to the discovery (January 11th, 1787) of two satellites of Uranus, *Oberon* and *Titania*; it was henceforward regularly employed. After several mishaps the 40-foot telescope (fig. 82) was successfully constructed. On the first evening on which it was employed (August 28th, 1789) a sixth satellite of Saturn (*Enceladus*) was detected, and on September 17th a much fainter seventh satellite (*Mimas*). Both satellites were found to be nearer to the planet than any of the five hitherto discovered, Mimas being the nearer of the two (cf. fig. 91).

Although for the detection of extremely faint objects such as these satellites the great telescope was unequalled, for many kinds of work and for all but the very clearest evenings a smaller instrument was as good, and being less unwieldy was much more used. The mirror of the great telescope deteriorated to some extent, and after 1811, Herschel's hand being then no longer equal to the delicate task of repolishing it, the telescope ceased to be used though it was left standing till 1839, when it was dismounted and closed up.

256. From the time of his establishment at Slough till he began to lose his powers through old age the story of Herschel's life is little but a record of the work he did. It was his practice to employ in observing the whole of every suitable night; his daylight hours were devoted to interpreting his observations and to writing the papers in which he embodied his results. His sister was nearly always present as his assistant when he was observing, and also did a good deal of cataloguing, indexing, and similar work for him. After leaving Bath she also did some observing on her own account, though only when her brother was away or for some other reason did not require her services; she specialised on comets, and succeeded from first to last in discovering no less than eight. To form any adequate idea of the discomfort and even danger attending the nights spent in observing, it is necessary to realise that the great telescopes used were erected in the open air, that for both the Newtonian and Herschelian forms of reflectors the observer has to be near the upper end of the telescope, and therefore at a considerable height above the ground. In the 40-foot, for example, ladders 50 feet in length were used to reach the platform on which the observer was stationed. Moreover from the nature of the case satisfactory observations could not be taken in the

presence either of the moon or of artificial light. It is not therefore surprising that Caroline Herschel's journals contain a good many expressions of anxiety for her brother's

FIG. 82.—Herschel's forty-foot telescope.

welfare on these occasions, and it is perhaps rather a matter of wonder that so few serious accidents occurred.

In addition to doing his real work Herschel had to

receive a large number of visitors who came to Slough out of curiosity or genuine scientific interest to see the great man and his wonderful telescopes. In 1801 he went to Paris, where he made Laplace's acquaintance and also saw Napoleon, whose astronomical knowledge he rated much below that of George III., while "his general air was something like affecting to know more than he did know."

In the spring of 1807 he had a serious illness; and from that time onwards his health remained delicate, and a larger proportion of his time was in consequence given to indoor work. The last of the great series of papers presented to the Royal Society appeared in 1818, when he was almost 80, though three years later he communicated a list of double stars to the newly founded Royal Astronomical Society. His last observation was taken almost at the same time, and he died rather more than a year afterwards (August 21st, 1822), when he was nearly 84.

He left one son, John, who became an astronomer only less distinguished than his father (chapter XIII., §§ 306–8). Caroline Herschel after her beloved brother's death returned to Hanover, chiefly to be near other members of her family; here she executed one important piece of work by cataloguing in a convenient form her brother's lists of nebulae, and for the remaining 26 years of her long life her chief interest seems to have been in the prosperous astronomical career of her nephew John.

257. The incidental references to Herschel's work that have been made in describing his career have shewn him chiefly as the constructor of giant telescopes far surpassing in power any that had hitherto been used, and as the diligent and careful observer of whatever could be seen with them in the skies. Sun and moon, planets and fixed stars, were all passed in review, and their peculiarities noted and described. But this merely descriptive work was in Herschel's eyes for the most part means to an end, for, as he said in 1811, "a knowledge of the construction of the heavens has always been the ultimate object of my observations."

Astronomy had for many centuries been concerned almost wholly with the positions of the various heavenly bodies on the celestial sphere, that is with their directions.

Coppernicus and his successors had found that the apparent motions on the celestial sphere of the members of the solar system could only be satisfactorily explained by taking into account their actual motions in space, so that the solar system came to be effectively regarded as consisting of bodies at different distances from the earth and separated from one another by so many miles. But with the fixed stars the case was quite different: for, with the unimportant exception of the proper motions of a few stars (chapter x., § 203), all their known apparent motions were explicable as the result of the motion of the earth; and the relative or actual distances of the stars scarcely entered into consideration. Although the belief in a real celestial sphere to which the stars were attached scarcely survived the onslaughts of Tycho Brahe and Galilei, and any astronomer of note in the latter part of the 17th or in the 18th century would, if asked, have unhesitatingly declared the stars to be at different distances from the earth, this was in effect a mere pious opinion which had no appreciable effect on astronomical work.

The geometrical conception of the stars as represented by points on a celestial sphere was in fact sufficient for ordinary astronomical purposes, and the attention of great observing astronomers such as Flamsteed, Bradley, and Lacaille was directed almost entirely towards ascertaining the positions of these points with the utmost accuracy or towards observing the motions of the solar system. Moreover the group of problems which Newton's work suggested naturally concentrated the attention of eighteenth-century astronomers on the solar system, though even from this point of view the construction of star catalogues had considerable value as providing reference points which could be used for fixing the positions of the members of the solar system.

Almost the only exception to this general tendency consisted in the attempts—hitherto unsuccessful—to find the parallaxes and hence the distances of some of the fixed stars, a problem which, though originally suggested by the Coppernican controversy, had been recognised as possessing great intrinsic interest.

Herschel therefore struck out an entirely new path when

he began to study the sidereal system *per se* and the mutual relations of its members. From this point of view the sun, with its attendant planets, became one of an innumerable host of stars, which happened to have received a fictitious importance from the accident that we inhabited one member of its system.

258. A complete knowledge of the positions in space of the stars would of course follow from the measurement of the parallax (chapter VI., § 129 and chapter X., § 207) of each. The failure of such astronomers as Bradley to get the parallax of any one star was enough to shew the hopelessness of this general undertaking, and, although Herschel did make an attack on the parallax problem (§ 263), he saw that the question of stellar distribution in space, if to be answered at all, required some simpler if less reliable method capable of application on a large scale.

Accordingly he devised (1784) his method of **star-gauging**. The most superficial view of the sky shews that the stars visible to the naked eye are very unequally distributed on the celestial sphere; the same is true when the fainter stars visible in a telescope are taken into account. If two portions of the sky of the same apparent or angular magnitude are compared, it may be found that the first contains many times as many stars as the second. If we realise that the stars are not actually on a sphere but are scattered through space at different distances from us, we can explain this inequality of distribution on the *sky* as due to either a real inequality of distribution in *space*, or to a difference in the distance to which the sidereal system extends in the directions in which the two sets of stars lie. The first region on the sky may correspond to a region of space in which the stars are really clustered together, or may represent a direction in which the sidereal system extends to a greater distance, so that the accumulation of layer after layer of stars lying behind one another produces the apparent density of distribution. In the same way, if we are standing in a wood and the wood appears less thick in one direction than in another, it may be because the trees are really more thinly planted there or because in that direction the edge of the wood is nearer.

In the absence of any *a priori* knowledge of the actual

clustering of the stars in space, Herschel chose the former of these two hypotheses; that is, he treated the apparent density of the stars on any particular part of the sky as a measure of the depth to which the sidereal systems extended in that direction, and interpreted from this point of view the results of a vast series of observations. He used a 20-foot telescope so arranged that he could see with it a circular portion of the sky 15′ in diameter (one-quarter the area of the sun or full moon), turned the telescope to different parts of the sky, and counted the stars visible in each case. To avoid accidental irregularities he usually took the average of several neighbouring fields, and published in 1785 the results of gauges thus made in 683 * regions,

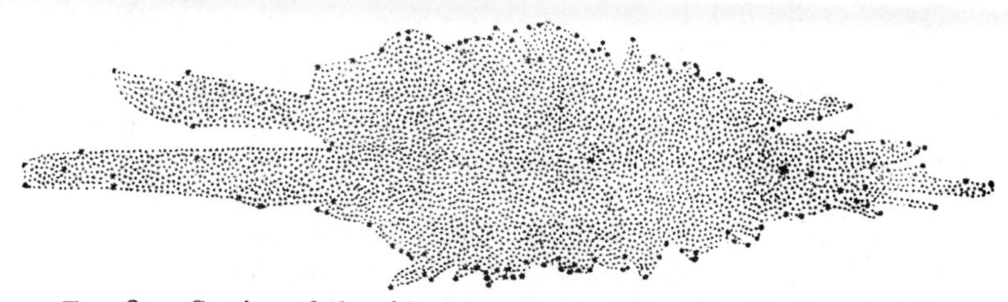

Fig. 83.—Section of the sidereal system. From Herschel's paper in the *Philosophical Transactions.*

while he subsequently added 400 others which he did not think it necessary to publish. Whereas in some parts of the sky he could see on an average only one star at a time, in others nearly 600 were visible, and he estimated that on one occasion about 116,000 stars passed through the field of view of his telescope in a quarter of an hour. The general result was, as rough naked-eye observation suggests, that stars are most plentiful in and near the Milky Way and least so in the parts of the sky most remote from it. Now the Milky Way forms on the sky an ill-defined band never deviating much from a great circle (sometimes called the **galactic** circle); so that on Herschel's hypothesis the space occupied by the stars is shaped roughly like a disc or grindstone, of which according to

* In his paper of 1817 Herschel gives the number as 863, but a reference to the original paper of 1785 shews that this must be a printer's error.

his figures the diameter is about five times the thickness. Further, the Milky Way is during part of its length divided into two branches, the space between the two branches being comparatively free of stars. Corresponding to this subdivision there has therefore to be assumed a cleft in the "grindstone."

This "grindstone" theory of the universe had been suggested in 1750 by *Thomas Wright* (1711–1786) in his *Theory of the Universe*, and again by Kant five years later; but neither had attempted, like Herschel, to collect numerical data and to work out consistently and in detail the consequences of the fundamental hypothesis.

That the assumption of uniform distribution of stars in space could not be true in detail was evident to Herschel from the beginning. A star cluster, for example, in which many thousands of faint stars are collected together in a very small space on the sky, would have to be interpreted as representing a long projection or spike full of stars, extending far beyond the limits of the adjoining portions of the sidereal system, and pointing directly away from the position occupied by the solar system. In the same way certain regions in the sky which are found to be bare of stars would have to be regarded as tunnels through the stellar system. That even one or two such spikes or tunnels should exist would be improbable enough, but as star clusters were known in considerable numbers before Herschel began his work, and were discovered by him in hundreds, it was impossible to explain their existence on this hypothesis, and it became necessary to assume that a star cluster occupied a region of space in which stars were really closer together than elsewhere.

Moreover further study of the arrangement of the stars, particularly of those in the Milky Way, led Herschel gradually to the belief that his original assumption was a wider departure from the truth than he had at first supposed; and in 1811, nearly 30 years after he had begun star-gauging, he admitted a definite change of opinion:—

"I must freely confess that by continuing my sweeps of the heavens my opinion of the arrangement of the stars . . . has undergone a gradual change. . . . For instance, an equal scattering of the stars may be admitted in certain calculations; but when we examine the Milky Way, or the closely compressed clusters of stars of which my catalogues have recorded so many instances, this supposed equality of scattering must be given up."

The method of star-gauging was intended primarily to give information as to the limits of the sidereal system—or the visible portions of it. Side by side with this method Herschel constantly made use of the brightness of a star as a probable test of nearness. If two stars give out actually the same amount of light, then that one which is nearer to us will appear the brighter; and on the assumption that no light is absorbed or stopped in its passage through space, the apparent brightness of the two stars will be inversely as the *square* of their respective distances. Hence, if we receive nine times as much light from one star as from another, and if it is assumed that this difference is merely due to difference of distance, then the first star is three times as far off as the second, and so on.

That the stars as a whole give out the same amount of light, so that the difference in their apparent brightness is due to distance only, is an assumption of the same general character as that of equal distribution. There must necessarily be many exceptions, but, in default of more exact knowledge, it affords a rough-and-ready method of estimating with some degree of probability relative distances of stars.

To apply this method it was necessary to have some means of comparing the amount of light received from different stars. This Herschel effected by using telescopes of different sizes. If the same star is observed with two reflecting telescopes of the same construction but of different sizes, then the light transmitted by the telescope to the eye is proportional to the area of the mirror which collects the light, and hence to the square of the diameter of the mirror. Hence the apparent brightness of a star as viewed through a telescope is proportional on the one hand to the inverse square of the distance, and on the other to the square of the diameter of the mirror of the telescope; hence the distance of the star is, as it were, exactly counterbalanced by the diameter of the mirror of the telescope. For example, if one star viewed in a telescope with an eight-inch mirror and another viewed in the great telescope with a four-foot mirror appear equally bright, then the second star is—on the fundamental assumption—six times as far off.

In the same way the size of the mirror necessary to make a star just visible was used by Herschel as a measure of the distance of the star, and it was in this sense that he constantly referred to the "space-penetrating power" of his telescope. On this assumption he estimated the faintest stars visible to the naked eye to be about twelve times as remote as one of the brightest stars, such as Arcturus, while Arcturus if removed to 900 times its present distance would just be visible in the 20-foot telescope which he commonly used, and the 40-foot would penetrate about twice as far into space.

Towards the end of his life (1817) Herschel made an attempt to compare statistically his two assumptions of uniform distribution in space and of uniform actual brightness, by counting the number of stars of each degree of apparent brightness and comparing them with the numbers that would result from uniform distribution in space if apparent brightness depended only on distance. The inquiry only extended as far as stars visible to the naked eye and to the brighter of the telescopic stars, and indicated the existence of an excess of the fainter stars of these classes, so that either these stars are more closely packed in space than the brighter ones, or they are in reality smaller or less luminous than the others; but no definite conclusions as to the arrangement of the stars were drawn.

259. Intimately connected with the structure of the sidereal system was the question of the distribution and nature of **nebulae** (cf. figs. 100, 102, facing pp. 397, 400) and **star clusters** (cf. fig. 104, facing p. 405). When Herschel began his work rather more than 100 such bodies were known, which had been discovered for the most part by the French observers Lacaille (chapter x., § 223) and *Charles Messier* (1730–1817). Messier may be said to have been a comet-hunter by profession; finding himself liable to mistake nebulae for comets, he put on record (1781) the positions of 103 of the former. Herschel's discoveries—carried out much more systematically and with more powerful instrumental appliances—were on a far larger scale. In 1786 he presented to the Royal Society a catalogue of 1,000

new nebulae and clusters, three years later a second catalogue of the same extent, and in 1802 a third comprising 500. Each nebula was carefully observed, its general appearance as well as its position being noted and described, and to obtain a general idea of the distribution of nebulae on the sky the positions were marked on a star map. The differences in brightness and in apparent structure led to a division into eight classes; and at quite an early stage of his work (1786) he gave a graphic account of the extraordinary varieties in form which he had noted:—

"I have seen double and treble nebulae, variously arranged; large ones with small, seeming attendants; narrow but much extended, lucid nebulae or bright dashes; some of the shape of a fan, resembling an electric brush, issuing from a lucid point; others of the cometic shape, with a seeming nucleus in the center; or like cloudy stars, surrounded with a nebulous atmosphere; a different sort again contain a nebulosity of the milky kind, like that wonderful inexplicable phenomenon about θ Orionis; while others shine with a fainter mottled kind of light, which denotes their being resolvable into stars."

260. But much the most interesting problem in classification was that of the relation between nebulae and star clusters. The Pleiades, for example, appear to ordinary eyes as a group of six stars close together, but many short-sighted people only see there a portion of the sky which is a little brighter than the adjacent region; again, the nebulous patch of light, as it appears to the ordinary eye, known as *Praesepe* (in the Crab), is resolved by the smallest telescope into a cluster of faint stars. In the same way there are other objects which in a small telescope appear cloudy or nebulous, but viewed in an instrument of greater power are seen to be star clusters. In particular Herschel found that many objects which to Messier were purely nebulous appeared in his own great telescopes to be undoubted clusters, though others still remained nebulous. Thus in his own words:—

"Nebulae can be selected so that an insensible gradation shall take place from a coarse cluster like the Pleiades down to a milky nebulosity like that in Orion, every intermediate step being represented."

These facts suggested obviously the inference that the difference between nebulae and star clusters was merely a question of the power of the telescope employed, and accordingly Herschel's next sentence is :—

"This tends to confirm the hypothesis that all are composed of stars more or less remote."

The idea was not new, having at any rate been suggested, rather on speculative than on scientific grounds, in 1755 by Kant, who had further suggested that a single nebula or star cluster is an assemblage of stars comparable in magnitude and structure with the whole of those which constitute the Milky Way and the other separate stars which we see. From this point of view the sun is one star in a cluster, and every nebula which we see is a system of the same order. This "island universe" theory of nebulae, as it has been called, was also at first accepted by Herschel, so that he was able once to tell Miss Burney that he had discovered 1,500 new universes.

Herschel, however, was one of those investigators who hold theories lightly, and as early as 1791 further observation had convinced him that these views were untenable, and that some nebulae at least were essentially distinct from star clusters. The particular object which he quotes in support of his change of view was a certain nebulous star—that is, a body resembling an ordinary star but surrounded by a circular halo gradually diminishing in brightness.

"Cast your eye," he says, "on this cloudy star, and the result will be no less decisive. . . . Your judgement, I may venture to say, will be, that the *nebulosity about the star is not of a starry nature*."

If the nebulosity were due to an aggregate of stars so far off as to be separately indistinguishable, then the central body would have to be a star of almost incomparably greater dimensions than an ordinary star; if, on the other hand, the central body were of dimensions comparable with those of an ordinary star, the nebulosity must be due to something other than a star cluster. In either case the object presented features markedly different from those of a star cluster of the recognised kind; and of the two alternative explanations Herschel chose the latter, considering the nebulosity to be "a shining fluid, of a nature totally unknown to us." One exception to his earlier views being thus admitted, others naturally followed by analogy, and henceforward he recognised nebulae of the "shining fluid" class as essentially different from star clusters, though it might be impossible in many cases to say to which class a particular body belonged.

The evidence accumulated by Herschel as to the distribution of nebulae also shewed that, whatever their nature, they could not be independent of the general sidereal system, as on the "island universe" theory. In the first place observation soon shewed him that an individual nebula or cluster was usually surrounded by a region of the sky comparatively free from stars; this was so commonly the case that it became his habit while sweeping for nebulae, after such a bare region had passed through the field of his telescope, to warn his sister to be ready to take down observations of nebulae. Moreover, as the position of a large number of nebulae came to be known and charted, it was seen that, whereas clusters were common near the Milky Way, nebulae which appeared incapable of resolution into clusters were scarce there, and shewed on the contrary a decided tendency to be crowded together in the regions of the sky most remote from the Milky Way—that is, round the poles of the galactic circle (§ 258). If nebulae were external systems, there would of course be no reason why their distribution on the sky should shew any connection either with the scarcity of stars generally or with the position of the Milky Way.

It is, however, rather remarkable that Herschel did not in this respect fully appreciate the consequences of his own observations, and up to the end of his life seems to have considered that some nebulae and clusters were external "universes," though many were part of our own system.

261. As early as 1789 Herschel had thrown out the idea that the different kinds of nebulae and clusters were objects of the same kind at different stages of development, some "clustering power" being at work converting a diffused nebula into a brighter and more condensed

body; so that condensation could be regarded as a sign of "age." And he goes on:—

"This method of viewing the heavens seems to throw them into a new kind of light. They are now seen to resemble a luxuriant garden, which contains the greatest variety of productions, in different flourishing beds; and one advantage we may at least reap from it is, that we can, as it were, extend the range of our experience to an immense duration. For, to continue the simile I have borrowed from the vegetable kingdom, is it not almost the same thing, whether we live successively to witness the germination, blooming, foliage, fecundity, fading, withering and corruption of a plant, or whether a vast number of specimens, selected from every stage through which the plant passes in the course of its existence, be brought at once to our view?"

His change of opinion in 1791 as to the nature of nebulae led to a corresponding modification of his views of this process of condensation. Of the star already referred to (§ 260) he remarked that its nebulous envelope "was more fit to produce a star by its condensation than to depend upon the star for its existence." In 1811 and 1814 he published a complete theory of a possible process whereby the shining fluid constituting a diffused nebula might gradually condense—the denser portions of it being centres of attraction—first into a denser nebula or compressed star cluster, then into one or more nebulous stars, lastly into a single star or group of stars. Every supposed stage in this process was abundantly illustrated from the records of actual nebulae and clusters which he had observed.

In the latter paper he also for the first time recognised that the clusters in and near the Milky Way really belonged to it, and were not independent systems that happened to lie in the same direction as seen by us.

262. On another allied point Herschel also changed his mind towards the end of his life. When he first used his great 20-foot telescope to explore the Milky Way, he thought that he had succeeded in completely resolving its faint cloudy light into component stars, and had thus penetrated to the end of the Milky Way; but afterwards he was convinced that this was not the case, but that there remained cloudy portions which—whether on account of their remote-

ness or for other reasons—his telescopes were unable to resolve into stars (cf. fig. 104, facing p. 405).

In both these respects therefore the structure of the Milky Way appeared to him finally less simple than at first.

263. One of the most notable of Herschel's discoveries was a bye-product of an inquiry of an entirely different character. Just as Bradley in trying to find the parallax of a star discovered aberration and nutation (chapter x., § 207), so also the same problem in Herschel's hands led to the discovery of double stars. He proposed to employ Galilei's differential or double-star method (chapter vi., § 129), in which the minute shift of a star's position, due to the earth's motion round the sun, is to be detected not by measuring its angular distance from standard points on the celestial sphere such as the pole or the zenith, but by observing the variations in its distance from some star close to it, which from its faintness or for some other reason might be supposed much further off and therefore less affected by the earth's motion.

With this object in view Herschel set to work to find pairs of stars close enough together to be suitable for his purpose, and, with his usual eagerness to see and to record all that could be seen, gathered in an extensive harvest of such objects. The limit of distance between the two members of a pair beyond which he did not think it worth while to go was 2′, an interval imperceptible to the naked eye except in cases of quite abnormally acute sight. In other words, the two stars—even if bright enough to be visible—would always appear as *one* to the ordinary eye. A first catalogue of such pairs, each forming what may be called a **double star,** was published early in 1782 and contained 269, of which 227 were new discoveries; a second catalogue of 434 was presented to the Royal Society at the end of 1784; and his last paper, sent to the Royal Astronomical Society in 1821 and published in the first volume of its memoirs, contained a list of 145 more. In addition to the position of each double star the angular distance between the two members, the direction of the line joining them, and the brightness of each were noted. In some cases also curious contrasts in the colour of the two components were observed. There were also not a few cases in which not merely two, but three, four, or more stars were found close enough to one another to be reckoned as forming a multiple star.

Herschel had begun with the idea that a double star was due to a merely accidental coincidence in the direction of two stars which had no connection with one another and one of which might be many times as remote as the other. It had, however, been pointed out by Michell (chapter x., § 219), as early as 1767, that even the few double stars then known afforded examples of coincidences which were very improbable as the result of mere random distribution of stars. A special case may be taken to make the argument clearer, though Michell's actual reasoning was not put into a numerical form. The bright star Castor (in the Twins) had for some time been known to consist of two stars, α and β, rather less than 5″ apart. Altogether there are about 50 stars of the same order of brightness as α, and 400 like β. Neither set of stars shews any particular tendency to be distributed in any special way over the celestial sphere. So that the question of probabilities becomes: if there are 50 stars of one sort and 400 of another distributed at random over the whole celestial sphere, the two distributions having no connection with one another, what is the chance that one of the first set of stars should be within 5″ of one of the second set? The chance is about the same as that, if 50 grains of wheat and 400 of barley are scattered at random in a field of 100 acres, one grain of wheat should be found within half an inch of a grain of barley. The odds against such a possibility are clearly very great and can be shewn to be more than 300,000 to one. These are the odds against the existence—without some real connection between the members—of a *single* double star like Castor; but when Herschel began to discover double stars by the hundred the improbability was enormously increased. In his first paper Herschel gave as his opinion that "it is much too soon to form any theories of small stars revolving round large ones," a remark shewing that the idea had been considered; and in 1784 Michell returned to the subject, and expressed the opinion that the odds in favour of a physical relation between the members of Herschel's newly discovered double stars were "beyond arithmetic."

264. Twenty years after the publication of his first catalogue Herschel was of Michell's opinion, but was now able to support it by evidence of an entirely novel and much more direct character. A series of observations of Castor, presented in two papers published in the *Philosophical Transactions* in 1803 and 1804, which were fortunately supplemented by an observation of Bradley's in 1759, had shewn a progressive alteration in the direction of the line joining its two components, of such a character as to leave no doubt that the two stars were revolving round one another; and there were five other cases in which a similar motion was observed. In these six cases it was thus shewn that the double star was really formed by a connected pair of stars near enough to influence one another's motion. A double star of this kind is called a **binary star** or a **physical double star,** as distinguished from a merely optical double star, the two members of which have no connection with one another. In three cases, including Castor, the observations were enough to enable the period of a complete revolution of one star round another, assumed to go on at a uniform rate, to be at any rate roughly estimated, the results given by Herschel being 342 years for Castor,* 375 and 1,200 years for the other two. It was an obvious inference that the motion of revolution observed in a binary star was due to the mutual gravitation of its members, though Herschel's data were not enough to determine with any precision the law of the motion, and it was not till five years after his death that the first attempt was made to shew that the orbit of a binary star was such as would follow from, or at any rate would be consistent with, the mutual gravitation of its members (chapter XIII., § 309: cf. also fig. 101). This may be regarded as the first direct evidence of the extension of the law of gravitation to regions outside the solar system.

Although only a few double stars were thus definitely shewn to be binary, there was no reason why many others

* The motion of Castor has become slower since Herschel's time, and the present estimate of the period is about 1,000 years, but it is by no means certain.

should not be so also, their motion not having been rapid enough to be clearly noticeable during the quarter of a century or so over which Herschel's observations extended; and this probability entirely destroyed the utility of double stars for the particular purpose for which Herschel had originally sought them. For if a double star is binary, then the two members are approximately at the same distance from the earth and therefore equally affected by the earth's motion, whereas for the purpose of finding the parallax it is essential that one should be much more remote than the other. But the discovery which he had made appeared to him far more interesting than that which he had attempted but failed to make; in his own picturesque language, he had, like Saul, gone out to seek his father's asses and had found a kingdom.

265. It had been known since Halley's time (chapter X., § 203) that certain stars had proper motions on the celestial sphere, relative to the general body of stars. The conviction, that had been gradually strengthening among astronomers, that the sun is only one of the fixed stars, suggested the possibility that the sun, like other stars, might have a motion in space. Thomas Wright, Lambert, and others had speculated on the subject, and Tobias Mayer (chapter X., §§ 225–6) had shewn how to look for such a motion.

If a single star appears to move, then by the principle of relative motion (chapter IV., § 77) this may be explained equally well by a motion of the star or by a motion of the observer, or by a combination of the two; and since in this problem the internal motions of the solar system may be ignored, this motion of the observer may be identified with that of the sun. When the proper motions of several stars are observed, a motion of the sun only is in general inadequate to explain them, but they may be regarded as due either solely to the motions in space of the stars or to combinations of these with some motion of the sun. If now the stars be regarded as motionless and the sun be moving towards a particular point on the celestial sphere, then by an obvious effect of perspective the stars near that point will appear to recede from it and one another on the celestial sphere, while those in the opposite region will approach one another, the magnitude of these changes depending on the rapidity of the sun's motion and on the nearness of the stars in question. The effect is exactly of the same nature as that produced when, on looking along a street at night, two lamps on opposite sides of the street at some distance from us appear close together, but as we walk down the street towards them they appear to become more and more separated from one another. In the figure, for example, L and L′ as seen from B appear farther apart than when seen from A.

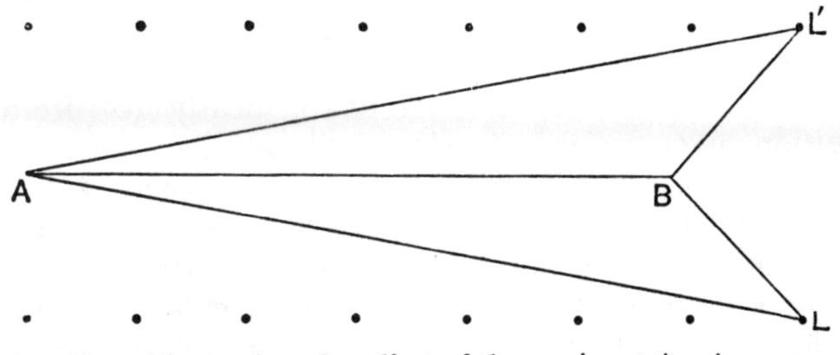

FIG. 84.—Illustrating the effect of the sun's motion in space.

If the observed proper motions of stars examined are not of this character, they cannot be explained as due *merely* to the motion of the sun; but if they shew some tendency to move in this way, then the observations can be most simply explained by regarding the sun as in motion, and by assuming that the discrepancies between the effects resulting from the assumed motion of the sun and the observed proper motions are due to the motions in space of the several stars.

From the few proper motions which Mayer had at his command he was, however, unable to derive any indication of a motion of the sun.

Herschel used the proper motions, published by Maskelyne and Lalande, of 14 stars (13 if the double star Castor be counted as only one), and with extraordinary insight detected in them a certain uniformity of motion of the kind already described, such as would result from a motion of the sun. The point on the celestial sphere towards which the sun was assumed to be moving, the **apex** as he called it, was taken to be the point marked by the star λ in the constellation Hercules. A motion of the sun in this direction would, he found, produce in the 14 stars apparent motions which were in the majority of cases in general agreement with those observed.* This result was published in 1783, and a few months later *Pierre Prévost* (1751–1839) deduced a very similar result from Tobias Mayer's collection of proper motions. More than 20 years later (1805) Herschel took up the question again, using six of the brightest stars in a collection of the proper motions of 36 published by Maskelyne in 1790, which were much more reliable than any earlier ones, and employing more elaborate processes of calculation; again the apex was placed in the constellation Hercules, though at a distance of nearly 30° from the position given in 1783. Herschel's results were avowedly to a large extent speculative, and were received by contemporary astronomers with a large measure of distrust; but a number of far more elaborate modern investigations of the same subject have confirmed the general correctness of his work, the earlier of his two estimates appearing, however, to be the more accurate. He also made some attempts in the same papers and in a third (published in 1806) to estimate the speed as well as the direction of the sun's motion; but the work necessarily involved so many assumptions as to the probable distances of the stars—which were quite unknown—that it is not worth while to quote results more definite than the statement made in the paper of 1783, that "We may in a general way estimate that the solar motion can certainly not be less than that which the earth has in her annual orbit."

266. The question of the comparative brightness of stars was, as we have seen (§ 258), of importance in connection with Herschel's attempts to estimate their relative distances from the earth and their arrangement in space; it also presented itself in connection with inquiries into the variability of the light of stars. Two remarkable cases of variability had been for some time known. A star in the Whale (*o Ceti* or *Mira*) had been found to be at times

* More precisely, counting motions in right ascension and in declination separately, he had 27 observed motions to deal with (one of the stars having no motion in declination); 22 agreed in sign with those which would result from the assumed motion of the sun.

invisible to the naked eye and at other times to be conspicuous; a Dutch astronomer, *Phocylides Holwarda* (1618–1651), first clearly recognised its variable character (1639), and *Ismaël Boulliau* or *Bullialdus* (1605–1694) in 1667 fixed its period at about eleven months, though it was found that its fluctuations were irregular both in amount and in period. Its variations formed the subject of the first paper published by Herschel in the *Philosophical Transactions* (1780). An equally remarkable variable star is that known as *Algol* (or *β Persei*), the fluctuations of which were found to be performed with almost absolute regularity. Its variability had been noted by *Geminiano Montanari* (1632–1687) in 1669, but the regularity of its changes was first detected in 1783 by *John Goodricke* (1764–1786), who was soon able to fix its period at very nearly 2 days 20 hours 49 minutes. Algol, when faintest, gives about one-quarter as much light as when brightest, the change from the first state to the second being effected in about ten hours; whereas Mira varies its light several hundredfold, but accomplishes its changes much more slowly.

At the beginning of Herschel's career these and three or four others of less interest were the only stars definitely recognised as variable, though a few others were added soon afterwards. Several records also existed of so-called "new" stars, which had suddenly been noticed in places where no star had previously been observed, and which for the most part rapidly became inconspicuous again (cf. chapter II., § 42; chapter V., § 100; chapter VII., § 138); such stars might evidently be regarded as variable stars, the times of greatest brightness occurring quite irregularly or at long intervals. Moreover various records of the brightness of stars by earlier astronomers left little doubt that a good many must have varied sensibly in brightness. For example, a small star in the Great Bear (close to the middle star of the "tail") was among the Arabs a noted test of keen sight, but is perfectly visible even in our duller climate to persons with ordinary eyesight; and Castor, which appeared the brighter of the two Twins to Bayer when he published his Atlas (1603), was in the 18th century (as now) less bright than Pollux.

Herschel made a good many definite measurements of the amounts of light emitted by stars of various magnitudes, but was not able to carry out any extensive or systematic measurements on this plan. With a view to the future detection of such changes of brightness as have just been mentioned, he devised and carried out on a large scale the extremely simple method of **sequences.** If a group of stars are observed and their order of brightness noted at two different times, then any alteration in the order will shew that the brightness of one or more has changed. So that if a number of stars are observed in sets in such a way that each star is recorded as being less bright than certain stars near it and brighter than certain other stars, materials are thereby provided for detecting at any future time any marked amount of variation of brightness. Herschel prepared on this plan, at various times between 1796 and 1799, four catalogues of comparative brightness based on naked eye observations and comprising altogether about 3,000 stars. In the course of the work a good many cases of slight variability were noticed; but the most interesting discovery of this kind was that of the variability of the well-known star α *Herculis*, announced in 1796. The period was estimated at 60 days, and the star thus seemed to form a connecting link between the known variables which like Algol had periods of a very few days and those (of which Mira was the best known) with periods of some hundreds of days. As usual, Herschel was not content with a mere record of observations, but attempted to explain the observed facts by the supposition that a variable star had a rotation and that its surface was of unequal brightness.

267. The novelty of Herschel's work on the fixed stars, and the very general character of the results obtained, have caused this part of his researches to overshadow in some respects his other contributions to astronomy.

Though it was no part of his plan to contribute to that precise knowledge of the motions of the bodies of the solar system which absorbed the best energies of most of the astronomers of the 18th century—whether they were observers or mathematicians—he was a careful and successful observer of the bodies themselves.

His discoveries of Uranus, of two of its satellites, and of two new satellites of Saturn have been already mentioned in connection with his life (§§ 253, 255). He believed himself to have seen also (1798) four other satellites of Uranus, but their existence was never satisfactorily verified; and the second pair of satellites now known to belong to Uranus, which were discovered by Lassell in 1847 (chapter XIII., § 295), do not agree in position and motion with any of Herschel's four. It is therefore highly probable that they were mere optical illusions due to defects of his mirror, though it is not impossible that he may have caught glimpses of one or other of Lassell's satellites and misinterpreted the observations.

Saturn was a favourite object of study with Herschel from the very beginning of his astronomical career, and seven papers on the subject were published by him between 1790 and 1806. He noticed and measured the deviation of the planet's form from a sphere (1790); he observed various markings on the surface of the planet itself, and seems to have seen the inner ring, now known from its appearance as the crape ring (chapter XIII., § 295), though he did not recognise its nature. By observations of some markings at some distance from the equator he discovered (1790) that Saturn rotated on an axis, and fixed the period of rotation at about 10 h. 16 m. (a period differing only by about 2 minutes from modern estimates), and by similar observations of the ring (1790) concluded that it rotated in about $10\frac{1}{2}$ hours, the axis of rotation being in each case perpendicular to the plane of the ring. The satellite Japetus, discovered by Cassini in 1671 (chapter VIII., § 160), had long been recognised as variable in brightness, the light emitted being several times as much at one time as at another. Herschel found that these variations were not only perfectly regular, but recurred at an interval equal to that of the satellite's period of rotation round its primary (1792), a conclusion which Cassini had thought of but rejected as inconsistent with his observations. This peculiarity was obviously capable of being explained by supposing that different portions of Japetus had unequal power of reflecting light, and that like our moon it turned on its axis once in every revolution, in such a way as always to present the same face towards its primary, and in consequence each face in turn to an observer on the earth. It was natural to conjecture that such an arrangement was general among satellites, and

Herschel obtained (1797) some evidence of variability in the satellites of Jupiter, which appeared to him to support this hypothesis.

Herschel's observations of other planets were less numerous and important. He rightly rejected the supposed observations by Schroeter (§ 271) of vast mountains on Venus, and was only able to detect some indistinct markings from which the planet's rotation on an axis could be somewhat doubtfully inferred. He frequently observed the familiar bright bands on Jupiter commonly called belts, which he was the first to interpret (1793) as bands of cloud. On Mars he noted the periodic diminution of the white caps on the two poles, and observed how in these and other respects Mars was of all planets the one most like the earth.

268. Herschel made also a number of careful observations on the sun, and based on them a famous theory of its structure. He confirmed the existence of various features of the solar surface which had been noted by the earlier telescopists such as Galilei, Scheiner, and Hevel, and added to them in some points of detail. Since Galilei's time a good many suggestions as to the nature of spots had been thrown out by various observers, such as that they were clouds, mountain-tops, volcanic products, etc., but none of these had been supported by any serious evidence. Herschel's observations of the appearances of spots suggested to him that they were depressions in the surface of the sun, a view which derived support from occasional observations of a spot when passing over the edge of the sun as a distinct depression or notch there. Upon this somewhat slender basis of fact he constructed (1795) an elaborate theory of the nature of the sun, which attracted very general notice by its ingenuity and picturesqueness and commanded general assent in the astronomical world for more than half a century. The interior of the sun was supposed to be a cold dark solid body, surrounded by two cloud-layers, of which the outer was the **photosphere** or ordinary surface of the sun, intensely hot and luminous, and the inner served as a fire-screen to protect the interior. The umbra (chapter VI., § 124) of a spot was the dark interior seen through an opening in the clouds, and the penumbra corresponded

to the inner cloud-layer rendered luminous by light from above.

"The sun viewed in this light appears to be nothing else than a very eminent, large, and lucid planet, evidently the first or, in strictness of speaking, the only primary one of our system; . . . it is most probably also inhabited, like the rest of the planets, by beings whose organs are adapted to the peculiar circumstances of that vast globe."

That spots were depressions had been suggested more than twenty years before (1774) by *Alexander Wilson* of Glasgow (1714–1786), and supported by evidence different from any adduced by Herschel and in some ways more conclusive. Wilson noticed, first in the case of a large spot seen in 1769, and afterwards in other cases, that as the sun's rotation carries a spot across its disc from one edge to another, its appearance changes exactly as it would do in accordance with ordinary laws of perspective if the spot were a saucer-shaped depression, of which the bottom formed the umbra and the sloping sides the penumbra, since the penumbra appears narrowest on the side nearest the centre of the sun and widest on the side nearest the edge. Hence Wilson inferred, like Herschel, but with less confidence, that the body of the sun is dark. In the paper referred to Herschel shews no signs of being acquainted with Wilson's work, but in a second paper (1801), which contained also a valuable series of observations of the detailed markings on the solar surface, he refers to Wilson's "geometrical proof" of the depression of the umbra of a spot.

Although it is easy to see now that Herschel's theory was a rash generalisation from slight data, it nevertheless explained—with fair success—most of the observations made up to that time.

Modern knowledge of heat, which was not accessible to Herschel, shews us the fundamental impossibility of the continued existence of a body with a cold interior and merely a shallow ring of hot and luminous material round it; and the theory in this form is therefore purely of historic interest (cf. also chapter XIII., §§ 298, 303).

269. Another suggestive idea of Herschel's was the analogy between the sun and a variable star, the known

variation in the number of spots and possibly of other markings on the sun suggesting to him the probability of a certain variability in the total amount of solar light and heat emitted. The terrestrial influence of this he tried to measure—in the absence of precise meteorological data—with characteristic ingenuity by the price of wheat, and some evidence was adduced to shew that at times when sun-spots had been noted to be scarce—corresponding according to Herschel's view to periods of diminished solar activity—wheat had been dear and the weather presumably colder. In reality, however, the data were insufficient to establish any definite conclusions.

270. In addition to carrying out the astronomical researches already sketched, and a few others of less importance, Herschel spent some time, chiefly towards the end of his life, in working at light and heat; but the results obtained, though of considerable value, belong rather to physics than to astronomy, and need not be dealt with here.

271. It is natural to associate Herschel's wonderful series of discoveries with his possession of telescopes of unusual power and with his formulation of a new programme of astronomical inquiry; and these were certainly essential elements. It is, however, significant, as shewing how important other considerations were, that though a great number of his telescopes were supplied to other astronomers, and though his astronomical programme when once suggested was open to all the world to adopt, hardly any of his contemporaries executed any considerable amount of work comparable in scope to his own.

Almost the only astronomer of the period whose work deserves mention beside Herschel's, though very inferior to it both in extent and in originality, was *Johann Hieronymus Schroeter* (1745–1816).

Holding an official position at Lilienthal, near Bremen, he devoted his leisure during some thirty years to a scrutiny of the planets and of the moon, and to a lesser extent of other bodies.

As has been seen in the case of Venus (§ 267), his results were not always reliable, but notwithstanding some errors he added considerably to our knowledge of the appearances presented by the various planets, and in particular studied the visible features of the moon with a minuteness and accuracy far exceeding that of any of his predecessors, and made some attempt to deduce from his observations data as to its physical condition. His two volumes on the moon (*Selenotopographische Fragmente*, 1791 and 1802), and other minor writings, are a storehouse of valuable detail, to which later workers have been largely indebted.

CHAPTER XIII.

THE NINETEENTH CENTURY.

"The greater the sphere of our knowledge, the larger is the surface of its contact with the infinity of our ignorance."

272. THE last three chapters have contained some account of progress made in three branches of astronomy which, though they overlap and exercise an important influence on one another, are to a large extent studied by different men and by different methods, and have different aims. The difference is perhaps best realised by thinking of the work of a great master in each department, Bradley, Laplace, and Herschel. So great is the difference that Delambre in his standard history of astronomy all but ignores the work of the great school of mathematical astronomers who were his contemporaries and immediate predecessors, not from any want of appreciation of their importance, but because he regards their work as belonging rather to mathematics than to astronomy; while Bessel (§ 277), in saying that the function of astronomy is "to assign the places on the sky where sun, moon, planets, comets, and stars have been, are, and will be," excludes from its scope nearly everything towards which Herschel's energies were directed.

Current modern practice is, however, more liberal in its use of language than either Delambre or Bessel, and finds it convenient to recognise all three of the subjects or groups of subjects referred to as integral parts of one science.

The mutual relation of gravitational astronomy and what has been for convenience called observational astronomy has been already referred to (chapter x., § 196). It should, however, be noticed that the latter term has in this book hitherto been used chiefly for only one part of the astronomical work which concerns itself primarily with observation. Observing played at least as large a part in Herschel's work as in Bradley's, but the aims of the two men were in many ways different. Bradley was interested chiefly in ascertaining as accurately as possible the apparent positions of the fixed stars on the celestial sphere, and the positions and motions of the bodies of the solar system, the former undertaking being in great part subsidiary to the latter. Herschel, on the other hand, though certain of his researches, *e.g.* into the parallax of the fixed stars and into the motions of the satellites of Uranus, were precisely like some of Bradley's, was far more concerned with questions of the appearances, mutual relations, and structure of the celestial bodies in themselves. This latter branch of astronomy may conveniently be called **descriptive astronomy**, though the name is not altogether appropriate to inquiries into the physical structure and chemical constitution of celestial bodies which are often put under this head, and which play an important part in the astronomy of the present day.

273. Gravitational astronomy and exact observational astronomy have made steady progress during the nineteenth century, but neither has been revolutionised, and the advances made have been to a great extent of such a nature as to be barely intelligible, still less interesting, to those who are not experts. The account of them to be given in this chapter must therefore necessarily be of the slightest character, and deal either with general tendencies or with isolated results of a less technical character than the rest.

Descriptive astronomy, on the other hand, which can be regarded as being almost as much the creation of Herschel as gravitational astronomy is of Newton, has not only been greatly developed on the lines laid down by its founder, but has received—chiefly through the invention of spectrum analysis (§ 299)—extensions into regions not only unthought of but barely imaginable a century ago. Most of the results of descriptive astronomy—unlike those of the older branches of the subject—are readily intelligible and fairly interesting to those who have but little knowledge of the subject; in particular they are as yet to a considerable extent independent of the mathematical ideas and language which dominate so much of astronomy and render it unattractive or inaccessible to many. Moreover, not only can descriptive astronomy be appreciated and studied, but its progress can materially be assisted, by observers who have neither knowledge of higher mathematics nor any elaborate instrumental equipment.

Accordingly, while the successors of Laplace and Bradley have been for the most part astronomers by profession, attached to public observatories or to universities, an immense mass of valuable descriptive work has been done by amateurs who, like Herschel in the earlier part of his career, have had to devote a large part of their energies to professional work of other kinds, and who, though in some cases provided with the best of instruments, have in many others been furnished with only a slender instrumental outfit. For these and other reasons one of the most notable features of nineteenth century astronomy has been a great development, particularly in this country and in the United States, of general interest in the subject, and the establishment of a large number of private observatories devoted almost entirely to the study of special branches of descriptive astronomy. The nineteenth century has accordingly witnessed the acquisition of an unprecedented amount of detailed astronomical knowledge. But the wealth of material thus accumulated has outrun our powers of interpretation, and in a number of cases our knowledge of some particular department of descriptive astronomy consists, on the one hand of an immense series of careful observations, and on the other of one or more highly speculative theories, seldom capable of explaining more than a small portion of the observed facts.

In dealing with the progress of modern descriptive astronomy the proverbial difficulty of seeing the wood on account of the trees is therefore unusually great. To give an account within the limits of a single chapter of even the most important facts added to our knowledge would be a hopeless endeavour ; fortunately it would also be superfluous, as they are to be found in many easily accessible textbooks on astronomy, or in treatises on special parts of the subject. All that can be attempted is to give some account of the chief lines on which progress has been made, and to indicate some general conclusions which seem to be established on a tolerably secure basis.

274. The progress of exact observation has of course been based very largely on instrumental advances. Not only have great improvements been made in the extremely delicate work of making large lenses, but the graduated circles and other parts of the mounting of a telescope upon which accuracy of measurement depends can also be constructed with far greater exactitude and certainty than at the beginning of the century. New methods of mounting telescopes and of making and recording observations have also been introduced, all contributing to greater accuracy. For certain special problems photography is found to present great advantages as compared with eye-observations, though its most important applications have so far been to descriptive astronomy.

275. The necessity for making allowance for various known sources of errors in observation, and for diminishing as far as possible the effect of errors due to unknown causes, had been recognised even by Tycho Brahe (chapter v., § 110), and had played an important part in the work of Flamsteed and Bradley (chapter x., §§ 198, 218). Some further important steps in this direction were taken in the earlier part of this century. The method of **least squares**, established independently by two great mathematicians, *Adrien Marie Legendre* (1752–1833) of Paris and *Carl Friedrich Gauss* (1777–1855) of Göttingen,* was a systematic method of combining observations, which gave slightly different results, in such a way as to be as near the truth as possible. Any ordinary physical measurement, *e.g.* of a length, however carefully executed, is necessarily imperfect; if the same measurement is made several times, even under almost identical conditions, the results will in general differ slightly; and the question arises of combining these so as to get the most satisfactory result. The common practice in this simple case has long been to take the arithmetical mean or average of the different results. But astronomers have constantly

* The method was published by Legendre in 1806 and by Gauss in 1809, but it was invented and used by the latter more than 20 years earlier.

to deal with more complicated cases in which *two* or more unknown quantities have to be determined from observations of different quantities, as, for example, when the elements of the orbit of a planet (chapter XI., § 236) have to be found from observations of the planet's position at different times. The method of least squares gives a rule for dealing with such cases, which was a generalisation of the ordinary rule of averages for the case of a single unknown quantity; and it was elaborated in such a way as to provide for combining observations of different value, such as observations taken by observers of unequal skill or with different instruments, or under more or less favourable conditions as to weather, etc. It also gives a simple means of testing, by means of their mutual consistency, the value of a series of observations, and comparing their probable accuracy with that of some other series executed under different conditions. The method of least squares and the special case of the "average" can be deduced from a certain assumption as to the general character of the causes which produce the error in question; but the assumption itself cannot be justified *a priori*; on the other hand, the satisfactory results obtained from the application of the rule to a great variety of problems in astronomy and in physics has shewn that in a large number of cases unknown causes of error must be approximately of the type considered. The method is therefore very widely used in astronomy and physics wherever it is worth while to take trouble to secure the utmost attainable accuracy.

276. Legendre's other contributions to science were almost entirely to branches of mathematics scarcely affecting astronomy. Gauss, on the other hand, was for nearly half a century head of the observatory of Göttingen, and though his most brilliant and important work was in pure mathematics, while he carried out some researches of first-rate importance in magnetism and other branches of physics, he also made some further contributions of importance to astronomy. These were for the most part processes of calculation of various kinds required for utilising astronomical observations, the best known being a method of calculating the orbit of a planet from three complete

observations of its position, which was published in his *Theoria Motus* (1809). As we have seen (chapter XI., § 236), the complete determination of a planet's orbit depends on six independent elements: any complete observation of the planet's position in the sky, at any time, gives two quantities, *e.g.* the right ascension and declination (chapter II., § 33); hence three complete observations give six equations and are theoretically adequate to determine the elements of the orbit; but it had not hitherto been found necessary to deal with the problem in this form. The orbits of all the planets but Uranus had been worked out gradually by the use of a series of observations extending over centuries; and it was feasible to use observations taken at particular times so chosen that certain elements could be determined without any accurate knowledge of the others; even Uranus had been under observation for a considerable time before its path was determined with anything like accuracy; and in the case of comets not only was a considerable series of observations generally available, but the problem was simplified by the fact that the orbit could be taken to be nearly or quite a parabola instead of an ellipse (chapter IX., § 190). The discovery of the new planet Ceres on January 1st, 1801 (§ 294), and its loss when it had only been observed for a few weeks, presented virtually a new problem in the calculation of an orbit. Gauss applied his new methods—including that of least squares—to the observations available, and with complete success, the planet being rediscovered at the end of the year nearly in the position indicated by his calculations.

277. The theory of the "reduction" of observations (chapter X., § 218) was first systematised and very much improved by *Friedrich Wilhelm Bessel* (1784–1846), who was for more than thirty years the director of the new Prussian observatory at Königsberg. His first great work was the reduction and publication of Bradley's Greenwich observations (chapter X., § 218). This undertaking involved an elaborate study of such disturbing causes as precession, aberration, and refraction, as well as of the errors of Bradley's instruments. Allowance was made for these on a uniform and systematic plan, and the result was the publication in 1818,

under the title *Fundamenta Astronomiae*, of a catalogue of the places of 3,222 stars as they were in 1755. A special problem dealt with in the course of the work was that of refraction. Although the complete theoretical solution was then as now unattainable, Bessel succeeded in constructing a table of refractions which agreed very closely with observation and was presented in such a form that the necessary correction for a star in almost any position could be obtained with very little trouble. His general methods of reduction—published finally in his *Tabulae Regiomontanae* (1830)—also had the great advantage of arranging the necessary calculations in such a way that they could be performed with very little labour and by an almost mechanical process, such as could easily be carried out by a moderately skilled assistant. In addition to editing Bradley's observations, Bessel undertook a fresh series of observations of his own, executed between the years 1821 and 1833, upon which were based two new catalogues, containing about 62,000 stars, which appeared after his death.

278. The most memorable of Bessel's special pieces of

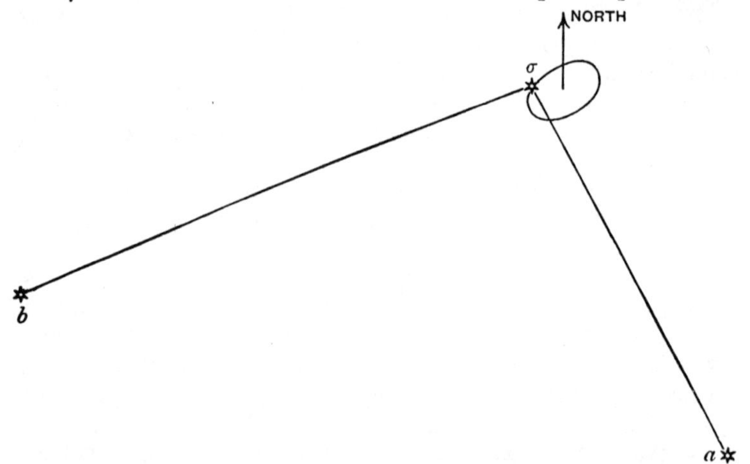

FIG. 85.—61 *Cygni* and the two neighbouring stars used by Bessel.

work was the first definite detection of the parallax of a fixed star. He abandoned the test of brightness as an indication of nearness, and selected a star (61 *Cygni*) which was barely visible to the naked eye but was remarkable for its large proper motion (about 5″ per annum); evidently if a star is moving at an assigned rate (in miles per hour) through space, the nearer to the observer it is the more rapid does its motion appear to be, so that apparent rapidity of motion, like brightness, is a probable but by no means infallible indication of nearness. A modification of Galilei's differential method (chapter VI., § 129, and chapter XII., § 263) being adopted, the angular distance of 61 *Cygni* from two neighbouring stars, the faintness and immovability of which suggested their great distance in space, was measured at frequent intervals during a year. From the changes in these distances $\sigma\ a$, $\sigma\ b$ (in fig. 85), the size of the small ellipse described by σ could be calculated. The result, announced at the end of 1838, was that the star had an annual parallax of about $\frac{1}{3}''$ (chapter VIII., § 161), *i.e.* that the star was at such distance that the greatest angular distance of the earth from the sun viewed from the star (the angle S σ E in fig. 86, where S is the sun and E the earth) was this insignificant angle.* The result was confirmed, with slight alterations, by a fresh investigation of Bessel's in 1839–40, but later work seems to shew that the parallax is a little less than $\frac{1}{2}''$.† With this latter estimate, the apparent size of the earth's path round the sun as seen from the star is the same as that of a halfpenny

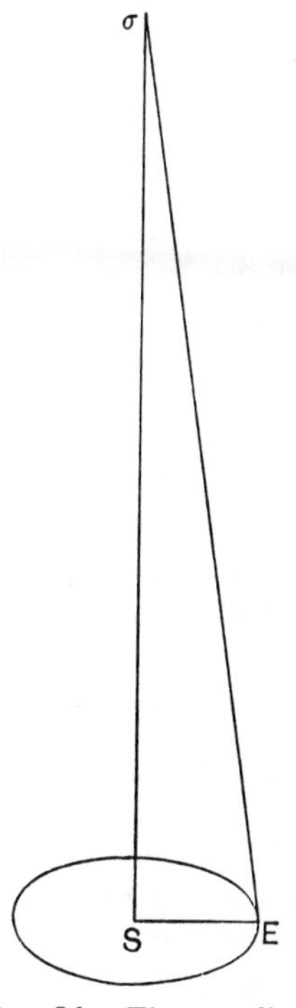

FIG. 86.—The parallax of 61 *Cygni*.

* The figure has to be enormously exaggerated, the angle S σ E as shewn there being about 10°, and therefore about 100,000 times too great.

† Sir R. S. Ball and the late Professor Pritchard (§ 279) have obtained respectively ·47″ and ·43″; the mean of these, ·45″, may be provisionally accepted as not very far from the truth.

at a distance of rather more than three miles. In other words, the distance of the star is about 400,000 times the distance of the sun, which is itself about 93,000,000 miles. A mile is evidently a very small unit by which to measure such a vast distance; and the practice of expressing such distances by means of the time required by light to perform the journey is often convenient. Travelling at the rate of 186,000 miles *per second* (§ 283), light takes rather more than six years to reach us from 61 *Cygni.*

279. Bessel's solution of the great problem which had baffled astronomers ever since the time of Coppernicus was immediately followed by two others. Early in 1839 *Thomas Henderson* (1798–1844) announced a parallax of nearly 1″ for the bright star α *Centauri* which he had observed at the Cape, and in the following year *Friedrich Georg Wilhelm Struve* (1793–1864) obtained from observations made at Pulkowa a parallax of $\frac{1}{4}''$ for *Vega*; later work has reduced these numbers to $\frac{3}{4}''$ and $\frac{1}{10}''$ respectively.

A number of other parallax determinations have subsequently been made. An interesting variation in method was made by the late Professor *Charles Pritchard* (1808–1893) of Oxford by *photographing* the star to be examined and its companions, and subsequently measuring the distances on the photograph, instead of measuring the angular distances directly with a micrometer.

At the present time some 50 stars have been ascertained with some reasonable degree of probability to have measurable, if rather uncertain, parallaxes; α *Centauri* still holds its own as the nearest star, the light-journey from it being about four years. A considerable number of other stars have been examined with negative or highly uncertain results, indicating that their parallaxes are too small to be measured with our present means, and that their distances are correspondingly great.

280. A number of star catalogues and star maps—too numerous to mention separately—have been constructed during this century, marking steady progress in our knowledge of the position of the stars, and providing fresh materials for ascertaining, by comparison of the state of the sky at different epochs, such quantities as the proper motions of the stars and the amount of precession. Among

the most important is the great catalogue of 324,198 stars in the northern hemisphere known as the Bonn *Durchmusterung*, published in 1859–62 by Bessel's pupil *Friedrich Wilhelm August Argelander* (1799–1875); this was extended (1875–85) so as to include 133,659 stars in a portion of the southern hemisphere by *Eduard Schönfeld* (1828–1891); and more recently Dr. *Gill* has executed at the Cape photographic observations of the remainder of the southern hemisphere, the reduction to the form of a catalogue (the first instalment of which was published in 1896) having been performed by Professor *Kapteyn* of Groningen. The star places determined in these catalogues do not profess to be the most accurate attainable, and for many purposes it is important to know with the utmost accuracy the positions of a smaller number of stars. The greatest undertaking of this kind, set on foot by the German Astronomical Society in 1867, aims at the construction, by the co-operation of a number of observatories, of catalogues of about 130,000 of the stars contained in the "approximate" catalogues of Argelander and Schönfeld; nearly half of the work has now been published.

The greatest scheme for a survey of the sky yet attempted is the photographic chart, together with a less extensive catalogue to be based on it, the construction of which was decided on at an international congress held at Paris in 1887. The whole sky has been divided between 18 observatories in all parts of the world, from Helsingfors in the north to Melbourne in the south, and each of these is now taking photographs with virtually identical instruments. It is estimated that the complete chart, which is intended to include stars of the 14th magnitude,* will contain about 20,000,000 stars, 2,000,000 of which will be catalogued also.

281. One other great problem—that of the distance of the sun—may conveniently be discussed under the head of observational astronomy.

The transits of Venus (chapter x., §§ 202, 227) which occurred in 1874 and 1882 were both extensively observed,

* An average star of the 14th magnitude is 10,000 times fainter than one of the 4th magnitude, which again is about 150 times less bright than Sirius. See § 316.

the old methods of time-observation being supplemented by photography and by direct micrometric measurements of the positions of Venus while transiting.

The method of finding the distance of the sun by means of observation of Mars in opposition (chapter VIII., § 161) has been employed on several occasions with considerable success, notably by Dr. Gill at Ascension in 1877. A method originally used by Flamsteed, but revived in 1857 by *Sir George Biddell Airy* (1801–1892), the late Astronomer Royal, was adopted on this occasion. For the determination of the parallax of a planet observations have to be made from two different positions at a known distance apart; commonly these are taken to be at two different observatories, as far as possible removed from one another in latitude. Airy pointed out that the same object could be attained if only one observatory were used, but observations taken at an interval of some hours, as the rotation of the earth on its axis would in that time produce a known displacement of the observer's position and so provide the necessary base line. The apparent shift of the planet's position could be most easily ascertained by measuring (with the micrometer) its distances from neighbouring fixed stars. This method (known as the **diurnal method**) has the great advantage, among others, of being simple in application, a single observer and instrument being all that is needed.

The diurnal method has also been applied with great success to certain of the minor planets (§ 294). Revolving as they do between Mars and Jupiter, they are all farther off from us than the former; but there is the compensating advantage that as a minor planet, unlike Mars, is, as a rule, too small to shew any appreciable disc, its angular distance from a neighbouring star is more easily measured. The employment of the minor planets in this way was first suggested by Professor *Galle* of Berlin in 1872, and recent observations of the minor planets *Victoria*, *Sappho*, and *Iris* in 1888–89, made at a number of observatories under the general direction of Dr. Gill, have led to some of the most satisfactory determinations of the sun's distance.

282. It was known to the mathematical astronomers of the 18th century that the distance of the sun could be obtained from a knowledge of various perturbations of

members of the solar system; and Laplace had deduced a value of the solar parallax from lunar theory. Improvements in gravitational astronomy and in observation of the planets and moon during the present century have added considerably to the value of these methods. A certain irregularity in the moon's motion known as the **parallactic inequality,** and another in the motion of the sun, called the **lunar equation,** due to the displacement of the earth by the attraction of the moon, alike depend on the ratio of the distances of the sun and moon from the earth; if the amount of either of these inequalities can be observed, the distance of the sun can therefore be deduced, that of the moon being known with great accuracy. It was by a virtual application of the first of these methods that Hansen (§ 286) in 1854, in the course of an elaborate investigation of the lunar theory, ascertained that the current value of the sun's distance was decidedly too large, and Leverrier (§ 288) confirmed the correction by the second method in 1858.

Again, certain changes in the orbits of our two neighbours, Venus and Mars, are known to depend upon the ratio of the masses of the sun and earth, and can hence be connected, by gravitational principles, with the quantity sought. Leverrier pointed out in 1861 that the motions of Venus and of Mars, like that of the moon, were inconsistent with the received estimate of the sun's distance, and he subsequently worked out the method more completely and deduced (1872) values of the parallax. The displacements to be observed are very minute, and their accurate determination is by no means easy, but they are both secular (chapter XI., § 242), so that in the course of time they will be capable of very exact measurement. Leverrier's method, which is even now a valuable one, must therefore almost inevitably outstrip all the others which are at present known; it is difficult to imagine, for example, that the transits of Venus due in 2004 and 2012 will have any value for the purpose of the determination of the sun's distance.

283. One other method, in two slightly different forms, has become available during this century. The displacement of a star by aberration (chapter X., § 210) depends

upon the ratio of the velocity of light to that of the earth in its orbit round the sun; and observations of Jupiter's satellites after the manner of Roemer (chapter VIII., § 162) give the **light-equation**, or time occupied by light in travelling from the sun to the earth. Either of these astronomical quantities—of which aberration is the more accurately known—can be used to determine the velocity of light when the dimensions of the solar system are known, or *vice versa*. No independent method of determining the velocity of light was known until 1849, when *Hippolyte Fizeau* (1819–1896) invented and successfully carried out a laboratory method.

New methods have been devised since, and three comparatively recent series of experiments, by M. *Cornu* in France (1874 and 1876) and by Dr. *Michelson* (1879) and Professor *Newcomb* (1880–82) in the United States, agreeing closely with one another, combine to fix the velocity of light at very nearly 186,300 miles (299,800 kilometres) per second; the solar parallax resulting from this by means of aberration is very nearly 8″·8.*

284. Encke's value of the sun's parallax, 8″·571, deduced from the transits of Venus (chapter X., § 227) in 1761 and 1769, and published in 1835, corresponding to a distance of about 95,000,000 miles, was generally accepted till past the middle of the century. Then the gravitational methods of Hansen and Leverrier, the earlier determinations of the velocity of light, and the observations made at the opposition of Mars in 1862, all pointed to a considerably larger value of the parallax; a fresh examination of the 18th century observations shewed that larger values than Encke's could easily be deduced from them; and for some time—from about 1860 onwards—a parallax of nearly 8″·95, corresponding to a distance of rather more than 91,000,000 miles, was in common use. Various small errors in the new methods were, however, detected, and the most probable value of the parallax has again increased. Three of the most reliable methods, the diurnal method as applied to Mars in 1877, the same applied to the minor planets in 1888–89, and

* Newcomb's velocity of light and Nyrén's constant of aberration (20″·4921) give 8″·794; Struve's constant of aberration (20″·445), Loewy's (20″·447), and Hall's (20″·454) each give 8″·81.

aberration, unite in giving values not differing from 8″·80 by more than two or three hundredths of a second. The results of the last transits of Venus, the publication and discussion of which have been spread over a good many years, point to a somewhat larger value of the parallax. Most astronomers appear to agree that a parallax of 8″·8, corresponding to a distance of rather less than 93,000,000 miles, represents fairly the available data.

285. The minute accuracy of modern observations is well illustrated by the recent discovery of a variation in the latitude of several observatories. Observations taken at Berlin in 1884–85 indicated a minute variation in the latitude; special series of observations to verify this were set on foot in several European observatories, and subsequently at Honolulu and at Cordoba. A periodic alteration in latitude amounting to about ½″ emerged as the result. Latitude being defined (chapter x., § 221) as the angle which the vertical at any place makes with the equator, which is the same as the elevation of the pole above the horizon, is consequently altered by any change in the equator, and therefore by an alteration in the position of the earth's poles or the ends of the axis about which it rotates.

Dr. *S. C. Chandler* succeeded (1891 and subsequently) in shewing that the observations in question could be in great part explained by supposing the earth's axis to undergo a minute change of position in such a way that either pole of the earth describes a circuit round its mean position in about 427 days, never deviating more than some 30 feet from it. It is well known from dynamical theory that a rotating body such as the earth can be displaced in this manner, but that if the earth were perfectly rigid the period should be 306 days instead of 427. The discrepancy between the two numbers has been ingeniously used as a test of the extent to which the earth is capable of yielding—like an elastic solid—to the various forces which tend to strain it.

286. All the great problems of gravitational astronomy have been rediscussed since Laplace's time, and further steps taken towards their solution.

Laplace's treatment of the lunar theory was first developed by *Marie Charles Theodore Damoiseau* (1768–1846), whose

Tables de la Lune (1824 and 1828) were for some time in general use.

Some special problems of both lunar and planetary theory were dealt with by *Siméon Denis Poisson* (1781–1840), who is, however, better known as a writer on other branches of mathematical physics than as an astronomer. A very elaborate and detailed theory of the moon, investigated by the general methods of Laplace, was published by *Giovanni Antonio Amadeo Plana* (1781–1869) in 1832, but unaccompanied by tables. A general treatment of both lunar and planetary theories, the most complete that had appeared up to that time, by *Philippe Gustave Doulcet de Pontécoulant* (1795–1874), appeared in 1846, with the title *Théorie Analytique du Système du Monde*; and an incomplete lunar theory similar to his was published by *John William Lubbock* (1803–1865) in 1830–34.

A great advance in lunar theory was made by *Peter Andreas Hansen* (1795–1874) of Gotha, who published in 1838 and 1862–64 the treatises commonly known respectively as the *Fundamenta* * and the *Darlegung*,† and produced in 1857 tables of the moon's motion of such accuracy that the discrepancies between the tables and observations in the century 1750–1850 were never greater than 1″ or 2″. These tables were at once used for the calculation of the *Nautical Almanac* and other periodicals of the same kind, and with some modifications have remained in use up to the present day.

A completely new lunar theory—of great mathematical interest and of equal complexity—was published by *Charles Delaunay* (1816–1872) in 1860 and 1867. Unfortunately the author died before he was able to work out the corresponding tables.

Professor Newcomb of Washington (§ 283) has rendered valuable services to lunar theory—as to other branches of astronomy—by a number of delicate and intricate calculations, the best known being his comparison of Hansen's tables with observation and consequent corrections of the tables.

* *Fundamenta Nova Investigationis Orbitae Verae quam Luna perlustrat.*

† *Darlegung der theoretischen Berechnung der in den Mondtafeln angewandten Störungen.*

New methods of dealing with lunar theory were devised by the late Professor *John Couch Adams* of Cambridge (1819–1892), and similar methods have been developed by Dr. *G. W. Hill* of Washington; so far they have not been worked out in detail in such a way as to be available for the calculation of tables, and their interest seems to be at present mathematical rather than practical; but the necessary detailed work is now in progress, and these and allied methods may be expected to lead to a considerable diminution of the present excessive intricacy of lunar theory.

287. One special point in lunar theory may be worth mentioning. The secular acceleration of the moon's mean motion which had perplexed astronomers since its first discovery by Halley (chapter x., § 201) had, as we have seen (chapter xi., § 240), received an explanation in 1787 at the hands of Laplace. Adams, on going through the calculation, found that some quantities omitted by Laplace as unimportant had in reality a very sensible effect on the result, so that a certain quantity expressing the rate of increase of the moon's motion came out to be between 5″ and 6″, instead of being about 10″, as Laplace had found and as observation required. The correction was disputed at first by several of the leading experts, but was confirmed independently by Delaunay and is now accepted. The moon appears in consequence to have a certain very minute increase in speed for which the theory of gravitation affords no explanation. An ingenious though by no means certain explanation was suggested by Delaunay in 1865. It had been noticed by Kant that **tidal friction**—that is, the friction set up between the solid earth and the ocean as the result of the tidal motion of the latter—would have the effect of checking to some extent the rotation of the earth; but as the effect seemed to be excessively minute and incapable of precise calculation it was generally ignored. An attempt to calculate its amount was, however, made in 1853 by *William Ferrel*, who also pointed out that, as the period of the earth's rotation—the day—is our fundamental unit of time, a reduction of the earth's rate of rotation involves the lengthening of our unit of time, and consequently produces an apparent increase of speed in all other motions

measured in terms of this unit. Delaunay, working independently, arrived at like conclusions, and shewed that tidal friction might thus be capable of producing just such an alteration in the moon's motion as had to be explained; if this explanation were accepted the observed motion of the moon would give a measure of the effect of tidal friction. The minuteness of the quantities involved is shewn by the fact that an alteration in the earth's rotation equivalent to the lengthening of the day by $\frac{1}{10}$ second in 10,000 years is sufficient to explain the acceleration in question. Moreover it is by no means certain that the usual estimate of the amount of this acceleration—based as it is in part on ancient eclipse observations—is correct, and even then a part of it may conceivably be due to some indirect effect of gravitation even more obscure than that detected by Laplace, or to some other cause hitherto unsuspected.

288. Most of the writers on lunar theory already mentioned have also made contributions to various parts of planetary theory, but some of the most important advances in planetary theory made since the death of Laplace have been due to the French mathematician *Urbain Jean Joseph Leverrier* (1811–1877), whose methods of determining the distance of the sun have been already referred to (§ 282). His first important astronomical paper (1839) was a discussion of the stability (chapter XI., § 245) of the system formed by the sun and the three largest and most distant planets then known, Jupiter, Saturn, and Uranus. Subsequently he worked out afresh the theory of the motion of the sun and of each of the principal planets, and constructed tables of them, which at once superseded earlier ones, and are now used as the basis of the chief planetary calculations in the *Nautical Almanac* and most other astronomical almanacs. Leverrier failed to obtain a satisfactory agreement between observation and theory in the case of Mercury, a planet which has always given great trouble to astronomers, and was inclined to explain the discrepancies as due to the influence either of a planet revolving between Mercury and the sun or of a number of smaller bodies analogous to the minor planets (§ 294).

Researches of a more abstract character, connecting planetary theory with some of the most recent advances in pure mathematics, have been carried out by *Hugo Gyldén* (1841–1896), while one of the most eminent pure mathematicians of the day, M. *Henri Poincaré* of Paris, has recently turned his attention to astronomy, and is engaged in investigations which, though they have at present but little bearing on practical astronomy, seem likely to throw important light on some of the general problems of celestial mechanics.

289. One memorable triumph of gravitational astronomy, the discovery of Neptune, has been described so often and so fully elsewhere * that a very brief account will suffice here. Soon after the discovery of Uranus (chapter XII., § 253) it was found that the planet had evidently been observed, though not recognised as a planet, as early as 1690, and on several occasions afterwards.

When the first attempts were made to compute its orbit carefully, it was found impossible satisfactorily to reconcile the earlier with the later observations, and in Bouvard's tables (chapter XI., § 247, note) published in 1821 the earlier observations were rejected. But even this drastic measure did not cure the evil; discrepancies between the observed and calculated places soon appeared and increased year by year. Several explanations were proposed, and more than one astronomer threw out the suggestion that the irregularities might be due to the attraction of a hitherto unknown planet. The first serious attempt to deduce from the irregularities in the motion of Uranus the position of this hypothetical body was made by Adams immediately after taking his degree (1843). By October 1845 he had succeeded in constructing an orbit for the new planet, and in assigning for it a position differing (as we now know) by less than 2° (four times the diameter of the full moon) from its actual position. No telescopic search for it was, however, undertaken. Meanwhile, Leverrier had independently taken up the inquiry, and by August 31st, 1846, he, like Adams, had succeeded in determining the orbit and the position of the disturbing body. On the 23rd of the follow-

* *E.g.* in Grant's *History of Physical Astronomy*, Herschel's *Outlines of Astronomy*, Miss Clerke's *History of Astronomy in the Nineteenth Century*, and the memoir by Dr. Glaisher prefixed to the first volume of Adams's *Collected Papers*.

ing month Dr. Galle of the Berlin Observatory received from Leverrier a request to search for it, and on the same evening found close to the position given by Leverrier a strange body shewing a small planetary disc, which was soon recognised as a new planet, known now as Neptune.

It may be worth while noticing that the error in the motion of Uranus which led to this remarkable discovery never exceeded 2′, a quantity imperceptible to the ordinary eye; so that if two stars were side by side in the sky, one in the true position of Uranus and one in the calculated position as given by Bouvard's tables, an observer of ordinary eyesight would see one star only.

290. The lunar tables of Hansen and Professor Newcomb, and the planetary and solar tables of Leverrier, Professor Newcomb, and Dr. Hill, represent the motions of the bodies dealt with much more accurately than the corresponding tables based on Laplace's work, just as these were in turn much more accurate than those of Euler, Clairaut, and Halley. But the agreement between theory and observation is by no means perfect, and the discrepancies are in many cases greater than can be explained as being due to the necessary imperfections in our observations.

The two most striking cases are perhaps those of Mercury and the moon. Leverrier's explanation of the irregularities of the former (§ 288) has never been fully justified or generally accepted; and the position of the moon as given in the *Nautical Almanac* and in similar publications is calculated by means of certain corrections to Hansen's tables which were deduced by Professor Newcomb from observation and have no justification in the theory of gravitation.

291. The calculation of the paths of comets has become of some importance during this century owing to the discovery of a number of comets revolving round the sun in comparatively short periods. Halley's comet (chapter XI., § 231) reappeared duly in 1835, passing through its perihelion within a few days of the times predicted by three independent calculators; and it may be confidently expected again about 1910. Four other comets are now known which, like Halley's, revolve in elongated elliptic orbits, completing a revolution in between 70 and 80 years;

two of these have been seen at two returns, that known as Olbers's comet in 1815 and 1887, and the Pons-Brooks comet in 1812 and 1884. Fourteen other comets with periods varying between $3\frac{1}{3}$ years (Encke's) and 14 years (Tuttle's), have been seen at more than one return; about a dozen more have periods estimated at less than a century; and 20 or 30 others move in orbits that are decidedly elliptic, though their periods are longer and consequently not known

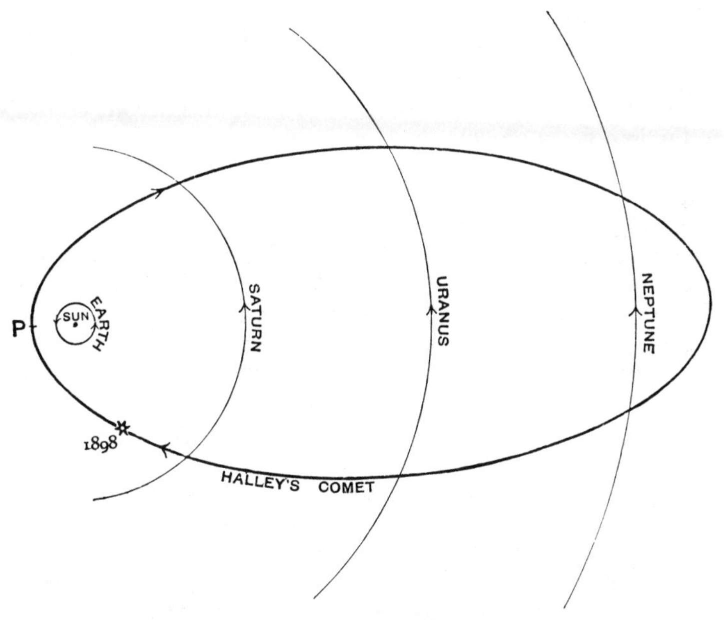

FIG. 87.—The path of Halley's comet.

with much certainty. Altogether the paths of about 230 or 240 comets have been computed, though many are highly uncertain.

292. In the theory of the tides the first important advance made after the publication of the *Mécanique Céleste* was the collection of actual tidal observations on a large scale, their interpretation, and their comparison with the results of theory. The pioneers in this direction were Lubbock (§ 286), who presented a series of papers on the subject

to the Royal Society in 1830–37, and *William Whewell* (1794–1866), whose papers on the subject appeared between 1833 and 1851. Airy (§ 281), then Astronomer Royal, also published in 1845 an important treatise dealing with the whole subject, and discussing in detail the theory of tides in bodies of water of limited extent and special form. The analysis of tidal observations, a large number of which taken from all parts of the world are now available, has subsequently been carried much further by new methods due to Lord *Kelvin* and Professor *G. H. Darwin.* A large quantity of information is thus available as to the way in which tides actually vary in different places and according to different positions of the sun and moon.

Of late years a good deal of attention has been paid to the effect of the attraction of the sun and moon in producing alterations—analogous to oceanic tides—in the earth itself. No body is perfectly rigid, and the forces in question must therefore produce some tidal effect. The problem was first investigated by Lord Kelvin in 1863, subsequently by Professor Darwin and others. Although definite numerical results are hardly attainable as yet, the work so far carried out points to the comparative smallness of these bodily tides and the consequent great rigidity of the earth, a result of interest in connection with geological inquiries into the nature of the interior of the earth.

Some speculations connected with tidal friction are referred to elsewhere (§ 320).

293. The series of propositions as to the stability of the solar system established by Lagrange and Laplace (chapter XI., §§ 244, 245), regarded as abstract propositions mathematically deducible from certain definite assumptions, have been confirmed and extended by later mathematicians such as Poisson and Leverrier; but their claim to give information as to the condition of the actual solar system at an indefinitely distant future time receives much less assent now than formerly. The general trend of scientific thought has been towards the fuller recognition of the merely approximate and probable character of even the best ascertained portions of our knowledge; "exact," "always," and "certain" are words which are disappearing from the scientific vocabulary, except as convenient abbreviations.

Propositions which profess to be—or are commonly interpreted as being—"exact" and valid throughout all future time are consequently regarded with considerable distrust, unless they are clearly mere abstractions.

In the case of the particular propositions in question the progress of astronomy and physics has thrown a good deal of emphasis on some of the points in which the assumptions required by Lagrange and Laplace are not satisfied by the actual solar system.

It was assumed for the purposes of the stability theorems that the bodies of the solar system are perfectly rigid; in other words, the motions relative to one another of the parts of any one body were ignored. Both the ordinary tides of the ocean and the bodily tides to which modern research has called attention were therefore left out of account. Tidal friction, though at present very minute in amount (§ 287), differs essentially from the perturbations which form the main subject-matter of gravitational astronomy, inasmuch as its action is irreversible. The stability theorems shewed in effect that the ordinary perturbations produced effects which sooner or later compensated one another, so that if a particular motion was accelerated at one time it would be retarded at another; but this is not the case with tidal friction. Tidal action between the earth and the moon, for example, gradually lengthens both the day and the month, and increases the distance between the earth and the moon. Solar tidal action has a similar though smaller effect on the sun and earth. The effect in each case—as far as we can measure it at all—seems to be minute almost beyond imagination, but there is no compensating action tending at any time to reverse the process. And on the whole the energy of the bodies concerned is thereby lessened. Again, modern theories of light and electricity require space to be filled with an "ether" capable of transmitting certain waves; and although there is no direct evidence that it in any way affects the motions of earth or planets, it is difficult to imagine a medium so different from all known forms of ordinary matter as to offer *no* resistance to a body moving through it. Such resistance would have the effect of slowly bringing the members of the solar system nearer to the sun, and gradually diminishing their times of revolution round it. This is again an irreversible tendency for which we know of no compensation.

In fact, from the point of view which Lagrange and Laplace occupied, the solar system appeared like a clock which, though not going quite regularly, but occasionally gaining and occasionally losing, nevertheless required no winding up; whereas modern research emphasises the analogy to a clock which after all is running down, though at an excessively slow rate. Modern study of the sun's heat (§ 319) also indicates an irreversible tendency towards the "running down" of the solar system in another way.

294. Our account of modern descriptive astronomy may conveniently begin with planetary discoveries.

The first day of the 19th century was marked by the discovery of a new planet, known as Ceres. It was seen by *Giuseppe Piazzi* (1746–1826) as a strange star in a region of the sky which he was engaged in mapping, and soon recognised by its motion as a planet. Its orbit—first calculated by Gauss (§ 276)—shewed it to belong to the space between Mars and Jupiter, which had been noted since the time of Kepler as abnormally large. That a planet should be found in this region was therefore no great surprise; but the discovery by *Heinrich Olbers* (1758–1840), scarcely a year later (March 1802), of a second body (*Pallas*), revolving at nearly the same distance from the sun, was wholly unexpected, and revealed an entirely new planetary arrangement. It was an obvious conjecture that if there was room for two planets there was room for more, and two fresh discoveries (*Juno* in 1804, *Vesta* in 1807) soon followed.

The new bodies were very much smaller than any of the other planets, and, so far from readily shewing a planetary disc like their neighbours Mars and Jupiter, were barely distinguishable in appearance from fixed stars, except in the most powerful telescopes of the time; hence the name **asteroid** (suggested by William Herschel) or **minor planet** has been generally employed to distinguish them from the other planets. Herschel attempted to measure their size, and estimated the diameter of the largest at under 200 miles (that of Mercury, the smallest of the ordinary planets, being 3000), but the problem was in reality

Fig. 88.—Photographic trail of a minor planet.

[*To face p.* 377.

too difficult even for his unrivalled powers of observation. The minor planets were also found to be remarkable for the great inclination and eccentricity of some of the orbits; the path of Pallas, for example, makes an angle of 35° with the ecliptic, and its eccentricity is $\frac{1}{4}$, so that its least distance from the sun is not much more than half its greatest distance. These characteristics suggested to Olbers that the minor planets were in reality fragments of a primeval planet of moderate dimensions which had been blown to pieces, and the theory, which fitted most of the facts then known, was received with great favour in an age when "catastrophes" were still in fashion as scientific explanations.

The four minor planets named were for nearly 40 years the only ones known; then a fifth was discovered in 1845 by *Karl Ludwig Hencke* (1793–1866) after 15 years of search. Two more were found in 1847, another in 1848, and the number has gone on steadily increasing ever since. The process of discovery has been very much facilitated by improvements in star maps, and latterly by the introduction of photography. In this last method, first used by Dr. *Max Wolf* of Heidelberg in 1891, a photographic plate is exposed for some hours; any planet present in the region of the sky photographed, having moved sensibly relatively to the stars in this period, is thus detected by the trail which its image leaves on the plate. The annexed figure shews (near the centre) the trail of the minor planet *Svea*, discovered by Dr. Wolf on March 21st, 1892.

At the end of 1897 no less than 432 minor planets were known, of which 92 had been discovered by a single observer, M. *Charlois* of Nice, and only nine less by Professor *Palisa* of Vienna.

The paths of the minor planets practically occupy the whole region between the paths of Mars and Jupiter, though few are near the boundaries; no orbit is more inclined to the ecliptic than that of Pallas, and the eccentricities range from almost zero up to about $\frac{1}{3}$.

Fig. 89 shews the orbits of the first two minor planets discovered, as well as of No. 323 (*Brucia*), which comes nearest to the sun, and of No. 361 (not yet named), which goes farthest from it. All the orbits are described in the standard, or west to east, direction. The most interesting characteristic in the distribution of the minor planets, first noted in 1866 by *Daniel Kirkwood* (1815–1895), is the existence of comparatively clear spaces in the regions where the disturbing action of Jupiter would by Lagrange's

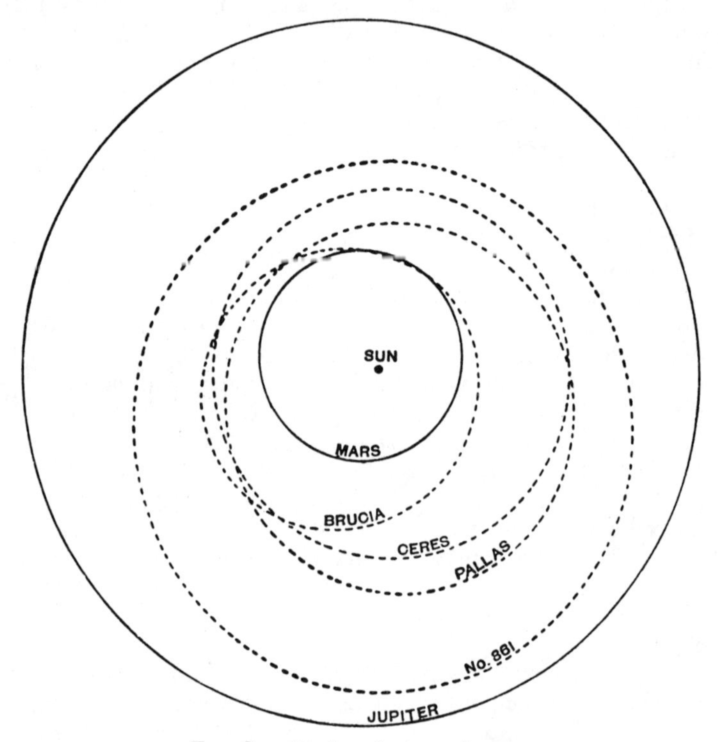

FIG. 89.—Paths of minor planets.

principle (chapter XI., § 243) be most effective : for instance, at a distance from the sun about five-eighths that of Jupiter, a planet would by Kepler's law revolve exactly *twice* as fast as Jupiter ; and accordingly there is a gap among the minor planets at about this distance.

Estimates of the sizes and masses of the minor planets are still very uncertain. The first direct measurements of any of the discs which seem reliable are those of Professor *E. E. Barnard*, made at the Lick Observatory in 1894 and 1895; according to these the three largest minor planets, Ceres, Pallas, and Vesta, have diameters of nearly 500 miles, about 300 and about 250 miles respectively. Their sizes compared with the moon are shewn on the diagram (fig 90). An alternative method—the only one available except for a few of the very largest

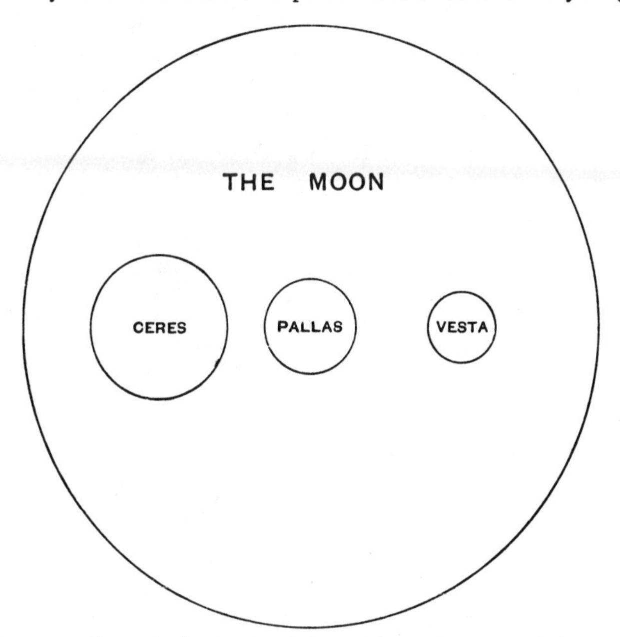

FIG. 90.—Comparative sizes of three minor planets and the moon.

of the minor planets—is to measure the amount of light received, and hence to deduce the size, on the assumption that the reflective power is the same as that of some known planet. This method gives diameters of about 300 miles for the brightest and of about a dozen miles for the faintest known.

Leverrier calculated from the perturbations of Mars that the total mass of all known or unknown bodies between Mars and Jupiter could not exceed a fourth that of the earth; but such knowledge of the sizes as we can derive from light-observations seems to indicate that the total mass of those at present known is many hundred times less than this limit.

295. Neptune and the minor planets are the only planets which have been discovered during this century, but several satellites have been added to our system.

Barely a fortnight after the discovery of Neptune (1846)

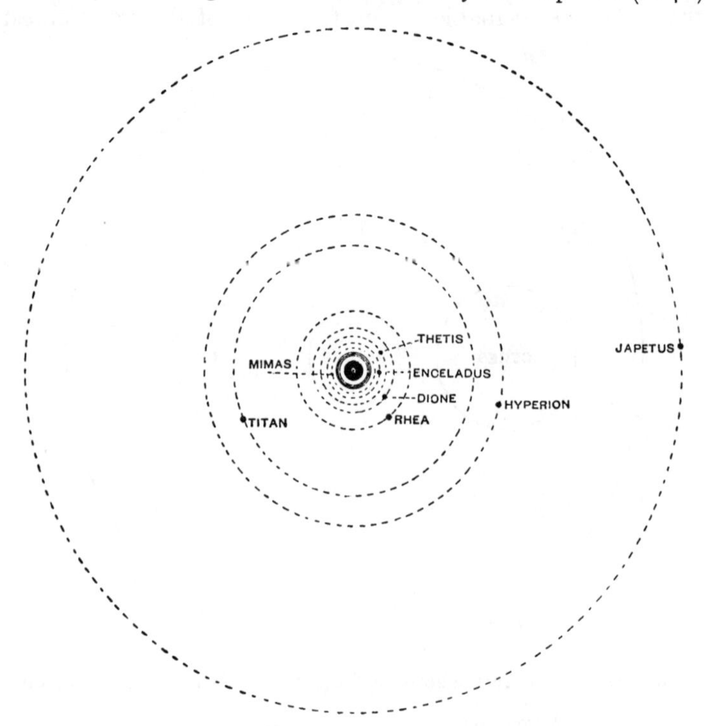

FIG. 91.—Saturn and its system.

a satellite was detected by *William Lassell* (1799–1880) at Liverpool. Like the satellites of Uranus, this revolves round its primary from east to west—that is, in the direction contrary to that of all the other known motions of the solar system (certain long-period comets not being counted).

Two years later (September 16th, 1848) *William Cranch Bond* (1789–1859) discovered, at the Harvard College

Observatory, an eighth satellite of Saturn, called *Hyperion*, which was detected independently by Lassell two days afterwards. In the following year Bond discovered that Saturn was accompanied by a third comparatively dark ring —now commonly known as the **crape ring**—lying immediately inside the bright rings (see fig. 95); and the discovery was made independently a fortnight later by

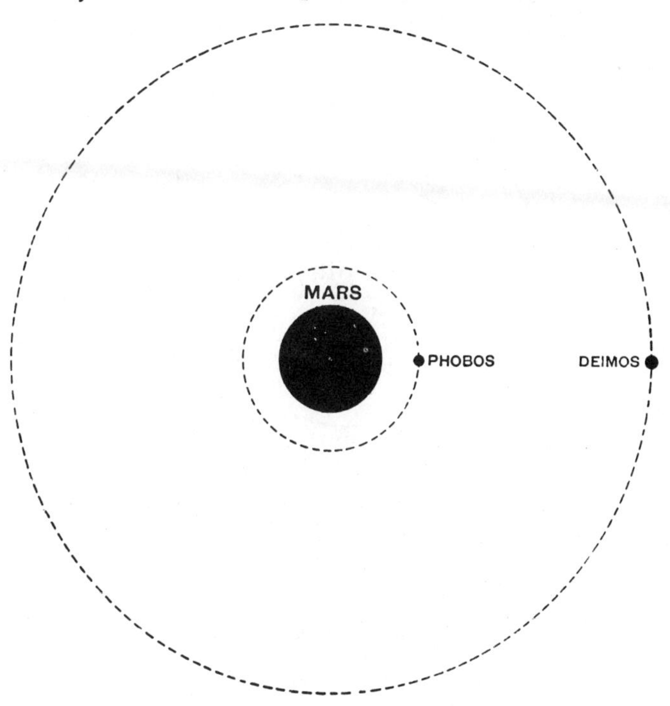

FIG. 92.—Mars and its satellites.

William Rutter Dawes (1799–1868) in England. Lassell discovered in 1851 two new satellites of Uranus, making a total of four belonging to that planet. The next discoveries were those of two satellites of Mars, known as *Deimos* and *Phobos*, by Professor *Asaph Hall* of Washington on August 11th and 17th, 1877. These are remarkable chiefly for their close proximity to Mars and their extremely rapid motion, the nearer one revolving more rapidly than

Mars rotates, so that to the Martians it must rise in the west and set in the east. Lastly, Jupiter's system received an addition after nearly three centuries by Professor Barnard's discovery at the Lick Observatory (September 9th, 1892) of

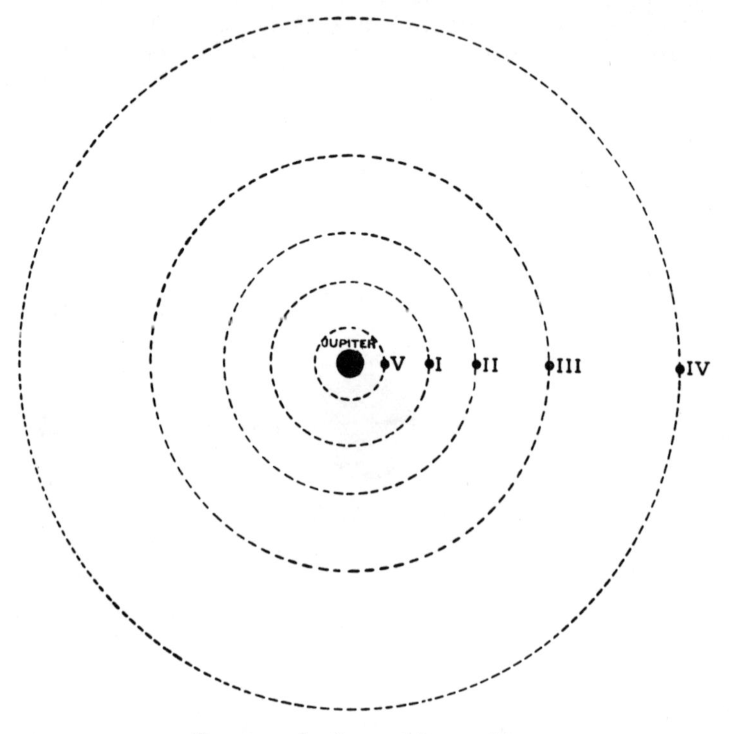

FIG. 93.—Jupiter and its satellites.

an extremely faint fifth satellite, a good deal nearer to Jupiter than the nearest of Galilei's satellites (chapter VI., § 121).

296. The surfaces of the various planets and satellites have been watched with the utmost care by an army of observers, but the observations have to a large extent remained without satisfactory interpretation, and little is known of the structure or physical condition of the bodies concerned.

Astronomers are naturally most familiar with the surface

Fig. 94.—The Apennines and adjoining regions of the moon. From a photograph taken at the Paris Observatory. [*To face p.* 383.

of our nearest neighbour, the moon. The visible half has been elaborately mapped, and the heights of the chief mountain ranges measured by means of their shadows. Modern knowledge has done much to dispel the view, held by the earlier telescopists and shared to some extent even by Herschel, that the moon closely resembles the earth and is suitable for inhabitants like ourselves. The dark spaces which were once taken to be seas and still bear that name are evidently covered with dry rock; and the craters with which the moon is covered are all—with one or two doubtful exceptions—extinct; the long dark lines known as **rills** and formerly taken for river-beds have clearly no water in them. The question of a lunar atmosphere is more difficult: if there is air its density must be very small, some hundredfold less than that of our atmosphere at the surface of the earth; but with this restriction there seems to be no bar to the existence of a lunar atmosphere of considerable extent, and it is difficult to explain certain observations without assuming the existence of some atmosphere.

297. Mars, being the nearest of the superior planets, is the most favourably situated for observation. The chief markings on its surface—provisionally interpreted as being land and water—are fairly permanent and therefore recognisable; several tolerably consistent maps of the surface have been constructed; and by observation of certain striking features the rotation period has been determined to a fraction of a second. Signor *Schiaparelli* of Milan detected at the opposition of 1877 a number of intersecting dark lines generally known as **canals**, and as the result of observations made during the opposition of 1881–82 announced that certain of them appeared doubled, two nearly parallel lines being then seen instead of one. These remarkable observations have been to a great extent confirmed by other observers, but remain unexplained.

The visible surfaces of Jupiter and Saturn appear to be layers of clouds; the low density of each planet (1·3 and 7 respectively, that of water being 1 and of the earth 5·5), the rapid changes on the surface, and other facts indicate that these planets are to a great extent in a fluid condition, and have a high temperature at a very moderate distance below the visible surface. The surface markings are in each case definite enough for the rotation periods to be fixed with some accuracy; though it is clear in the case of Jupiter, and probably also in that of Saturn, that—as with the sun (§ 298)—different parts of the surface move at different rates.

Laplace had shewn that Saturn's ring (or rings) could not be, as it appeared, a uniform solid body; he rashly inferred —without any complete investigation—that it might be an irregularly weighted solid body. The first important advance was made by *James Clerk Maxwell* (1831–1879), best known as a writer on electricity and other branches of physics. Maxwell shewed (1857) that the rings could neither be continuous solid bodies nor liquid, but that all the important dynamical conditions would be satisfied if they were made up of a very large number of small solid bodies revolving independently round the sun.* The theory thus suggested on mathematical grounds has received a good deal of support from telescopic evidence. The rings thus bear to Saturn a relation having some analogy to that which the minor planets bear to the sun; and Kirkwood pointed out in 1867 that Cassini's division between the two main rings can be explained by the perturbations due to certain of the satellites, just as the corresponding gaps in the minor planets can be explained by the action of Jupiter (§ 294).

The great distance of Uranus and Neptune naturally makes the study of them difficult, and next to nothing is known of the appearance or constitution of either; their rotation periods are wholly uncertain.

Mercury and Venus, being inferior planets, are never very far from the sun in the sky, and therefore also extremely difficult to observe satisfactorily. Various bright and dark markings on their surfaces have been recorded, but different observers give very different accounts of them. The rotation periods are also very uncertain, though a good many astronomers support the view put forward by Sig. Schiaparelli, in 1882 and 1890 for Mercury and Venus respectively, that each rotates in a time equal to its period of revolution round the sun, and thus always turns the same face towards the sun. Such a motion—which is analogous to that of the

* This had been suggested as a possibility by several earlier writers.

FIG. 95.—Saturn and its rings. From a drawing by Professor Barnard. [*To face p.* 384.

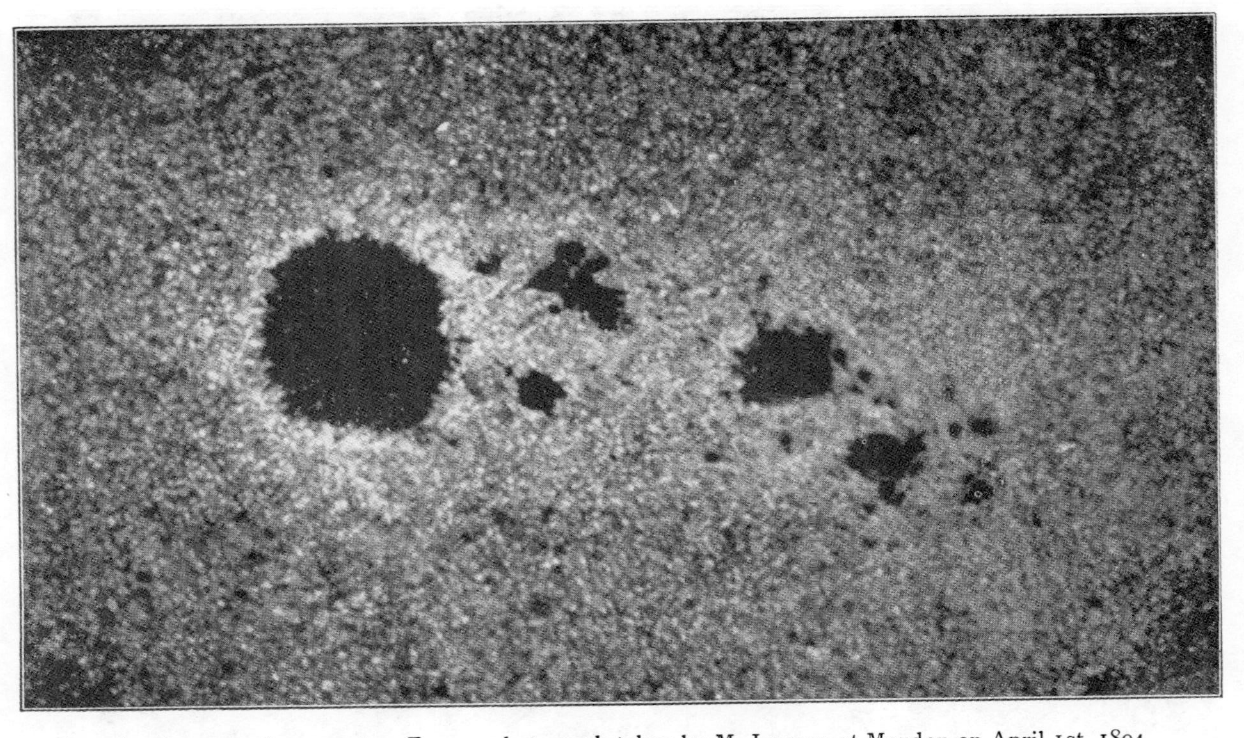

FIG. 96.—A group of sun-spots. From a photograph taken by M. Janssen at Meudon on April 1st, 1894.

[*To face p.* 385.

moon round the earth and of Japetus round Saturn (chapter XII., § 267)—could be easily explained as the result of tidal action at some past time when the planets were to a great extent fluid.

298. Telescopic study of the surface of the sun during the century has resulted in an immense accumulation of detailed knowledge of peculiarities of the various markings on the surface. The most interesting results of a general nature are connected with the distribution and periodicity of sun-spots. The earliest telescopists had noticed that the number of spots visible on the sun varied from time to time, but no law of variation was established till 1851, when *Heinrich Schwabe* of Dessau (1789–1875) published in Humboldt's *Cosmos* the results of observations of sun-spots carried out during the preceding quarter of a century, shewing that the number of spots visible increased and decreased in a tolerably regular way in a period of about ten years.

Earlier records and later observations have confirmed the general result, the period being now estimated as slightly over 11 years on the average, though subject to considerable fluctuations. A year later (1852) three independent investigators, Sir *Edward Sabine* (1788–1883) in England, *Rudolf Wolf* (1816–1893) and *Alfred Gautier* (1793–1881) in Switzerland, called attention to the remarkable similarity between the periodic variations of sun-spots and of various magnetic disturbances on the earth. Not only is the period the same, but it almost invariably happens that when spots are most numerous on the sun magnetic disturbances are most noticeable on the earth, and that similarly the times of scarcity of the two sets of phenomena coincide. This wholly unexpected and hitherto quite unexplained relationship has been confirmed by the occurrence on several occasions of decided magnetic disturbances simultaneously with rapid changes on the surface of the sun.

A long series of observations of the position of spots on the sun undertaken by *Richard Christopher Carrington* (1826–1875) led to the first clear recognition of the difference in the rate of rotation of the different parts of the surface of the sun, the period of rotation being fixed (1859) at about 25 days at the equator, and two and a half days longer half-way between the equator and the poles; while in addition spots were seen to have also independent "proper motions." Carrington also established (1858) the scarcity of spots in the immediate neighbourhood of the equator, and confirmed statistically their prevalence in the adjacent regions, and their great scarcity more than about 35° from the equator; and noticed further certain regular changes in the distribution of spots on the sun in the course of the 11-year cycle.

Wilson's theory (chapter XII., § 268) that spots are depressions was confirmed by an extensive series of photographs taken at Kew in 1858–72, shewing a large preponderance of cases of the perspective effect noticed by him; but, on the other hand, Mr. *F. Howlett*, who has watched the sun for some 35 years and made several thousand drawings of spots, considers (1894) that his observations are decidedly against Wilson's theory. Other observers are divided in opinion.

299. **Spectrum analysis,** which has played such an important part in recent astronomical work, is essentially a method of ascertaining the nature of a body by a process of sifting or analysing into different components the light received from it.

It was first clearly established by Newton, in 1665–66 (chapter IX., § 168), that ordinary white light, such as sunlight, is composite, and that by passing a beam of sunlight —with proper precautions—through a glass prism it can be decomposed into light of different colours; if the beam so decomposed is received on a screen, it produces a band of colours known as a **spectrum**, red being at one end and violet at the other.

Now according to modern theories light consists essentially of a series of disturbances or waves transmitted at extremely short but regular intervals from the luminous object to the eye, the medium through which the disturbances travel being called **ether**. The most important characteristic distinguishing different kinds of light is the interval of time or space between one wave and the next, which is generally expressed by means of **wave-length**, or the distance between any point of one wave and the corresponding point of the next. Differences in wave-length shew themselves most readily as differences of colour; so

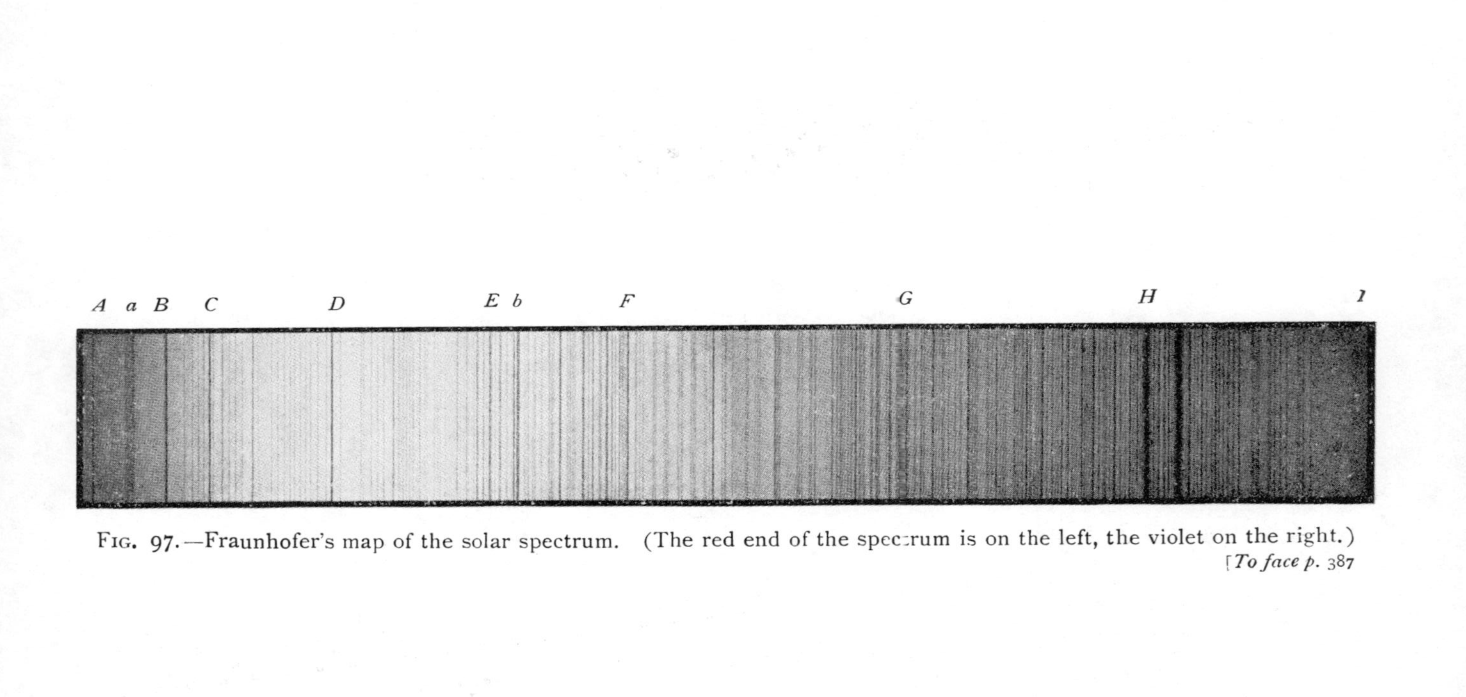

Fig. 97.—Fraunhofer's map of the solar spectrum. (The red end of the spectrum is on the left, the violet on the right.)

[*To face p.* 387

that light of a particular colour found at a particular part of the spectrum has a definite wave-length. At the extreme violet end of the spectrum, for example, the wave-length is about fifteen millionths of an inch, at the red end it is about twice as great; from which it follows (§ 283), from the known velocity of light, that when we look at the red end of a spectrum about 400 billion waves of light enter the eye per second, and twice that number when we look at the other end. Newton's experiment thus shews that a prism sorts out light of a composite nature according to the wave-length of the different kinds of light present. The same thing can be done by substituting for the prism a so-called **diffraction-grating**, and this is for many purposes superseding the prism. In general it is necessary, to ensure purity in the spectrum and to make it large enough, to admit light through a narrow slit, and to use certain lenses in combination with one or more prisms or a grating; and the arrangement is such that the spectrum is not thrown on to a screen, but either viewed directly by the eye or photographed. The whole apparatus is known as a **spectroscope.**

The solar spectrum appeared to Newton as a continuous band of colours; but in 1802 *William Hyde Wollaston* (1766–1828) observed certain dark lines running across the spectrum, which he took to be the boundaries of the natural colours. A few years later (1814–15) the great Munich optician *Joseph Fraunhofer* (1787–1826) examined the sun's spectrum much more carefully, and discovered about 600 such dark lines, the positions of 324 of which he mapped (see fig. 97). These dark lines are accordingly known as **Fraunhofer lines:** for purposes of identification Fraunhofer attached certain letters of the alphabet to a few of the most conspicuous; the rest are now generally known by the wave-length of the corresponding kind of light.

It was also gradually discovered that dark bands could be produced artificially in spectra by passing light through various coloured substances; and that, on the other hand, the spectra of certain flames were crossed by various *bright* lines.

Several attempts were made to explain and to connect these various observations, but the first satisfactory and tolerably complete explanation was given in 1859 by *Gustav*

Robert Kirchhoff (1824–1887) of Heidelberg, who at first worked in co-operation with the chemist Bunsen.

Kirchhoff shewed that a luminous solid or liquid—or, as we now know, a highly compressed gas—gives a continuous spectrum; whereas a substance in the gaseous state gives a spectrum consisting of bright lines (with or without a faint continuous spectrum), and these bright lines depend on the particular substance and are characteristic of it. Consequently the presence of a particular substance in the form of gas in a hot body can be inferred from the presence of its characteristic lines in the spectrum of the light. The *dark* lines in the solar spectrum were explained by the fundamental principle—often known as Kirchhoff's law—that a body's capacity for stopping or absorbing light of a particular wave-length is proportional to its power, under like conditions, of giving out the same light. If, in particular, light from a luminous solid or liquid body, giving a continuous spectrum, passes through a gas, the gas absorbs light of the same wave-length as that which it itself gives out: if the gas gives out more light of these particular wave-lengths than it absorbs, then the spectrum is crossed by the corresponding bright lines; but if it absorbs more than it gives out, then there is a deficiency of light of these wave-lengths and the corresponding parts of the spectrum appear dark—that is, the spectrum is crossed by dark lines in the same position as the bright lines in the spectrum of the gas alone. Whether the gas absorbs more or less than it gives out is essentially a question of temperature, so that if light from a hot solid or liquid passes through a gas at a higher temperature a spectrum crossed by bright lines is the result, whereas if the gas is cooler than the body behind it dark lines are seen in the spectrum.

300. The presence of the Fraunhofer lines in the spectrum of the sun shews that sunlight comes from a hot solid or liquid body (or from a highly compressed gas), and that it has passed through cooler gases which have absorbed light of the wave-lengths corresponding to the dark lines. These gases must be either round the sun or in our atmosphere; and it is not difficult to shew that, although some of the Fraunhofer lines are due to our

atmosphere, the majority cannot be, and are therefore caused by gases in the atmosphere of the sun.

For example, the metal sodium when vaporised gives a spectrum characterised by two nearly coincident bright lines in the yellow part of the spectrum; these agree in position with a pair of dark lines (known as D) in the spectrum of the sun (see fig. 97); Kirchhoff inferred therefore that the atmosphere of the sun contains sodium. By comparison of the dark lines in the spectrum of the sun with the bright lines in the spectra of metals and other substances, their presence or absence in the solar atmosphere can accordingly be ascertained. In the case of iron—which has an extremely complicated spectrum—Kirchhoff succeeded in identifying 60 lines (since increased to more than 2,000) in its spectrum with dark lines in the spectrum of the sun. Some half-dozen other known elements were also identified by Kirchhoff in the sun.

The inquiry into solar chemistry thus started has since been prosecuted with great zeal. Improved methods and increased care have led to the construction of a series of maps of the solar spectrum, beginning with Kirchhoff's own, published in 1861–62, of constantly increasing complexity and accuracy. Knowledge of the spectra of the metals has also been greatly extended. At the present time between 30 and 40 elements have been identified in the sun, the most interesting besides those already mentioned being hydrogen, calcium, magnesium, and carbon.

The first spectroscopic work on the sun dealt only with the light received from the sun as a whole, but it was soon seen that by throwing an image of the sun on to the slit of the spectroscope by means of a telescope the spectrum of a particular part of the sun's surface, such as a spot or a facula, could be obtained; and an immense number of observations of this character have been made.

301. Observations of total eclipses of the sun have shewn that the bright surface of the sun as we ordinarily see it is not the whole, but that outside this there is an envelope of some kind too faint to be seen ordinarily but becoming visible when the intense light of the sun itself is cut off by the moon. A white halo of considerable extent round the eclipsed sun, now called the **corona**, is referred to by

Plutarch, and discussed by Kepler (chapter VII., § 145) Several 18th century astronomers noticed a red streak along some portion of the common edge of the sun and moon, and red spots or clouds here and there (cf. chapter X., § 205). But little serious attention was given to the subject till after the total solar eclipse of 1842. Observations made then and at the two following eclipses of 1851 and 1860, in the latter of which years photography was for the first time effectively employed, made it evident that the red streak represented a continuous envelope of some kind surrounding the sun, to which the name of **chromosphere** has been given, and that the red objects, generally known as **prominences,** were in general projecting parts of the chromosphere, though sometimes detached from it. At the eclipse of 1868 the spectrum of the prominences and the chromosphere was obtained, and found to be one of bright lines, shewing that they consisted of gas. Immediately afterwards M. *Janssen*, who was one of the observers of the eclipse, and Sir *J. Norman Lockyer* independently devised a method whereby it was possible to get the spectrum of a prominence at the edge of the sun's disc in ordinary daylight, without waiting for an eclipse; and a modification introduced by Sir *William Huggins* in the following year (1869) enabled the form of a prominence to be observed spectroscopically. Recently (1892) Professor *G. E. Hale* of Chicago has succeeded in obtaining by a photographic process a representation of the whole of the chromosphere and prominences, while the same method gives also photographs of faculae (chapter VIII., § 153) on the visible surface of the sun.

The most important lines ordinarily present in the spectrum of the chromosphere are those of hydrogen, two lines (H and K) which have been identified with some difficulty as belonging to calcium, and a yellow line the substance producing which, known as **helium,** has only recently (1895) been discovered on the earth. But the chromosphere when disturbed and many of the prominences give spectra containing a number of other lines.

The corona was for some time regarded as of the nature of an optical illusion produced in the atmosphere. That it is, at any rate in great part, an actual appendage of the sun was first established in 1869 by the American astronomers

FIG. 98.—The total solar eclipse of August 29th, 1886. From a drawing based on photographs by Dr. Schuster and Mr. Maunder.

[*To ace p.* 390.

Professor *Harkness* and Professor *C. A. Young*, who discovered a bright line—of unknown origin *—in its spectrum, thus shewing that it consists in part of glowing gas. Subsequent spectroscopic work shews that its light is partly reflected sunlight.

The corona has been carefully studied at every solar eclipse during the last 30 years, both with the spectroscope and with the telescope, supplemented by photography, and a number of ingenious theories of its constitution have been propounded; but our present knowledge of its nature hardly goes beyond Professor Young's description of it as "an inconceivably attenuated cloud of gas, fog, and dust, surrounding the sun, formed and shaped by solar forces."

302. The spectroscope also gives information as to certain motions taking place on the sun. It was pointed out in 1842 by *Christian Doppler* (1803–1853), though in an imperfect and partly erroneous way, that if a luminous body is approaching the observer, or *vice versa*, the waves of light are as it were crowded together and reach the eye at shorter intervals than if the body were at rest, and that the character of the light is thereby changed. The colour and the position in the spectrum both depend on the interval between one wave and the next, so that if a body giving out light of a particular wave-length, *e.g.* the blue light corresponding to the F line of hydrogen, is approaching the observer rapidly, the line in the spectrum appears slightly on one side of its usual position, being displaced towards the violet end of the spectrum; whereas if the body is receding the line is, in the same way, displaced in the opposite direction. This result is usually known as **Doppler's principle.** The effect produced can easily be expressed numerically. If, for example, the body is approaching with a speed equal to $\frac{1}{1000}$ that of light, then 1001 waves enter the eye or the spectroscope in the same time in which there would otherwise only be 1000; and there is in consequence a virtual shortening of the wave-length in the ratio of 1001 to 1000. So that if it is found that a line in the spectrum of a body is displaced from its ordinary position in such

* The discovery of a terrestrial substance with this line in its spectrum has been announced while this book has been passing through the press.

a way that its wave-length is apparently decreased by $\frac{1}{1000}$ part, it may be inferred that the body is approaching with the speed just named, or about 186 miles per second, and if the wave-length appears increased by the same amount (the line being displaced towards the red end of the spectrum) the body is receding at the same rate.

Some of the earliest observations of the prominences by Sir J. N. Lockyer (1868), and of spots and other features of the sun by the same and other observers, shewed displacements and distortions of the lines in the spectrum, which were soon seen to be capable of interpretation by this method, and pointed to the existence of violent disturbances in the atmosphere of the sun, velocities as great as 300 miles per second being not unknown. The method has received an interesting confirmation from observations of the spectrum of opposite edges of the sun's disc, of which one is approaching and the other receding owing to the rotation of the sun. Professor *Dunér* of Upsala has by this process ascertained (1887–89) the rate of rotation of the surface of the sun beyond the regions where spots exist, and therefore outside the limits of observations such as Carrington's (§ 298).

303. The spectroscope tells us that the atmosphere of the sun contains iron and other metals in the form of vapour; and the photosphere, which gives the continuous part of the solar spectrum, is certainly hotter. Moreover everything that we know of the way in which heat is communicated from one part of a body to another shews that the outer regions of the sun, from which heat and light are radiating on a very large scale, must be the coolest parts, and that the temperature in all probability rises very rapidly towards the interior. These facts, coupled with the low density of the sun (about a fourth that of the earth) and the violently disturbed condition of the surface, indicate that the bulk of the interior of the sun is an intensely hot and highly compressed mass of gas. Outside this come in order, their respective boundaries and mutual relations being, however, very uncertain, first the photosphere, generally regarded as a cloud-layer, then the **reversing stratum** which produces most of the Fraunhofer lines, then the chromosphere and prominences, and finally the corona. Sun-spots, faculae, and

FIG. 99.—The great comet of 1882 (ii) on November 7th. From a photograph by Dr. Gill.

[*To face p.* 393.

prominences have been explained in a variety of different ways as joint results of solar disturbances of various kinds; but no detailed theory that has been given explains satisfactorily more than a fraction of the observed facts or commands more than a very limited amount of assent among astronomical experts.

304. More than 200 comets have been seen during the present century; not only have the motions of most of them been observed and their orbits computed (§ 291), but in a large number of cases the appearance and structure of the comet have been carefully observed telescopically, while latterly spectrum analysis and photography have also been employed.

Independent lines of inquiry point to the extremely unsubstantial character of a comet, with the possible exception of the bright central part or **nucleus**, which is nearly always present. More than once, as in 1767 (chapter XI., § 248), a comet has passed close to some member of the solar system, and has never been ascertained to affect its motion. The mass of a comet is therefore very small, but its bulk or volume, on the other hand, is in general very great, the tail often being millions of miles in length; so that the density must be extremely small. Again, stars have often been observed shining through a comet's tail (as shewn in fig. 99), and even through the head at no great distance from the nucleus, their brightness being only slightly, if at all, affected. Twice at least (1819, 1861) the earth has passed through a comet's tail, but we were so little affected that the fact was only discovered by calculations made after the event. The early observation (chapter III., § 69) that a comet's tail points away from the sun has been abundantly verified; and from this it follows that very rapid changes in the position of the tail must occur in some cases. For example, the comet of 1843 passed very close to the sun at such a rate that in about two hours it had passed from one side of the sun to the opposite; it was then much too near the sun to be seen, but if it followed the ordinary law its tail, which was unusually long, must have entirely reversed its direction within this short time. It is difficult to avoid the inference that the tail is not a permanent part of the comet, but is a stream of matter driven off from it in some way by the action of the sun, and in this respect comparable with the smoke issuing from a chimney. This view is confirmed by the fact that the tail is only developed when the comet approaches the sun, a comet when at a great distance from the sun appearing usually as an indistinct patch of nebulous light, with perhaps a brighter spot representing the nucleus. Again, if the tail be formed by an outpouring of matter from the comet, which only takes place when the comet is near the sun, the more often a comet approaches the sun the more must it waste away; and we find accordingly that the short-period comets, which return to the neighbourhood of the sun at frequent intervals (§ 291), are inconspicuous bodies. The same theory is supported by the shape of the tail. In some cases it is straight, but more commonly it is curved to some extent, and the curvature is then always *backwards* in relation to the comet's motion. Now by ordinary dynamical principles matter shot off from the head of the comet while it is revolving round the sun would tend, as it were, to lag behind more and more the farther it receded from the head, and an apparent backward curvature of the tail—less or greater according to the speed with which the particles forming the tail were repelled—would be the result. Variations in curvature of the tails of different comets, and the existence of two or more differently curved tails of the same comet, are thus readily explained by supposing them made of different materials, repelled from the comet's head at different speeds.

The first application of the spectroscope to the study of comets was made in 1864 by *Giambattista Donati* (1826–1873), best known as the discoverer of the magnificent comet of 1858. A spectrum of three bright bands, wider than the ordinary "lines," was obtained, but they were not then identified. Four years later Sir William Huggins obtained a similar spectrum, and identified it with that of a compound of carbon and hydrogen. Nearly every comet examined since then has shewn in its spectrum bright bands indicating the presence of the same or some other hydrocarbon, but in a few cases other substances have also been detected. A comet is therefore in part at least self-luminous, and some of the light which it sends us is that of a glowing gas. It also shines to a considerable extent by reflected sunlight; there is nearly always a continuous spectrum, and in a few cases—first in 1881—the spectrum has been distinct enough to shew the Fraunhofer lines crossing it. But the continuous spectrum seems also to be due in part to solid or liquid matter in the comet itself, which is hot enough to be self-luminous.

305. The work of the last 30 or 40 years has established a remarkable relation between comets and the minute bodies which are seen in the form of **meteors** or **shooting stars.** Only a few of the more important links in the chain of evidence can, however, be mentioned. Showers of shooting stars, the occurrence of which has been known from quite early times, have been shewn to be due to the passage of the earth through a swarm of bodies revolving in elliptic orbits round the sun. The paths of four such swarms were ascertained with some precision in 1866–67, and found in each case to agree closely with the paths of known comets. And since then a considerable number of other cases of resemblance or identity between the paths of meteor swarms and of comets have been detected. One of the four comets just referred to, known as Biela's, with a period of between six and seven years, was duly seen on several successive returns, but in 1845–46 was observed first to become somewhat distorted in shape, and afterwards to have divided into two distinct comets; at the next return (1852) the pair were again seen; but since then nothing has been seen of either portion. At the end of November in each year the earth almost crosses the path of this comet, and on two occasions (1872, and 1885) it did so nearly at the time when the comet was due at the same spot; if, as seemed likely, the comet had gone to pieces since its last appearance, there seemed a good chance of falling in with some of its remains, and this expectation was fulfilled by the occurrence on both occasions of a meteor shower much more brilliant than that usually observed at the same date.

Biela's comet is not the only comet which has shewn signs of breaking up; Brooks's comet of 1889, which is probably identical with Lexell's (chapter XI., § 248), was found to be accompanied by three smaller companions; as this comet has more than once passed extremely close to Jupiter, a plausible explanation of its breaking up is at once given in the attractive force of the planet. Moreover certain systems of comets, the members of which revolve in the same orbit but separated by considerable intervals of time, have also been discovered. Tebbutt's comet of 1881 moves in practically the same path as one seen in 1807, and the great comet of 1880, the great comet of 1882 (shewn in fig. 99), and a third which appeared in 1887, all move in paths closely resembling that of the comet of 1843, while that of 1668 is more doubtfully connected with the same system. And it is difficult to avoid regarding the members of a system as fragments of an earlier comet, which has passed through the stages in which we have actually seen the comets of Biela and Brooks.

Evidence of such different kinds points to an intimate connection between comets and meteors, though it is perhaps still premature to state confidently that meteors are fragments of decayed comets, or that conversely comets are swarms of meteors.

306. Each of the great problems of sidereal astronomy which Herschel formulated and attempted to solve has been elaborately studied by the astronomers of the 19th century. The multiplication of observatories, improvements in telescopes, and the introduction of photography—to mention only three obvious factors of progress—have added enormously to the extent and accuracy of our knowledge of the stars, while the invention of spectrum analysis has thrown an entirely new light on several important problems.

William Herschel's most direct successor was his son *John Frederick William* (1792–1871), who was not only an astronomer, but also made contributions of importance to pure mathematics, to physics, to the nascent art of photography, and to the philosophy of scientific discovery. He began his astronomical career about 1816 by re-measuring, first alone, then in conjunction with *James South* (1785–1867), a number of his father's double stars. The first result of this work was a catalogue, with detailed measurements, of some hundred double and multiple stars (published in 1824), which formed a valuable third term of comparison with his father's observations of 1781–82 and 1802–03, and confirmed in several cases the slow motions of revolution the beginnings of which had been observed before. A great survey of nebulae followed, resulting in a catalogue

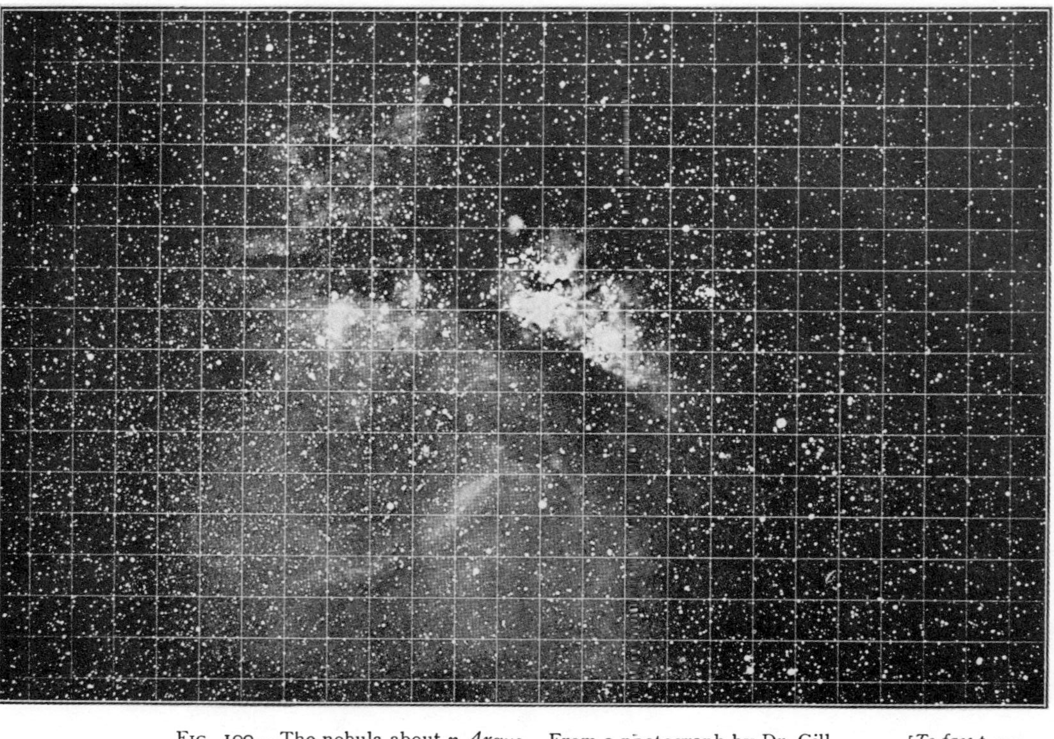

FIG. 100.—The nebula about η *Argus*. From a photograph by Dr. Gill.

[*To face p.* 397.

(1833) of about 2500, of which some 500 were new and 2000 were his father's, a few being due to other observers; incidentally more than 3000 pairs of stars close enough together to be worth recording as double stars were observed.

307. Then followed his well-known expedition to the Cape of Good Hope (1833–1838), where he "swept" the southern skies in very much the same way in which his father had explored the regions visible in our latitude. Some 1200 double and multiple stars, and a rather larger number of new nebulae, were discovered and studied, while about 500 known nebulae were re-observed; star-gauging on William Herschel's lines was also carried out on an extensive scale. A number of special observations of interest were made almost incidentally during this survey: the remarkable variable star η *Argus* and the nebula surrounding it (a modern photograph of which is reproduced in fig. 100), the wonderful collections of nebulae clusters and stars, known as the *Nubeculae* or *Magellanic Clouds*, and Halley's comet were studied in turn; and the two faintest satellites of Saturn then known (chapter XII., § 255) were seen again for the first time since the death of their discoverer.

An important investigation of a somewhat different character—that of the amount of heat received from the sun—was also carried out (1837) during Herschel's residence at the Cape; and the result agreed satisfactorily with that of an independent inquiry made at the same time in France by *Claude Servais Mathias Pouillet* (1791–1868). In both cases the heat received on a given area of the earth in a given time from direct sunshine was measured; and allowance being made for the heat stopped in the atmosphere as the sun's rays passed through it, an estimate was formed of the total amount of heat received annually by the earth from the sun, and hence of the total amount radiated by the sun in all directions, an insignificant fraction of which (one part in 2,000,000,000) is alone intercepted by the earth. But the allowance for the heat intercepted in our atmosphere was necessarily uncertain, and later work, in particular that of Dr. *S. P. Langley* in 1880–81, shews that it was very much under-estimated by both Herschel and Pouillet. According to Herschel's results, the heat received annually from the sun—including that intercepted in the atmosphere—would be sufficient to melt a shell of ice 120 feet thick covering the whole earth; according to Dr. Langley, the thickness would be about 160 feet.*

308. With his return to England in 1838 Herschel's career as an observer came to an end; but the working out of the results of his Cape observations, the arrangement and cataloguing of his own and his father's discoveries, provided occupation for many years. A magnificent volume on the *Results of Astronomical Observations made during the years* 1834–8 *at the Cape of Good Hope* appeared in 1847; and a catalogue of all known nebulae and clusters, amounting to 5,079, was presented to the Royal Society in 1864, while a corresponding catalogue of more than 10,000 double and multiple stars was never finished, though the materials collected for it were published posthumously in 1879. John Herschel's great catalogue of nebulae has since been revised and enlarged by Dr. *Dreyer*, the result being a list of 7,840 nebulae and clusters known up to the end of 1887; and a supplementary list of discoveries made in 1888–94 published by the same writer contains 1,529 entries, so that the total number now known is between 9,000 and 10,000, of which more than half have been discovered by the two Herschels.

309. Double stars have been discovered and studied by a number of astronomers besides the Herschels. One of the most indefatigable workers at this subject was the elder Struve (§ 279), who was successively director of the two Russian observatories of Dorpat and Pulkowa. He observed altogether some 2,640 double and multiple stars, measuring in each case with care the length and direction of the line joining the two components, and noting other peculiarities, such as contrasts in colour between the members of a pair. He paid attention only to double stars the two components of which were not more than 32″ apart, thus rejecting a good many which William Herschel would have noticed; as the number of known doubles rapidly increased, it was clearly necessary to concentrate attention on those which might with some reasonable degree of

* Observations made on Mont Blanc under the direction of M. Janssen in 1897 indicate a slightly larger number than Dr. Langley's.

probability turn out to be genuine binaries (chapter XII., § 264).

In addition to a number of minor papers Struve published three separate books on the subject in 1827, 1837, and 1852.* A comparison of his own earlier and later observations, and of both with Herschel's earlier ones, shewed about 100 cases of change of relative positions of two members of a pair, which indicated more or less clearly a motion of revolution, and further results of a like character have been obtained

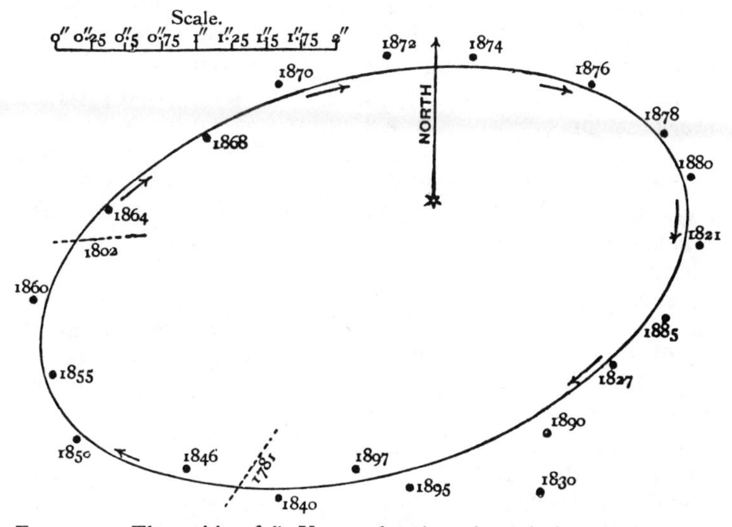

FIG. 101.—The orbit of ξ *Ursae*, shewing the relative positions of the two components at various times between 1781 and 1897. (The observations of 1781 and 1802 were only enough to determine the direction of the line joining the two components, not its length.)

from a comparison of Struve's observations with those of later observers.

William Herschel's observations of binary systems (chapter XII., § 264) only sufficed to shew that a motion of revolution of some kind appeared to be taking place; it was an obvious conjecture that the two members of a pair

* *Catalogus novus stellarum duplicium*, *Stellarum duplicium et multiplicium mensurae micrometricae*, and *Stellarum fixarum imprimis duplicium et multiplicium positiones mediae pro epocha* 1830.

attracted one another according to the law of gravitation, so that the motion of revolution was to some extent analogous to that of a planet round the sun; if this were the case, then each star of a pair should describe an ellipse (or conceivably some other conic) round the other, or each round the common centre of gravity, in accordance with Kepler's laws, and the apparent path as seen on the sky should be of this nature but in general foreshortened by being projected on to the celestial sphere. The first attempt to shew that this was actually the case was made by *Felix Savary* (1797–1841) in 1827, the star being ξ *Ursae*, which was found to be revolving in a period of about 60 years.

Many thousand double stars have been discovered by the Herschels, Struve, and a number of other observers, including several living astronomers, among whom Professor *S. W. Burnham* of Chicago, who has discovered some 1300, holds a leading place. Among these stars there are about 300 which we have fair reason to regard as binary, but not more than 40 or 50 of the orbits can be regarded as at all satisfactorily known. One of the most satisfactory is that of Savary's star ξ *Ursae*, which is shewn in fig. 101. Apart from the binaries discovered by the spectroscopic method (§ 314), which form to some extent a distinct class, the periods of revolution which have been computed range between about ten years and several centuries, the longer periods being for the most part decidedly uncertain.

310. William Herschel's telescopes represented for some time the utmost that could be done in the construction of reflectors; the first advance was made by Lord *Rosse* (1800–1867), who—after a number of less successful experiments—finally constructed (1845), at Parsonstown in Ireland, a reflecting telescope nearly 60 feet in length, with a mirror which was six feet across, and had consequently a "light-grasp" more than double that of Herschel's greatest telescope. Lord Rosse used the new instrument in the first instance to re-examine a number of known nebulae, and in the course of the next few years discovered a variety of new features, notably the spiral form of certain nebulae (fig. 102), and the resolution into apparent star clusters of a number of nebulae which Herschel had been unable to resolve

Fig. 102.—Spiral nebulae. From drawings by Lord Rosse.

[*To face p.* 400.

and had accordingly put into "the shining fluid" class (chapter XII., § 260). This last discovery, being exactly analogous to Herschel's experience when he first began to examine nebulae hitherto only observed with inferior telescopes, naturally led to a revival of the view that nebulae are indistinguishable from clusters of stars, though many of the arguments from probability urged by Herschel and others were in reality unaffected by the new discoveries.

311. The question of the status of nebulae in its simplest form may be said to have been settled by the first application of spectrum analysis. Fraunhofer (§ 299) had seen as early as 1823 that stars had spectra characterised like that of the sun by dark lines, and more complete investigations made soon after Kirchhoff's discoveries by several astronomers, in particular by Sir William Huggins and by the eminent Jesuit astronomer *Angelo Secchi* (1818–1878), confirmed this result as regards nearly all stars observed.

The first spectrum of a nebula was obtained by Sir William Huggins in 1864, and was seen to consist of three *bright* lines; by 1868 he had examined 70, and found in about one-third of the cases, including that of the Orion nebula, a similar spectrum of bright lines. In these cases therefore the luminous part of the nebula is gaseous, and Herschel's suggestion of a "shining fluid" was confirmed in the most satisfactory way. In nearly all cases three bright lines are seen, one of which is a hydrogen line, while the other two have not been identified, and in the case of a few of the brighter nebulae some other lines have also been seen. On the other hand, a considerable number of nebulae, including many of those which appear capable of telescopic resolution into star clusters, give a continuous spectrum, so that there is no clear spectroscopic evidence to distinguish them from clusters of stars, since the dark lines seen usually in the spectra of the latter could hardly be expected to be visible in the case of such faint objects as nebulae.

312. Stars have been classified, first by Secchi (1863), afterwards in slightly different ways by others, according to the general arrangement of the dark lines in their spectra; and some attempts have been made to base on these

differences inferences as to the relative "ages," or at any rate the stages of development, of different stars.

Many of the dark lines in the spectra of stars have been identified, first by Sir William Huggins in 1864, with the lines of known terrestrial elements, such as hydrogen, iron, sodium, calcium; so that a certain identity between the materials of which our own earth is made and that of bodies so remote as the fixed stars is thus established.

In addition to the classes of stars already mentioned, the spectroscope has shewn the existence of an extremely interesting if rather perplexing class of stars, falling into several subdivisions, which seem to form a connecting link between ordinary stars and nebulae, for, though indistinguishable telescopically from ordinary stars, their spectra shew *bright* lines either periodically or regularly. A good many stars of this class are variable, and several "new" stars which have appeared and faded away of late years have shewn similar characteristics.

313. The first application to the fixed stars of the spectroscopic method (§ 302) of determining motion towards or away from the observer was made by Sir William Huggins in 1868. A minute displacement from its usual position of a dark hydrogen line (F) in the spectrum of Sirius was detected, and interpreted as shewing that the star was receding from the solar system at a considerable speed. A number of other stars were similarly observed in the following year, and the work has been taken up since by a number of other observers, notably at Potsdam under the direction of Professor *H. C. Vogel*, and at Greenwich.

314. A very remarkable application of this method to binary stars has recently been made. If two stars are revolving round one another, their motions towards and away from the earth are changing regularly and are different; hence, if the light from both stars is received in the spectroscope, two spectra are formed—one for each star—the lines of which shift regularly relatively to one another. If a particular line, say the F line, common to the spectra of both stars, is observed when both stars are moving towards (or away from) the earth at the same rate—which happens twice in each revolution—only one line is seen; but when they are moving differently, if the spectroscope

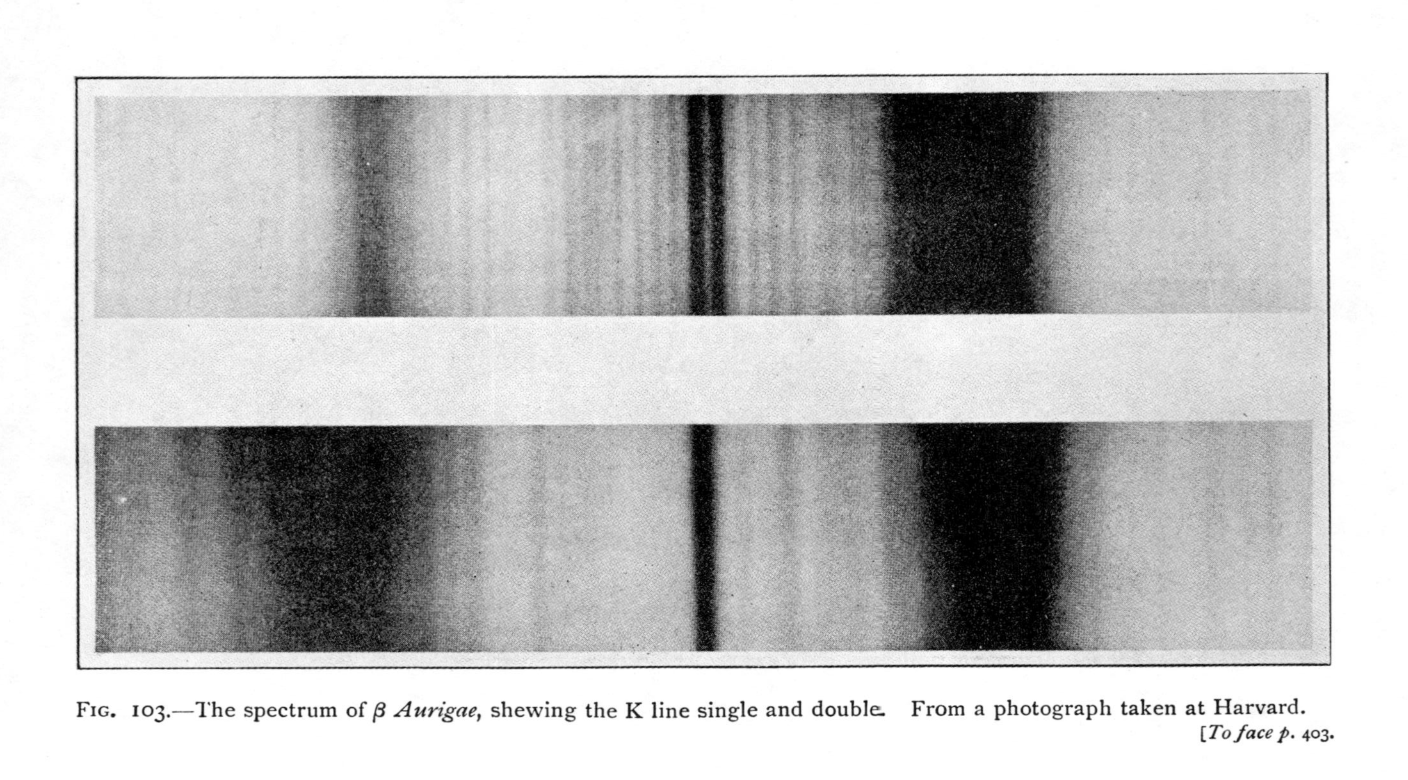

FIG. 103.—The spectrum of β *Aurigae*, shewing the K line single and double. From a photograph taken at Harvard.

[*To face p.* 403.

be powerful enough to detect the minute quantity involved, the line will appear doubled, one component being due to one star and one to the other. A periodic doubling of this kind was detected at the end of 1889 by Professor *E. C. Pickering* of Harvard in the case of ζ *Ursae*, which was thus for the first time shewn to be binary, and found to have the remarkably short period of only 104 days. This discovery was followed almost immediately by Professor Vogel's detection of a periodical shift in the position of the dark lines in the spectrum of the variable star Algol (chapter XII., § 266); but as in this case no doubling of the lines can be seen, the inference is that the companion star is nearly or quite dark, so that as the two revolve round one another the spectrum of the bright star shifts in the manner observed. Thus the eclipse-theory of Algol's variability received a striking verification.

A number of other cases of both classes of spectroscopic binary stars (as they may conveniently be called) have since been discovered. The upper part of fig. 103 shews the doubling of one of the lines in the spectrum of the double star β *Aurigae*; and the lower part shews the corresponding part of the spectrum at a time when the line appeared single.

315. Variable stars of different kinds have received a good deal of attention during this century, particularly during the last few years. About 400 stars are now clearly recognised as variable, while in a large number of other cases variability of light has been suspected; except, however, in a few cases, like that of Algol, the causes of variability are still extremely obscure.

316. The study of the relative brightness of stars—a branch of astronomy now generally known as stellar **photometry**—has also been carried on extensively during the century and has now been put on a scientific basis. The traditional classification of stars into magnitudes, according to their brightness, was almost wholly arbitrary, and decidedly uncertain. As soon as exact quantitative comparisons of stars of different brightness began to be carried out on a considerable scale, the need of a more precise system of classification became felt. John Herschel was one of the pioneers in this direction; he suggested a scale

capable of precise expression, and agreeing roughly, at any rate as far as naked-eye stars are concerned, with the current usages; while at the Cape he measured carefully the light of a large number of bright stars and classified them on this principle. According to the scale now generally adopted, first suggested in 1856 by *Norman Robert Pogson* (1829–1891), the light of a star of any magnitude bears a fixed ratio (which is taken to be 2·512...) to that of a star of the next magnitude. The number is so chosen that a star of the sixth magnitude—thus defined—is 100 times fainter than one of the first magnitude.* Stars of intermediate brightness have magnitudes expressed by fractions which can be at once calculated (according to a simple mathematical rule) when the ratio of the light received from the star to that received from a standard star has been observed.†

Most of the great star catalogues (§ 280) have included estimates of the magnitudes of stars. The most extensive and accurate series of measurements of star brightness have been those executed at Harvard and at Oxford under the superintendence of Professor E. C. Pickering and the late Professor Pritchard respectively. Both catalogues deal with stars visible to the naked eye; the Harvard catalogue (published in 1884) comprises 4,260 stars between the North Pole and 30° southern declination, and the *Uranometria Nova Oxoniensis* (1885), as it is called, only goes 10° south of the equator and includes 2,784 stars. Portions of more extensive catalogues dealing with fainter stars, in progress at Harvard and at Potsdam, have also been published.

* *I.e.* 2·512... is chosen as being the number the logarithm of which is ·4, so that $(2{\cdot}512...)^{5/2} = 10$.

† If L be the ratio of the light received from a star to that received from a standard first magnitude star, such as Aldebaran or Altair, then its magnitude *m* is given by the formula

$$L = \left(\frac{1}{2{\cdot}512}\right)^{m-1} = \left(\frac{1}{100}\right)^{\frac{m-1}{5}}, \text{ whence } m - 1 = -\frac{5}{2}\log L.$$

A star brighter than Aldebaran has a magnitude less than 1, while the magnitude of Sirius, which is about nine times as bright as Aldebaran, is a *negative* quantity, – 1·4, according to the Harvard photometry.

317. The great problem to which Herschel gave so much attention, that of the general arrangement of the stars and the structure of the system, if any, formed by them and the nebulae, has been affected in a variety of ways by the additions which have been made to our knowledge of the stars. But so far are we from any satisfactory solution of the problem that no modern theory can fairly claim to represent the facts now known to us as well as Herschel's earlier theory fitted the much scantier stock which he had at his command. In this as in so many cases an increase of knowledge has shewn the insufficiency of a previously accepted theory, but has not provided a successor. Detailed study of the form of the Milky Way (cf. fig. 104) and of its relation to the general body of stars has shewn the inadequacy of any simple arrangement of stars to represent its appearance; William Herschel's cloven grindstone, the ring which his son was inclined to substitute for it as the result of his Cape studies, and the more complicated forms which later writers have suggested, alike fail to account for its peculiarities. Again, such evidence as we have of the distance of the stars, when compared with their brightness, shews that there are large variations in their actual sizes as well as in their apparent sizes, and thus tells against the assumption of a certain uniformity which underlay much of Herschel's work. The "island universe" theory of nebulae, partially abandoned by Herschel after 1791 (chapter XII., § 260), but brought into credit again by Lord Rosse's discoveries (§ 310), scarcely survived the spectroscopic proof of the gaseous character of certain nebulae. Other evidence has pointed clearly to intimate relations between nebulae and stars generally; Herschel's observation that nebulae are densest in regions farthest from the Milky Way has been abundantly verified —as far as irresoluble nebulae are concerned—while obvious star clusters shew an equally clear preference for the neighbourhood of the Milky Way. In many cases again individual stars or groups seen on the sky in or near a nebula have been clearly shewn, either by their arrangement or in some cases by peculiarities of their spectra, to be really connected with the nebula, and not merely to be accidentally in the same direction. Stars which have bright lines

in their spectra (§ 312) form another link connecting nebulae with stars.

A good many converging lines of evidence thus point to a greater variety in the arrangement, size, and structure of the bodies with which the telescope makes us acquainted than seemed probable when sidereal astronomy was first seriously studied; they also indicate the probability that these bodies should be regarded as belonging to a single system, even if it be of almost inconceivable complexity, rather than to a number of perfectly distinct systems of a simpler type.

318. Laplace's nebular hypothesis (chapter XI., § 250) was published a little more than a century ago (1796), and has been greatly affected by progress in various departments of astronomical knowledge. Subsequent discoveries of planets and satellites (§§ 294, 295) have marred to some extent the uniformity and symmetry of the motions of the solar system on which Laplace laid so much stress; but it is not impossible to give reasonable explanations of the backward motions of the satellites of the two most distant planets, and of the large eccentricity and inclination of the paths of some of the minor planets, while apart from these exceptions the number of bodies the motions of which have the characteristics which Laplace pointed out has been considerably increased. The case for some sort of common origin of the bodies of the solar system has perhaps in this way gained as much as it has lost. Again, the telescopic evidence which Herschel adduced (chapter XII., § 261) in favour of the existence of certain processes of condensation in nebulae has been strengthened by later evidence of a similar character, and by the various pieces of evidence already referred to which connect nebulae with single stars and with clusters. The differences in the spectra of stars also receive their most satisfactory explanation as representing different stages of condensation of bodies of the same general character.

319. An entirely new contribution to the problem has resulted from certain discoveries as to the nature of heat, culminating in the recognition (about 1840–50) of heat as only one form of what physicists now call **energy**, which manifests itself also in the motion of bodies, in the

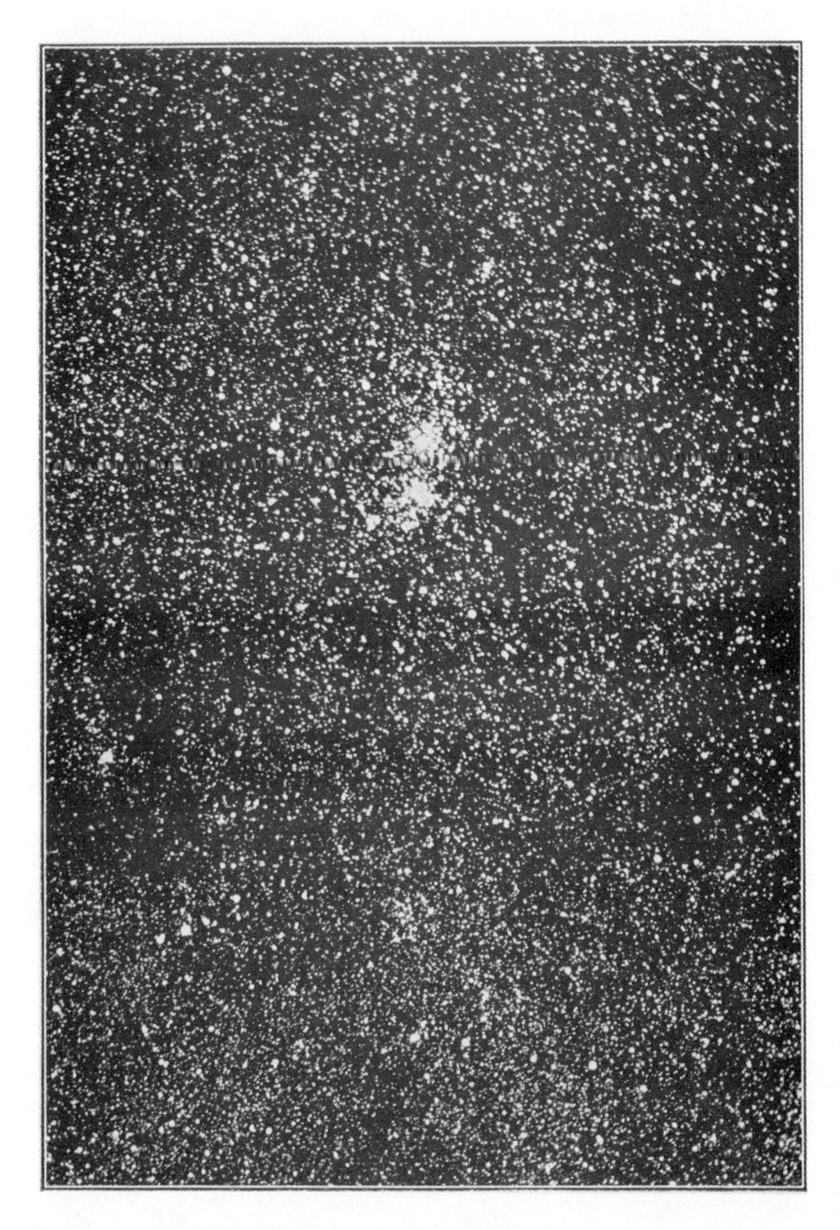

Fig. 104.—The Milky Way near the cluster in Perseus. From a photograph by Professor Barnard.

separation of bodies which attract one another, as well as in various electrical, chemical, and other ways. With this discovery was closely connected the general theory known as the **conservation of energy**, according to which energy, though capable of many transformations, can neither be increased nor decreased in quantity. A body which, like the sun, is giving out heat and light is accordingly thereby losing energy, and is like a machine doing work; either then it is receiving energy from some other source to compensate this loss or its store of energy is diminishing. But a body which goes on indefinitely giving out heat and light without having its store of energy replenished is exactly analogous to a machine which goes on working indefinitely without any motive power to drive it; and both are alike impossible.

The results obtained by John Herschel and Pouillet in 1836 (§ 307) called attention to the enormous expenditure of the sun in the form of heat, and astronomers thus had to face the problem of explaining how the sun was able to go on radiating heat and light in this way. Neither in the few thousand years of the past covered by historic records, nor in the enormously great periods of which geologists and biologists take account, is there any evidence of any important permanent alteration in the amount of heat and light received annually by the earth from the sun. Any theory of the sun's heat must therefore be able to account for the continual expenditure of heat at something like the present rate for an immense period of time. The obvious explanation of the sun as a furnace deriving its heat from combustion is found to be totally inadequate when put to the test of figures, as the sun could in this way be kept going at most for a few thousand years. The explanation now generally accepted was first given by the great German physicist *Hermann von Helmholtz* (1821–1894) in a popular lecture in 1854. The sun possesses an immense store of energy in the form of the mutual gravitation of its parts; if from any cause it shrinks, a certain amount of gravitational energy is necessarily lost and takes some other form. In the shrinkage of the sun we have therefore a possible source of energy. The precise amount of energy liberated by a definite amount of shrinkage of the sun depends upon the internal distribution of density in the sun, which is uncertain, but making any reasonable assumption as to this we find that the amount of shrinking required to supply the sun's expenditure of heat would only diminish the diameter by a few hundred feet annually, and would therefore be imperceptible with our present telescopic power for centuries, while no earlier records of the sun's size are accurate enough to shew it. It is easy to calculate on the same principles the amount of energy liberated by a body like the sun in shrinking from an indefinitely diffused condition to its present state, and from its present state to one of assigned greater density; the result being that we can in this way account for an expenditure of sun-heat at the present rate for a period to be counted in millions of years in either past or future time, while if the rate of expenditure was less in the remote past or becomes less in the future the time is extended to a corresponding extent.

No other cause that has been suggested is competent to account for more than a small fraction of the actual heat-expenditure of the sun; the gravitational theory satisfies all the requirements of astronomy proper, and goes at any rate some way towards meeting the demands of biology and geology.

If then we accept it as provisionally established, we are led to the conclusion that the sun was in the past larger and less condensed than now, and by going sufficiently far back into the past we find it in a condition not unlike the primitive nebula which Laplace presupposed, with the exception that it need not have been hot.

320. A new light has been thrown on the possible development of the earth and moon by Professor G. H. Darwin's study of the effects of tidal friction (cf. § 287 and §§ 292, 293). Since the tides increase the length of the day and month and gradually repel the moon from the earth, it follows that in the past the moon was nearer to the earth than now, and that tidal action was consequently much greater. Following out this clue, Professor Darwin found, by a series of elaborate calculations published in 1879–81, strong evidence of a past time when the moon was close to the earth, revolving round it in the same time

in which the earth rotated on its axis, which was then a little over two hours. The two bodies, in fact, were moving as if they were connected; it is difficult to avoid the probable inference that at an earlier stage the two really were one, and that the moon is in reality a fragment of the earth driven off from it by the too-rapid spinning of the earth, or otherwise.

Professor Darwin has also examined the possibility of explaining in a similar way the formation of the satellites of the other planets and of the planets themselves from the sun, but the circumstances of the moon-earth system turn out to be exceptional, and tidal influence has been less effective in other cases, though it gives a satisfactory explanation of certain peculiarities of the planets and their satellites. More recently (1892) Dr. *See* has applied a somewhat similar line of reasoning to explain by means of tidal action the development of double stars from an earlier nebulous condition.

Speaking generally, we may say that the outcome of the 19th century study of the problem of the early history of the solar system has been to discredit the details of Laplace's hypothesis in a variety of ways, but to establish on a firmer basis the general view that the solar system has been formed by some process of condensation out of an earlier very diffused mass bearing a general resemblance to one of the nebulae which the telescope shews us, and that stars other than the sun are not unlikely to have been formed in a somewhat similar way; and, further, the theory of tidal friction supplements this general but vague theory, by giving a rational account of a process which seems to have been the predominant factor in the development of the system formed by our own earth and moon, and to have had at any rate an important influence in a number of other cases.

AUTHORITIES AND BOOKS FOR STUDENTS.

I. GENERAL.

I HAVE made great use throughout of R. Wolf's *Geschichte der Astronomie*, and of the six volumes of Delambre's *Histoire de l'Astronomie* (*Ancienne*, 2 vols.; *du Moyen Age*, 1 vol.; *Moderne*, 2 vols.; *du Dixhuitième Siècle*, 1 vol.). I shall subsequently refer to these books simply as *Wolf* and *Delambre* respectively. I have used less often the astronomical sections of Whewell's *History of the Inductive Sciences* (referred to as *Whewell*), and I am indebted—chiefly for dates and references —to the histories of mathematics written respectively by Marie, W. W. R. Ball, and Cajori, to Poggendorff's *Handwörterbuch der Exacten Wissenschaften*, and to articles in various biographical dictionaries, encyclopaedias, and scientific journals. Of general treatises on astronomy Newcomb's *Popular Astronomy*, Young's *General Astronomy*, and Proctor's *Old and New Astronomy* have been the most useful for my purposes.

It is difficult to make a selection among the very large number of books on astronomy which are adapted to the general reader. For students who wish for an introductory account of astronomy the Astronomer Royal's *Primer of Astronomy* may be recommended; Young's *Elements of Astronomy* is a little more advanced, and Sir R. S. Ball's *Story of the Heavens*, Newcomb's *Popular Astronomy*, and Proctor's *Old and New Astronomy* enter into the subject in much greater detail. Young's *General Astronomy* may also be recommended to those who are not afraid of a little mathematics. There are also three modern English books dealing generally with the history of astronomy, in all of which the biographical element is much more prominent than in this book: viz. Sir R. S. Ball's *Great Astronomers*, Lodge's *Pioneers of Science*, and Morton's *Heroes of Science: Astronomers*.

II. Special Periods.

Chapters I. and II.—In addition to the general histories quoted above—especially *Wolf*—I have made most use of Tannery's *Recherches sur l'Histoire de l'Astronomie Ancienne* and of several biographical articles (chiefly by De Morgan) in Smith's *Dictionary of Classical Biography and Mythology*. Ideler's *Chronologische Untersuchungen*, Hankel's *Geschichte der Mathematik im Alterthum und Mittelalter*, G. C. Lewis's *Astronomy of the Ancients*, and Epping & Strassmaier's *Astronomisches aus Babylon* have also been used to some extent. Unfortunately my attention was only called to Susemihl's *Geschichte der Griechischen Litteratur in der Alexandriner Zeit* when most of my book was in proof, and I have consequently been able to make but little use of it.

I have in general made no attempt to consult the original Greek authorities, but I have made some use of translations of Aristarchus, of the *Almagest*, and of the astronomical writings of Plato and Aristotle.

Chapter III.—The account of Eastern astronomy is based chiefly on Delambre, and on Hankel's *Geschichte der Mathematik im Alterthum und Mittelalter*; to a less extent on Whewell. For the West I have made more use of Whewell, and have borrowed biographical material for the English writers from the *Dictionary of National Biography*. I have also consulted a good many of the original astronomical books referred to in the latter part of the chapter.

I know of no accessible book in English to which to refer students except *Whewell*.

Chapter IV.—For biographical material, for information as to the minor writings, and as to the history of the publication of the *De Revolutionibus* I have used little but Prowe's elaborate *Nicolaus Coppernicus*, and the documents printed in it. My account of the *De Revolutionibus* is taken from the book itself. The portrait is taken from Dandeleau's engraving of a picture in Lalande's possession. I have not been able to discover any portrait which was clearly made during Coppernicus's lifetime, but the close resemblance between several portraits dating from the 17th century and Dandeleau's seems to shew that the latter is substantially authentic.

There is a readable account of Coppernicus, as well as of several other astronomers, in Bertrand's *Fondateurs de l'Astronomie Moderne*; but I have not used the book as an authority.

Chapter V.—For the life of Tycho I have relied chiefly on Dreyer's *Tycho Brahe*, which has also been used as a guide to his scientific work; but I have made constant reference to the original writings: I have also made some use of Gassendi's *Vita*

Tychonis Brahe. The portrait is a reproduction of a picture in the possession of Dr. Crompton of Manchester, described by him in the *Memoirs of the Manchester Literary and Philosophical Society*, Vol. VI., Ser. III. For minor Continental writers I have used chiefly *Wolf* and *Delambre*, and for English writers, *Whewell*, various articles by De Morgan quoted by him, and articles in the *Dictionary of National Biography.*

Students will find in Dreyer's book all that they are likely to want to know about Tycho.

Chapter VI.—For Galilei's life I have used chiefly Karl von Gebler's *Galilei und die Römische Curie*, partly in the original German form and partly in the later English edition (translated by Mrs. Sturge). For the disputed questions connected with the trial I have relied as far as possible on the original documents preserved in the Vatican, which have been published by von Gebler and independently by L'Épinois in *Les Pièces du Procès de Galilée:* in the latter book some of the most important documents are reproduced in facsimile. For personal characteristics I have used the charming *Private Life of Galileo, compiled chiefly from his correspondence and that of his daughter Marie Céleste.* I have also read with great interest the estimate of Galilei's work contained in H. Martin's *Galilée*, and have probably borrowed from it to some extent. What I have said about Galilei's scientific work has been based almost entirely on study of his own books, either in the original or in translation : I have used freely the translations of the *Dialogue on the Two Chief Systems of the World* and of the *Letter to the Grand Duchess Christine* by Salusbury, that of the *Two New Sciences* by Weston (as well as that by Salusbury), and that of the *Sidereal Messenger* by Carlos. I have also made some use of various controversial tracts written by enemies of Galilei, which are to be found (together with his comments on them) in the magnificent national edition of his works now in course of publication ; and of the critical account of Galilei's contributions to dynamics contained in Mach's *Geschichte der Mechanik.*

Wolf and *Delambre* have only been used to a very small extent in this chapter, chiefly for the minor writers who are referred to.

The portrait is a reproduction of one by Sustermans in the Uffizi Gallery.

There is an excellent popular account of Galilei's life and work in the *Lives of Eminent Persons* published by the Society for the Diffusion of Useful Knowledge ; students who want fuller accounts of Galilei's life should read Gebler's book and the *Private Life*, which have been already quoted, and are strongly recommended to read at any rate parts of the *Dialogue on the Two Chief Systems of the World*, either in the original or in the picturesque old translation by Salusbury: there is also a modern German version of this book, as well as of the *Two New Sciences*, in Ostwald's series of *Klassiker der exakten Wissenschaften*.

Chapter VII.—For Kepler's life I have used chiefly *Wolf* and the life—or rather biographical material—given by Frisch in the last volume of his edition of Kepler's works, also to a small extent Breitschwerdt's *Johann Keppler.* For Kepler's scientific discoveries I have used chiefly his own writings, but I am indebted to some extent to *Wolf* and *Delambre*, especially for information with regard to his minor works. The portrait is a reproduction of one by Nordling given in Frisch's edition.

The *Lives of Eminent Persons*, already referred to, also contains an excellent popular account of Kepler's life and work.

Chapter VIII.—I have used chiefly *Wolf* and *Delambre*; for the English writers Gascoigne and Horrocks I have used *Whewell* and articles in the *Dict. Nat. Biog.* What I have said about the work of Huygens is taken directly from the books of his which are quoted in the text; and for special points I have consulted the *Principia* of Descartes, and a very few of Cassini's extensive writings.

There is no obvious book to recommend to students.

Chapter IX.—For the external events of Newton's life I have relied chiefly on Brewster's *Memoirs of Sir Isaac Newton*; and for the history of the growth of his ideas on the subject of gravitation I have made extensive use of W. W. R. Ball's *Essay on Newton's Principia*, and of the original documents contained in it. I have also made some use of the articles on Newton in the *Encyclopaedia Britannica* and the *Dictionary of National Biography*; as well as of Rigaud's *Correspondence of Scientific Men of the Seventeenth Century*, of Edleston's *Correspondence of Sir Isaac Newton and Prof. Cotes*, and of Baily's *Account of the Revd. John Flamsteed.* The portrait is a reproduction of one by Kneller.

Students are recommended to read Brewster's book, quoted above, or the abridged *Life of Sir Isaac Newton* by the same author. The Laws of Motion are discussed in most modern text-books of dynamics; the best treatment that I am acquainted with of the various difficulties connected with them is in an article by W. H. Macaulay in the *Bulletin of the American Mathematical Society*, Ser. II., Vol. III., No. 10, July 1897.

Chapter X.—For Flamsteed I have used chiefly Baily's *Account of the Revd. John Flamsteed*; for Bradley little but the *Miscellaneous Works and Correspondence of the Rev. James*

Bradley (edited by Rigaud), from which the portrait has been taken. My account of Halley's work is based to a considerable extent on his own writings; there is a good deal of biographical information about him in the books already quoted in connection with Newton and Flamsteed, and there is a useful article on him in the *Dictionary of National Biography*. I have made a good deal of use in this chapter of *Wolf* and *Delambre*, especially in dealing with Continental astronomers; and for special parts of the subject I have used Grant's *History of Physical Astronomy*, Todhunter's *History of the Mathematical Theories of Attraction and the Figure of the Earth*, and Poynting's *Density of the Earth*.

Chapter XI.—Most of the biographical material has been taken from *Wolf*, from articles in various encyclopaedias and biographical dictionaries, chiefly French, and from Delambre's *Éloge* of Lagrange. The two portraits are taken respectively from Serret's edition of the *Oeuvres de Lagrange* and from the Academy's edition of the *Oeuvres Complètes de Laplace*. Gautier's *Essai Historique sur le Problème des Trois Corps* and Grant's *History of Physical Astronomy* have been the books most used for my account of the scientific contributions of the various astronomers dealt with; I have also consulted various modern treatises on gravitational astronomy, especially Tisserand's *Mécanique Céleste*, Brown's *Lunar Theory*, and to a less extent Cheyne's *Planetary Theory* and Airy's *Gravitation*. For special points I have used Todhunter's *History*, already referred to. Of the original writings I have made a good deal of use of Laplace's *Mécanique Céleste* as well as of his *Système du Monde*; I have also consulted a certain number of his other writings and of those of Lagrange and Clairaut; but have made no systematic study of them.

Students who wish to know more about gravitational astronomy but have little knowledge of mathematics should try to read Airy's *Gravitation*; Herschel's *Outlines of Astronomy* and Grant's *History* (quoted above) also deal with the subject without employing mathematics, and are tolerably intelligible.

Chapter XII.—The account of Herschel's career is taken chiefly from Mrs. John Herschel's *Memoir of Caroline Herschel*, from Miss A. M. Clerke's *The Herschels and Modern Astronomy*, from the *Popular History of Astronomy in the Nineteenth Century* by the same author, and from Holden's *Sir William Herschel, his Life and Works*. The last three books and the *Synopsis and Subject Index to the Writings of Sir William Herschel* by Holden & Hastings have been my chief guides to Herschel's long series of papers; but nearly everything that I have said about his chief pieces of work is based on his own writings. I have made also some little use of Grant's *History* (already quoted), of *Wolf*, and of Miss Clerke's *System of the Stars*.

Students are recommended to read any or all of the first four books named above; the *Memoir* gives a charming picture of Herschel's personal life and especially of his relations with his sister. There is also a good critical account of Herschel's work on sidereal astronomy in Proctor's *Old and New Astronomy*.

Chapter XIII.—Except in the articles dealing with gravitational astronomy I have constantly used Miss Clerke's *History* (already quoted), a book which students are strongly recommended to read; and in dealing with the first half of the century I have been helped a good deal by Grant's *History*. But for the most part the materials for the chapter have been drawn from a great number of sources—consisting very largely of the original writings of the astronomers referred to—which it would be difficult and hardly worth while to enumerate; for the lives of astronomers (especially of English ones), as well as for recent astronomical history generally, I have been much helped by the obituary notices and the reports on the progress of astronomy which appear annually in the *Monthly Notices* of the Royal Astronomical Society.

I add the names of a few books which deal with special parts of modern astronomy in a non-technical way:—

The Sun, C. A. Young; *The Sun*, R. A. Proctor; *The Story of the Sun*, R. S. Ball; *The Sun's Place in Nature*, J. N. Lockyer.

The Moon, E. Neison; *The Moon*, T. G. Elger.

Saturn and its System, R. A. Proctor.

Mars, Percival Lowell.

The World of Comets, A. Guillemin (a well-illustrated but uncritical book, now rather out of date); *Remarkable Comets*, W. T. Lynn (a very small book full of useful information); *The Great Meteoritic Shower of November*, W. F. Denning.

The Tides and Kindred Phenomena in the Solar System, G. H. Darwin.

Remarkable Eclipses, W. T. Lynn (of the same character as his book on Comets.)

The System of the Stars, A. M. Clerke.

Spectrum Analysis, H. Schellen; *Spectrum Analysis*, H. E. Roscoe.

INDEX OF NAMES.

Roman figures refer to the chapters, Arabic to the articles. The numbers given in brackets after the name of an astronomer are the dates of birth and death. All dates are A.D. *unless otherwise stated. In cases in which an author's name occurs in several articles, the numbers of the articles in which the principal account of him or of his work is given are printed in clarendon type thus :* **286.** *The names of living astronomers are italicised.*

[*Roman figures refer to the chapters, Arabic to the articles.*]

[*Roman figures refer to the chapters, Arabic to the articles.*]

[*Roman figures refer to the chapters, Arabic to the articles.*]

[*Roman figures refer to the chapters, Arabic to the articles.*]

[*Roman figures refer to the chapters, Arabic to the articles.*]

[*Roman figures refer to the chapters, Arabic to the articles.*]

[Roman figures refer to the chapters, Arabic to the articles.]

GENERAL INDEX.

Roman figures refer to the chapters, Arabic to the articles. When several articles are given under one heading the numbers of the most important are printed in clarendon type, thus: **207**. *The names of books are printed in italics.*

[*Roman figures refer to the chapters, Arabic to the articles.*]

[*Roman figures refer to the chapters, Arabic to the articles.*]

[*Roman figures refer to the chapters, Arabic to the articles.*]

[*Roman figures refer to the chapters, Arabic to the articles.*]

[*Roman figures refer to the chapters, Arabic to the articles.*]

[*Roman figures refer to the chapters, Arabic to the articles.*]

[*Roman figures refer to the chapters, Arabic to the articles.*]

[*Roman figures refer to the chapters, Arabic to the articles.*]

[*Roman figures refer to the chapters, Arabic to the articles.*]

[*Roman figures refer to the chapters, Arabic to the articles.*]

[*Roman figures refer to the chapters Arabic to the articles.*]

[*Roman figures refer to the chapters, Arabic to the articles.*]

[*Roman figures refer to the chapters, Arabic to the articles.*]

[*Roman figures refer to the chapters, Arabic to the articles.*]

[Roman figures refer to the chapters, Arabic to the articles.]

A CATALOGUE OF SELECTED DOVER BOOKS IN ALL FIELDS OF INTEREST

A CATALOGUE OF SELECTED DOVER BOOKS IN ALL FIELDS OF INTEREST

AMERICA'S OLD MASTERS, James T. Flexner. Four men emerged unexpectedly from provincial 18th century America to leadership in European art: Benjamin West, J. S. Copley, C. R. Peale, Gilbert Stuart. Brilliant coverage of lives and contributions. Revised, 1967 edition. 69 plates. 365pp. of text.
21806-6 Paperbound $3.00

FIRST FLOWERS OF OUR WILDERNESS: AMERICAN PAINTING, THE COLONIAL PERIOD, James T. Flexner. Painters, and regional painting traditions from earliest Colonial times up to the emergence of Copley, West and Peale Sr., Foster, Gustavus Hesselius, Feke, John Smibert and many anonymous painters in the primitive manner. Engaging presentation, with 162 illustrations. xxii + 368pp.
22180-6 Paperbound $3.50

THE LIGHT OF DISTANT SKIES: AMERICAN PAINTING, 1760-1835, James T. Flexner. The great generation of early American painters goes to Europe to learn and to teach: West, Copley, Gilbert Stuart and others. Allston, Trumbull, Morse; also contemporary American painters—primitives, derivatives, academics—who remained in America. 102 illustrations. xiii + 306pp. 22179-2 Paperbound $3.00

A HISTORY OF THE RISE AND PROGRESS OF THE ARTS OF DESIGN IN THE UNITED STATES, William Dunlap. Much the richest mine of information on early American painters, sculptors, architects, engravers, miniaturists, etc. The only source of information for scores of artists, the major primary source for many others. Unabridged reprint of rare original 1834 edition, with new introduction by James T. Flexner, and 394 new illustrations. Edited by Rita Weiss. 6⅝ x 9⅝.
21695-0, 21696-9, 21697-7 Three volumes, Paperbound $13.50

EPOCHS OF CHINESE AND JAPANESE ART, Ernest F. Fenollosa. From primitive Chinese art to the 20th century, thorough history, explanation of every important art period and form, including Japanese woodcuts; main stress on China and Japan, but Tibet, Korea also included. Still unexcelled for its detailed, rich coverage of cultural background, aesthetic elements, diffusion studies, particularly of the historical period. 2nd, 1913 edition. 242 illustrations. lii + 439pp. of text.
20364-6, 20365-4 Two volumes, Paperbound $6.00

THE GENTLE ART OF MAKING ENEMIES, James A. M. Whistler. Greatest wit of his day deflates Oscar Wilde, Ruskin, Swinburne; strikes back at inane critics, exhibitions, art journalism; aesthetics of impressionist revolution in most striking form. Highly readable classic by great painter. Reproduction of edition designed by Whistler. Introduction by Alfred Werner. xxxvi + 334pp.
21875-9 Paperbound $2.50

THE RED FAIRY BOOK, Andrew Lang. Lang's color fairy books have long been children's favorites. This volume includes Rapunzel, Jack and the Bean-stalk and 35 other stories, familiar and unfamiliar. 4 plates, 93 illustrations x + 367pp.
21673-X Paperbound $2.50

THE BLUE FAIRY BOOK, Andrew Lang. Lang's tales come from all countries and all times. Here are 37 tales from Grimm, the Arabian Nights, Greek Mythology, and other fascinating sources. 8 plates, 130 illustrations. xi + 390pp.
21437-0 Paperbound $2.50

HOUSEHOLD STORIES BY THE BROTHERS GRIMM. Classic English-language edition of the well-known tales — Rumpelstiltskin, Snow White, Hansel and Gretel, The Twelve Brothers, Faithful John, Rapunzel, Tom Thumb (52 stories in all). Translated into simple, straightforward English by Lucy Crane. Ornamented with headpieces, vignettes, elaborate decorative initials and a dozen full-page illustrations by Walter Crane. x + 269pp. 21080-4 Paperbound $2.50

THE MERRY ADVENTURES OF ROBIN HOOD, Howard Pyle. The finest modern versions of the traditional ballads and tales about the great English outlaw. Howard Pyle's complete prose version, with every word, every illustration of the first edition. Do not confuse this facsimile of the original (1883) with modern editions that change text or illustrations. 23 plates plus many page decorations. xxii + 296pp.
22043-5 Paperbound $2.50

THE STORY OF KING ARTHUR AND HIS KNIGHTS, Howard Pyle. The finest children's version of the life of King Arthur; brilliantly retold by Pyle, with 48 of his most imaginative illustrations. xviii + 313pp. $6\frac{1}{8}$ x $9\frac{1}{4}$.
21445-1 Paperbound $2.50

THE WONDERFUL WIZARD OF OZ, L. Frank Baum. America's finest children's book in facsimile of first edition with all Denslow illustrations in full color. The edition a child should have. Introduction by Martin Gardner. 23 color plates, scores of drawings. iv + 267pp. 20691-2 Paperbound $2.25

THE MARVELOUS LAND OF OZ, L. Frank Baum. The second Oz book, every bit as imaginative as the Wizard. The hero is a boy named Tip, but the Scarecrow and the Tin Woodman are back, as is the Oz magic. 16 color plates, 120 drawings by John R. Neill. 287pp. 20692-0 Paperbound $2.50

THE MAGICAL MONARCH OF MO, L. Frank Baum. Remarkable adventures in a land even stranger than Oz. The best of Baum's books not in the Oz series. 15 color plates and dozens of drawings by Frank Verbeck. xviii + 237pp.
21892-9 Paperbound $2.00

THE BAD CHILD'S BOOK OF BEASTS, MORE BEASTS FOR WORSE CHILDREN, A MORAL ALPHABET, Hilaire Belloc. Three complete humor classics in one volume. Be kind to the frog, and do not call him names . . . and 28 other whimsical animals. Familiar favorites and some not so well known. Illustrated by Basil Blackwell. 156pp. (USO) 20749-8 Paperbound $1.25

A HISTORY OF COSTUME, Carl Köhler. Definitive history, based on surviving pieces of clothing primarily, and paintings, statues, etc. secondarily. Highly readable text, supplemented by 594 illustrations of costumes of the ancient Mediterranean peoples, Greece and Rome, the Teutonic prehistoric period; costumes of the Middle Ages, Renaissance, Baroque, 18th and 19th centuries. Clear, measured patterns are provided for many clothing articles. Approach is practical throughout. Enlarged by Emma von Sichart. 464pp. 21030-8 Paperbound $3.50

ORIENTAL RUGS, ANTIQUE AND MODERN, Walter A. Hawley. A complete and authoritative treatise on the Oriental rug—where they are made, by whom and how, designs and symbols, characteristics in detail of the six major groups, how to distinguish them and how to buy them. Detailed technical data is provided on periods, weaves, warps, wefts, textures, sides, ends and knots, although no technical background is required for an understanding. 11 color plates, 80 halftones, 4 maps. vi + 320pp. 6⅛ x 9⅛. 22366-3 Paperbound $5.00

TEN BOOKS ON ARCHITECTURE, Vitruvius. By any standards the most important book on architecture ever written. Early Roman discussion of aesthetics of building, construction methods, orders, sites, and every other aspect of architecture has inspired, instructed architecture for about 2,000 years. Stands behind Palladio, Michelangelo, Bramante, Wren, countless others. Definitive Morris H. Morgan translation. 68 illustrations. xii + 331pp. 20645-9 Paperbound $2.50

THE FOUR BOOKS OF ARCHITECTURE, Andrea Palladio. Translated into every major Western European language in the two centuries following its publication in 1570, this has been one of the most influential books in the history of architecture. Complete reprint of the 1738 Isaac Ware edition. New introduction by Adolf Placzek, Columbia Univ. 216 plates. xxii + 110pp. of text. 9½ x 12¾. 21308-0 Clothbound $10.00

STICKS AND STONES: A STUDY OF AMERICAN ARCHITECTURE AND CIVILIZATION, Lewis Mumford.One of the great classics of American cultural history. American architecture from the medieval-inspired earliest forms to the early 20th century; evolution of structure and style, and reciprocal influences on environment. 21 photographic illustrations. 238pp. 20202-X Paperbound $2.00

THE AMERICAN BUILDER'S COMPANION, Asher Benjamin. The most widely used early 19th century architectural style and source book, for colonial up into Greek Revival periods. Extensive development of geometry of carpentering, construction of sashes, frames, doors, stairs; plans and elevations of domestic and other buildings. Hundreds of thousands of houses were built according to this book, now invaluable to historians, architects, restorers, etc. 1827 edition. 59 plates. 114pp. 7⅞ x 10¾. 22236-5 Paperbound $3.00

DUTCH HOUSES IN THE HUDSON VALLEY BEFORE 1776, Helen Wilkinson Reynolds. The standard survey of the Dutch colonial house and outbuildings, with constructional features, decoration, and local history associated with individual homesteads. Introduction by Franklin D. Roosevelt. Map. 150 illustrations. 469pp. 6⅝ x 9¼. 21469-9 Paperbound $4.00

THE ARCHITECTURE OF COUNTRY HOUSES, Andrew J. Downing. Together with Vaux's *Villas and Cottages* this is the basic book for Hudson River Gothic architecture of the middle Victorian period. Full, sound discussions of general aspects of housing, architecture, style, decoration, furnishing, together with scores of detailed house plans, illustrations of specific buildings, accompanied by full text. Perhaps the most influential single American architectural book. 1850 edition. Introduction by J. Stewart Johnson. 321 figures, 34 architectural designs. xvi + 560pp.
22003-6 Paperbound $4.00

LOST EXAMPLES OF COLONIAL ARCHITECTURE, John Mead Howells. Full-page photographs of buildings that have disappeared or been so altered as to be denatured, including many designed by major early American architects. 245 plates. xvii + 248pp. 7⅞ x 10¾. 21143-6 Paperbound $3.50

DOMESTIC ARCHITECTURE OF THE AMERICAN COLONIES AND OF THE EARLY REPUBLIC, Fiske Kimball. Foremost architect and restorer of Williamsburg and Monticello covers nearly 200 homes between 1620-1825. Architectural details, construction, style features, special fixtures, floor plans, etc. Generally considered finest work in its area. 219 illustrations of houses, doorways, windows, capital mantels. xx + 314pp. 7⅞ x 10¾. 21743-4 Paperbound $4.00

EARLY AMERICAN ROOMS: 1650-1858, edited by Russell Hawes Kettell. Tour of 12 rooms, each representative of a different era in American history and each furnished, decorated, designed and occupied in the style of the era. 72 plans and elevations, 8-page color section, etc., show fabrics, wall papers, arrangements, etc. Full descriptive text. xvii + 200pp. of text. 8⅜ x 11¼.
21633-0 Paperbound $5.00

THE FITZWILLIAM VIRGINAL BOOK, edited by J. Fuller Maitland and W. B. Squire. Full modern printing of famous early 17th-century ms. volume of 300 works by Morley, Byrd, Bull, Gibbons, etc. For piano or other modern keyboard instrument; easy to read format. xxxvi + 938pp. 8⅜ x 11.
21068-5, 21069-3 Two volumes, Paperbound $10.00

KEYBOARD MUSIC, Johann Sebastian Bach. Bach Gesellschaft edition. A rich selection of Bach's masterpieces for the harpsichord: the six English Suites, six French Suites, the six Partitas (Clavierübung part I), the Goldberg Variations (Clavierübung part IV), the fifteen Two-Part Inventions and the fifteen Three-Part Sinfonias. Clearly reproduced on large sheets with ample margins; eminently playable. vi + 312pp. 8⅛ x 11. 22360-4 Paperbound $5.00

THE MUSIC OF BACH: AN INTRODUCTION, Charles Sanford Terry. A fine, nontechnical introduction to Bach's music, both instrumental and vocal. Covers organ music, chamber music, passion music, other types. Analyzes themes, developments, innovations. x + 114pp. 21075-8 Paperbound $1.25

BEETHOVEN AND HIS NINE SYMPHONIES, Sir George Grove. Noted British musicologist provides best history, analysis, commentary on symphonies. Very thorough, rigorously accurate; necessary to both advanced student and amateur music lover. 436 musical passages. vii + 407 pp. 20334-4 Paperbound $2.75

POEMS OF ANNE BRADSTREET, edited with an introduction by Robert Hutchinson. A new selection of poems by America's first poet and perhaps the first significant woman poet in the English language. 48 poems display her development in works of considerable variety—love poems, domestic poems, religious meditations, formal elegies, "quaternions," etc. Notes, bibliography. viii + 222pp.
22160-1 Paperbound $2.00

THREE GOTHIC NOVELS: THE CASTLE OF OTRANTO BY HORACE WALPOLE; VATHEK BY WILLIAM BECKFORD; THE VAMPYRE BY JOHN POLIDORI, WITH FRAGMENT OF A NOVEL BY LORD BYRON, edited by E. F. Bleiler. The first Gothic novel, by Walpole; the finest Oriental tale in English, by Beckford; powerful Romantic supernatural story in versions by Polidori and Byron. All extremely important in history of literature; all still exciting, packed with supernatural thrills, ghosts, haunted castles, magic, etc. xl + 291pp.
21232-7 Paperbound $2.00

THE BEST TALES OF HOFFMANN, E. T. A. Hoffmann. 10 of Hoffmann's most important stories, in modern re-editings of standard translations: Nutcracker and the King of Mice, Signor Formica, Automata, The Sandman, Rath Krespel, The Golden Flowerpot, Master Martin the Cooper, The Mines of Falun, The King's Betrothed, A New Year's Eve Adventure. 7 illustrations by Hoffmann. Edited by E. F. Bleiler. xxxix + 419pp. 21793-0 Paperbound $2.50

GHOST AND HORROR STORIES OF AMBROSE BIERCE, Ambrose Bierce. 23 strikingly modern stories of the horrors latent in the human mind: The Eyes of the Panther, The Damned Thing, An Occurrence at Owl Creek Bridge, An Inhabitant of Carcosa, etc., plus the dream-essay, Visions of the Night. Edited by E. F. Bleiler. xxii + 199pp. 20767-6 Paperbound $1.50

BEST GHOST STORIES OF J. S. LEFANU, J. Sheridan LeFanu. Finest stories by Victorian master often considered greatest supernatural writer of all. Carmilla, Green Tea, The Haunted Baronet, The Familiar, and 12 others. Most never before available in the U. S. A. Edited by E. F. Bleiler. 8 illustrations from Victorian publications. xvii + 467pp. 20415-4 Paperbound $2.50

THE TIME STREAM, THE GREATEST ADVENTURE, AND THE PURPLE SAPPHIRE—THREE SCIENCE FICTION NOVELS, John Taine (Eric Temple Bell). Great American mathematician was also foremost science fiction novelist of the 1920's. *The Time Stream,* one of all-time classics, uses concepts of circular time; *The Greatest Adventure,* incredibly ancient biological experiments from Antarctica threaten to escape; The *Purple Sapphire,* superscience, lost races in Central Tibet, survivors of the Great Race. 4 illustrations by Frank R. Paul. v + 532pp.
21180-0 Paperbound $3.00

SEVEN SCIENCE FICTION NOVELS, H. G. Wells. The standard collection of the great novels. Complete, unabridged. *First Men in the Moon, Island of Dr. Moreau, War of the Worlds, Food of the Gods, Invisible Man, Time Machine, In the Days of the Comet.* Not only science fiction fans, but every educated person owes it to himself to read these novels. 1015pp. 20264-X Clothbound $5.00

AMERICAN FOOD AND GAME FISHES, David S. Jordan and Barton W. Evermann. Definitive source of information, detailed and accurate enough to enable the sportsman and nature lover to identify conclusively some 1,000 species and sub-species of North American fish, sought for food or sport. Coverage of range, physiology, habits, life history, food value. Best methods of capture, interest to the angler, advice on bait, fly-fishing, etc. 338 drawings and photographs. l + 574pp. 6⅝ x 9⅜.
22383-1 Paperbound $4.50

THE FROG BOOK, Mary C. Dickerson. Complete with extensive finding keys, over 300 photographs, and an introduction to the general biology of frogs and toads, this is the classic non-technical study of Northeastern and Central species. 58 species; 290 photographs and 16 color plates. xvii + 253pp.
21973-9 Paperbound $4.00

THE MOTH BOOK: A GUIDE TO THE MOTHS OF NORTH AMERICA, William J. Holland. Classical study, eagerly sought after and used for the past 60 years. Clear identification manual to more than 2,000 different moths, largest manual in existence. General information about moths, capturing, mounting, classifying, etc., followed by species by species descriptions. 263 illustrations plus 48 color plates show almost every species, full size. 1968 edition, preface, nomenclature changes by A. E. Brower. xxiv + 479pp. of text. 6½ x 9¼.
21948-8 Paperbound $5.00

THE SEA-BEACH AT EBB-TIDE, Augusta Foote Arnold. Interested amateur can identify hundreds of marine plants and animals on coasts of North America; marine algae; seaweeds; squids; hermit crabs; horse shoe crabs; shrimps; corals; sea anemones; etc. Species descriptions cover: structure; food; reproductive cycle; size; shape; color; habitat; etc. Over 600 drawings. 85 plates. xii + 490pp.
21949-6 Paperbound $3.50

COMMON BIRD SONGS, Donald J. Borror. 33⅓ 12-inch record presents songs of 60 important birds of the eastern United States. A thorough, serious record which provides several examples for each bird, showing different types of song, individual variations, etc. Inestimable identification aid for birdwatcher. 32-page booklet gives text about birds and songs, with illustration for each bird.
21829-5 Record, book, album. Monaural. $2.75

FADS AND FALLACIES IN THE NAME OF SCIENCE, Martin Gardner. Fair, witty appraisal of cranks and quacks of science: Atlantis, Lemuria, hollow earth, flat earth, Velikovsky, orgone energy, Dianetics, flying saucers, Bridey Murphy, food fads, medical fads, perpetual motion, etc. Formerly "In the Name of Science." x + 363pp.
20394-8 Paperbound $2.00

HOAXES, Curtis D. MacDougall. Exhaustive, unbelievably rich account of great hoaxes: Locke's moon hoax, Shakespearean forgeries, sea serpents, Loch Ness monster, Cardiff giant, John Wilkes Booth's mummy, Disumbrationist school of art, dozens more; also journalism, psychology of hoaxing. 54 illustrations. xi + 338pp.
20465-0 Paperbound $2.75

EAST O' THE SUN AND WEST O' THE MOON, George W. Dasent. Considered the best of all translations of these Norwegian folk tales, this collection has been enjoyed by generations of children (and folklorists too). Includes True and Untrue, Why the Sea is Salt, East O' the Sun and West O' the Moon, Why the Bear is Stumpy-Tailed, Boots and the Troll, The Cock and the Hen, Rich Peter the Pedlar, and 52 more. The only edition with all 59 tales. 77 illustrations by Erik Werenskiold and Theodor Kittelsen. xv + 418pp. 22521-6 Paperbound $3.50

GOOPS AND HOW TO BE THEM, Gelett Burgess. Classic of tongue-in-cheek humor, masquerading as etiquette book. 87 verses, twice as many cartoons, show mischievous Goops as they demonstrate to children virtues of table manners, neatness, courtesy, etc. Favorite for generations. viii + 88pp. 6½ x 9¼.
22233-0 Paperbound $1.25

ALICE'S ADVENTURES UNDER GROUND, Lewis Carroll. The first version, quite different from the final *Alice in Wonderland,* printed out by Carroll himself with his own illustrations. Complete facsimile of the "million dollar" manuscript Carroll gave to Alice Liddell in 1864. Introduction by Martin Gardner. viii + 96pp. Title and dedication pages in color. 21482-6 Paperbound $1.25

THE BROWNIES, THEIR BOOK, Palmer Cox. Small as mice, cunning as foxes, exuberant and full of mischief, the Brownies go to the zoo, toy shop, seashore, circus, etc., in 24 verse adventures and 266 illustrations. Long a favorite, since their first appearance in St. Nicholas Magazine. xi + 144pp. 6⅝ x 9¼.
21265-3 Paperbound $1.75

SONGS OF CHILDHOOD, Walter De La Mare. Published (under the pseudonym Walter Ramal) when De La Mare was only 29, this charming collection has long been a favorite children's book. A facsimile of the first edition in paper, the 47 poems capture the simplicity of the nursery rhyme and the ballad, including such lyrics as I Met Eve, Tartary, The Silver Penny. vii + 106pp. 21972-0 Paperbound $1.25

THE COMPLETE NONSENSE OF EDWARD LEAR, Edward Lear. The finest 19th-century humorist-cartoonist in full: all nonsense limericks, zany alphabets, Owl and Pussy-cat, songs, nonsense botany, and more than 500 illustrations by Lear himself. Edited by Holbrook Jackson. xxix + 287pp. (USO) 20167-8 Paperbound $2.00

BILLY WHISKERS: THE AUTOBIOGRAPHY OF A GOAT, Frances Trego Montgomery. A favorite of children since the early 20th century, here are the escapades of that rambunctious, irresistible and mischievous goat—Billy Whiskers. Much in the spirit of *Peck's Bad Boy,* this is a book that children never tire of reading or hearing. All the original familiar illustrations by W. H. Fry are included: 6 color plates, 18 black and white drawings. 159pp. 22345-0 Paperbound $2.00

MOTHER GOOSE MELODIES. Faithful republication of the fabulously rare Munroe and Francis "copyright 1833" Boston edition—the most important Mother Goose collection, usually referred to as the "original." Familiar rhymes plus many rare ones, with wonderful old woodcut illustrations. Edited by E. F. Bleiler. 128pp. 4½ x 6⅜. 22577-1 Paperbound $1.25

THE PHILOSOPHY OF THE UPANISHADS, Paul Deussen. Clear, detailed statement of upanishadic system of thought, generally considered among best available. History of these works, full exposition of system emergent from them, parallel concepts in the West. Translated by A. S. Geden. xiv + 429pp.
21616-0 Paperbound $3.00

LANGUAGE, TRUTH AND LOGIC, Alfred J. Ayer. Famous, remarkably clear introduction to the Vienna and Cambridge schools of Logical Positivism; function of philosophy, elimination of metaphysical thought, nature of analysis, similar topics. "Wish I had written it myself," Bertrand Russell. 2nd, 1946 edition. 160pp.
20010-8 Paperbound $1.35

THE GUIDE FOR THE PERPLEXED, Moses Maimonides. Great classic of medieval Judaism, major attempt to reconcile revealed religion (Pentateuch, commentaries) and Aristotelian philosophy. Enormously important in all Western thought. Unabridged Friedländer translation. 50-page introduction. lix + 414pp.
(USO) 20351-4 Paperbound $2.50

OCCULT AND SUPERNATURAL PHENOMENA, D. H. Rawcliffe. Full, serious study of the most persistent delusions of mankind: crystal gazing, mediumistic trance, stigmata, lycanthropy, fire walking, dowsing, telepathy, ghosts, ESP, etc., and their relation to common forms of abnormal psychology. Formerly *Illusions and Delusions of the Supernatural and the Occult.* iii + 551pp. 20503-7 Paperbound $3.50

THE EGYPTIAN BOOK OF THE DEAD: THE PAPYRUS OF ANI, E. A. Wallis Budge. Full hieroglyphic text, interlinear transliteration of sounds, word for word translation, then smooth, connected translation; Theban recension. Basic work in Ancient Egyptian civilization; now even more significant than ever for historical importance, dilation of consciousness, etc. clvi + 377pp. 6½ x 9¼.
21866-X Paperbound $3.95

PSYCHOLOGY OF MUSIC, Carl E. Seashore. Basic, thorough survey of everything known about psychology of music up to 1940's; essential reading for psychologists, musicologists. Physical acoustics; auditory apparatus; relationship of physical sound to perceived sound; role of the mind in sorting, altering, suppressing, creating sound sensations; musical learning, testing for ability, absolute pitch, other topics. Records of Caruso, Menuhin analyzed. 88 figures. xix + 408pp.
21851-1 Paperbound $2.75

THE I CHING (THE BOOK OF CHANGES), translated by James Legge. Complete translated text plus appendices by Confucius, of perhaps the most penetrating divination book ever compiled. Indispensable to all study of early Oriental civilizations. 3 plates. xxiii + 448pp. 21062-6 Paperbound $3.00

THE UPANISHADS, translated by Max Müller. Twelve classical upanishads: Chandogya, Kena, Aitareya, Kaushitaki, Isa, Katha, Mundaka, Taittiriyaka, Brhadaranyaka, Svetasvatara, Prasna, Maitriyana. 160-page introduction, analysis by Prof. Müller. Total of 826pp. 20398-0, 20399-9 Two volumes, Paperbound $5.00

ALPHABETS AND ORNAMENTS, Ernst Lehner. Well-known pictorial source for decorative alphabets, script examples, cartouches, frames, decorative title pages, calligraphic initials, borders, similar material. 14th to 19th century, mostly European. Useful in almost any graphic arts designing, varied styles. 750 illustrations. 256pp. 7 x 10. 21905-4 Paperbound $4.00

PAINTING: A CREATIVE APPROACH, Norman Colquhoun. For the beginner simple guide provides an instructive approach to painting: major stumbling blocks for beginner; overcoming them, technical points; paints and pigments; oil painting; watercolor and other media and color. New section on "plastic" paints. Glossary. Formerly *Paint Your Own Pictures*. 221pp. 22000-1 Paperbound $1.75

THE ENJOYMENT AND USE OF COLOR, Walter Sargent. Explanation of the relations between colors themselves and between colors in nature and art, including hundreds of little-known facts about color values, intensities, effects of high and low illumination, complementary colors. Many practical hints for painters, references to great masters. 7 color plates, 29 illustrations. x + 274pp.
20944-X Paperbound $2.75

THE NOTEBOOKS OF LEONARDO DA VINCI, compiled and edited by Jean Paul Richter. 1566 extracts from original manuscripts reveal the full range of Leonardo's versatile genius: all his writings on painting, sculpture, architecture, anatomy, astronomy, geography, topography, physiology, mining, music, etc., in both Italian and English, with 186 plates of manuscript pages and more than 500 additional drawings. Includes studies for the Last Supper, the lost Sforza monument, and other works. Total of xlvii + 866pp. 7⅞ x 10¾.
22572-0, 22573-9 Two volumes, Paperbound $10.00

MONTGOMERY WARD CATALOGUE OF 1895. Tea gowns, yards of flannel and pillow-case lace, stereoscopes, books of gospel hymns, the New Improved Singer Sewing Machine, side saddles, milk skimmers, straight-edged razors, high-button shoes, spittoons, and on and on . . . listing some 25,000 items, practically all illustrated. Essential to the shoppers of the 1890's, it is our truest record of the spirit of the period. Unaltered reprint of Issue No. 57, Spring and Summer 1895. Introduction by Boris Emmet. Innumerable illustrations. xiii + 624pp. 8½ x 11⅝.
22377-9 Paperbound $6.95

THE CRYSTAL PALACE EXHIBITION ILLUSTRATED CATALOGUE (LONDON, 1851). One of the wonders of the modern world—the Crystal Palace Exhibition in which all the nations of the civilized world exhibited their achievements in the arts and sciences—presented in an equally important illustrated catalogue. More than 1700 items pictured with accompanying text—ceramics, textiles, cast-iron work, carpets, pianos, sleds, razors, wall-papers, billiard tables, beehives, silverware and hundreds of other artifacts—represent the focal point of Victorian culture in the Western World. Probably the largest collection of Victorian decorative art ever assembled—indispensable for antiquarians and designers. Unabridged republication of the Art-Journal Catalogue of the Great Exhibition of 1851, with all terminal essays. New introduction by John Gloag, F.S.A. xxxiv + 426pp. 9 x 12.
22503-8 Paperbound $4.50

EINSTEIN'S THEORY OF RELATIVITY, Max Born. Relativity theory analyzed, explained for intelligent layman or student with some physical, mathematical background. Includes Lorentz, Minkowski, and others. Excellent verbal account for teachers. Generally considered the finest non-technical account. vii + 376pp.
60769-0 Paperbound $2.75

PHYSICAL PRINCIPLES OF THE QUANTUM THEORY, Werner Heisenberg. Nobel Laureate discusses quantum theory, uncertainty principle, wave mechanics, work of Dirac, Schroedinger, Compton, Wilson, Einstein, etc. Middle, non-mathematical level for physicist, chemist not specializing in quantum; mathematical appendix for specialists. Translated by C. Eckart and F. Hoyt. 19 figures. viii + 184pp. 60113-7 Paperbound $2.00

PRINCIPLES OF QUANTUM MECHANICS, William V. Houston. For student with working knowledge of elementary mathematical physics; uses Schroedinger's wave mechanics. Evidence for quantum theory, postulates of quantum mechanics, applications in spectroscopy, collision problems, electrons, similar topics. 21 figures. 288pp. 60524-8 Paperbound $3.00

ATOMIC SPECTRA AND ATOMIC STRUCTURE, Gerhard Herzberg. One of the best introductions to atomic spectra and their relationship to structure; especially suited to specialists in other fields who require a comprehensive basic knowledge. Treatment is physical rather than mathematical. 2nd edition. Translated by J. W. T. Spinks. 80 illustrations. xiv + 257pp. 60115-3 Paperbound $2.00

ATOMIC PHYSICS: AN ATOMIC DESCRIPTION OF PHYSICAL PHENOMENA, Gaylord P. Harnwell and William E. Stephens. One of the best introductions to modern quantum ideas. Emphasis on the extension of classical physics into the realms of atomic phenomena and the evolution of quantum concepts. 156 problems. 173 figures and tables. xi + 401pp. 61584-7 Paperbound $3.00

ATOMS, MOLECULES AND QUANTA, Arthur E. Ruark and Harold C. Urey. 1964 edition of work that has been a favorite of students and teachers for 30 years. Origins and major experimental data of quantum theory, development of concepts of atomic and molecular structure prior to new mechanics, laws and basic ideas of quantum mechanics, wave mechanics, matrix mechanics, general theory of quantum dynamics. Very thorough, lucid presentation for advanced students. 230 figures. Total of xxiii + 810pp.
61106-X, 61107-8 Two volumes, Paperbound $6.00

INVESTIGATIONS ON THE THEORY OF THE BROWNIAN MOVEMENT, Albert Einstein. Five papers (1905-1908) investigating the dynamics of Brownian motion and evolving an elementary theory of interest to mathematicians, chemists and physical scientists. Notes by R. Fürth, the editor, discuss the history of study of Brownian movement, elucidate the text and analyze the significance of the papers. Translated by A. D. Cowper. 3 figures. iv + 122pp.
60304-0 Paperbound $1.50

LAST AND FIRST MEN AND STAR MAKER, TWO SCIENCE FICTION NOVELS, Olaf Stapledon. Greatest future histories in science fiction. In the first, human intelligence is the "hero," through strange paths of evolution, interplanetary invasions, incredible technologies, near extinctions and reemergences. Star Maker describes the quest of a band of star rovers for intelligence itself, through time and space: weird inhuman civilizations, crustacean minds, symbiotic worlds, etc. Complete, unabridged. v + 438pp. 21962-3 Paperbound $2.50

THREE PROPHETIC NOVELS, H. G. WELLS. Stages of a consistently planned future for mankind. *When the Sleeper Wakes,* and *A Story of the Days to Come,* anticipate *Brave New World* and *1984,* in the 21st Century; *The Time Machine,* only complete version in print, shows farther future and the end of mankind. All show Wells's greatest gifts as storyteller and novelist. Edited by E. F. Bleiler. x + 335pp. (USO) 20605-X Paperbound $2.50

THE DEVIL'S DICTIONARY, Ambrose Bierce. America's own Oscar Wilde—Ambrose Bierce—offers his barbed iconoclastic wisdom in over 1,000 definitions hailed by H. L. Mencken as "some of the most gorgeous witticisms in the English language." 145pp. 20487-1 Paperbound $1.25

MAX AND MORITZ, Wilhelm Busch. Great children's classic, father of comic strip, of two bad boys, Max and Moritz. Also Ker and Plunk (Plisch und Plumm), Cat and Mouse, Deceitful Henry, Ice-Peter, The Boy and the Pipe, and five other pieces. Original German, with English translation. Edited by H. Arthur Klein; translations by various hands and H. Arthur Klein. vi + 216pp. 20181-3 Paperbound $2.00

PIGS IS PIGS AND OTHER FAVORITES, Ellis Parker Butler. The title story is one of the best humor short stories, as Mike Flannery obfuscates biology and English. Also included, That Pup of Murchison's, The Great American Pie Company, and Perkins of Portland. 14 illustrations. v + 109pp. 21532-6 Paperbound $1.25

THE PETERKIN PAPERS, Lucretia P. Hale. It takes genius to be as stupidly mad as the Peterkins, as they decide to become wise, celebrate the "Fourth," keep a cow, and otherwise strain the resources of the Lady from Philadelphia. Basic book of American humor. 153 illustrations. 219pp. 20794-3 Paperbound $1.50

PERRAULT'S FAIRY TALES, translated by A. E. Johnson and S. R. Littlewood, with 34 full-page illustrations by Gustave Doré. All the original Perrault stories—Cinderella, Sleeping Beauty, Bluebeard, Little Red Riding Hood, Puss in Boots, Tom Thumb, etc.—with their witty verse morals and the magnificent illustrations of Doré. One of the five or six great books of European fairy tales. viii + 117pp. 8⅛ x 11. 22311-6 Paperbound $2.00

OLD HUNGARIAN FAIRY TALES, Baroness Orczy. Favorites translated and adapted by author of the *Scarlet Pimpernel.* Eight fairy tales include "The Suitors of Princess Fire-Fly," "The Twin Hunchbacks," "Mr. Cuttlefish's Love Story," and "The Enchanted Cat." This little volume of magic and adventure will captivate children as it has for generations. 90 drawings by Montagu Barstow. 96pp. (USO) 22293-4 Paperbound $1.95

ADVENTURES OF AN AFRICAN SLAVER, Theodore Canot. Edited by Brantz Mayer. A detailed portrayal of slavery and the slave trade, 1820-1840. Canot, an established trader along the African coast, describes the slave economy of the African kingdoms, the treatment of captured negroes, the extensive journeys in the interior to gather slaves, slave revolts and their suppression, harems, bribes, and much more. Full and unabridged republication of 1854 edition. Introduction by Malcom Cowley. 16 illustrations. xvii + 448pp. 22456-2 Paperbound $3.50

MY BONDAGE AND MY FREEDOM, Frederick Douglass. Born and brought up in slavery, Douglass witnessed its horrors and experienced its cruelties, but went on to become one of the most outspoken forces in the American anti-slavery movement. Considered the best of his autobiographies, this book graphically describes the inhuman treatment of slaves, its effects on slave owners and slave families, and how Douglass's determination led him to a new life. Unaltered reprint of 1st (1855) edition. xxxii + 464pp. 22457-0 Paperbound $2.50

THE INDIANS' BOOK, recorded and edited by Natalie Curtis. Lore, music, narratives, dozens of drawings by Indians themselves from an authoritative and important survey of native culture among Plains, Southwestern, Lake and Pueblo Indians. Standard work in popular ethnomusicology. 149 songs in full notation. 23 drawings, 23 photos. xxxi + 584pp. 6⅝ x 9⅜. 21939-9 Paperbound $4.50

DICTIONARY OF AMERICAN PORTRAITS, edited by Hayward and Blanche Cirker. 4024 portraits of 4000 most important Americans, colonial days to 1905 (with a few important categories, like Presidents, to present). Pioneers, explorers, colonial figures, U. S. officials, politicians, writers, military and naval men, scientists, inventors, manufacturers, jurists, actors, historians, educators, notorious figures, Indian chiefs, etc. All authentic contemporary likenesses. The only work of its kind in existence; supplements all biographical sources for libraries. Indispensable to anyone working with American history. 8,000-item classified index, finding lists, other aids. xiv + 756pp. 9¼ x 12¾. 21823-6 Clothbound $30.00

TRITTON'S GUIDE TO BETTER WINE AND BEER MAKING FOR BEGINNERS, S. M. Tritton. All you need to know to make family-sized quantities of over 100 types of grape, fruit, herb and vegetable wines; as well as beers, mead, cider, etc. Complete recipes, advice as to equipment, procedures such as fermenting, bottling, and storing wines. Recipes given in British, U. S., and metric measures. Accompanying booklet lists sources in U. S. A. where ingredients may be bought, and additional information. 11 illustrations. 157pp. 5⅝ x 8⅛. (USO) 22090-7 Clothbound $3.50

GARDENING WITH HERBS FOR FLAVOR AND FRAGRANCE, Helen M. Fox. How to grow herbs in your own garden, how to use them in your cooking (over 55 recipes included), legends and myths associated with each species, uses in medicine, perfumes, etc.—these are elements of one of the few books written especially for American herb fanciers. Guides you step-by-step from soil preparation to harvesting and storage for each type of herb. 12 drawings by Louise Mansfield. xiv + 334pp. 22540-2 Paperbound $2.50

PLANETS, STARS AND GALAXIES: DESCRIPTIVE ASTRONOMY FOR BEGINNERS, A. E. Fanning. Comprehensive introductory survey of astronomy: the sun, solar system, stars, galaxies, universe, cosmology; up-to-date, including quasars, radio stars, etc. Preface by Prof. Donald Menzel. 24pp. of photographs. 189pp. 5¼ x 8¼.
21680-2 Paperbound $1.50

TEACH YOURSELF CALCULUS, P. Abbott. With a good background in algebra and trig, you can teach yourself calculus with this book. Simple, straightforward introduction to functions of all kinds, integration, differentiation, series, etc. "Students who are beginning to study calculus method will derive great help from this book." Faraday House Journal. 308pp. 20683-1 Clothbound $2.00

TEACH YOURSELF TRIGONOMETRY, P. Abbott. Geometrical foundations, indices and logarithms, ratios, angles, circular measure, etc. are presented in this sound, easy-to-use text. Excellent for the beginner or as a brush up, this text carries the student through the solution of triangles. 204pp. 20682-3 Clothbound $2.00

TEACH YOURSELF ANATOMY, David LeVay. Accurate, inclusive, profusely illustrated account of structure, skeleton, abdomen, muscles, nervous system, glands, brain, reproductive organs, evolution. "Quite the best and most readable account,' *Medical Officer.* 12 color plates. 164 figures. 311pp. 4¾ x 7.
21651-9 Clothbound $2.50

TEACH YOURSELF PHYSIOLOGY, David LeVay. Anatomical, biochemical bases; digestive, nervous, endocrine systems; metabolism; respiration; muscle; excretion; temperature control; reproduction. "Good elementary exposition," *The Lancet.* 6 color plates. 44 illustrations. 208pp. 4¼ x 7. 21658-6 Clothbound $2.50

THE FRIENDLY STARS, Martha Evans Martin. Classic has taught naked-eye observation of stars, planets to hundreds of thousands, still not surpassed for charm, lucidity, adequacy. Completely updated by Professor Donald H. Menzel, Harvard Observatory. 25 illustrations. 16 x 30 chart. x + 147pp. 21099-5 Paperbound $1.25

MUSIC OF THE SPHERES: THE MATERIAL UNIVERSE FROM ATOM TO QUASAR, SIMPLY EXPLAINED, Guy Murchie. Extremely broad, brilliantly written popular account begins with the solar system and reaches to dividing line between matter and nonmatter; latest understandings presented with exceptional clarity. Volume One: Planets, stars, galaxies, cosmology, geology, celestial mechanics, latest astronomical discoveries; Volume Two: Matter, atoms, waves, radiation, relativity, chemical action, heat, nuclear energy, quantum theory, music, light, color, probability, antimatter, antigravity, and similar topics. 319 figures. 1967 (second) edition. Total of xx + 644pp. 21809-0, 21810-4 Two volumes, Paperbound $5.00

OLD-TIME SCHOOLS AND SCHOOL BOOKS, Clifton Johnson. Illustrations and rhymes from early primers, abundant quotations from early textbooks, many anecdotes of school life enliven this study of elementary schools from Puritans to middle 19th century. Introduction by Carl Withers. 234 illustrations. xxxiii + 381pp.
21031-6 Paperbound $2.50

MATHEMATICAL PUZZLES FOR BEGINNERS AND ENTHUSIASTS, Geoffrey Mott-Smith. 189 puzzles from easy to difficult—involving arithmetic, logic, algebra, properties of digits, probability, etc.—for enjoyment and mental stimulus. Explanation of mathematical principles behind the puzzles. 135 illustrations. viii + 248pp.
20198-8 Paperbound $1.75

PAPER FOLDING FOR BEGINNERS, William D. Murray and Francis J. Rigney. Easiest book on the market, clearest instructions on making interesting, beautiful origami. Sail boats, cups, roosters, frogs that move legs, bonbon boxes, standing birds, etc. 40 projects; more than 275 diagrams and photographs. 94pp.
20713-7 Paperbound $1.00

TRICKS AND GAMES ON THE POOL TABLE, Fred Herrmann. 79 tricks and games—some solitaires, some for two or more players, some competitive games—to entertain you between formal games. Mystifying shots and throws, unusual caroms, tricks involving such props as cork, coins, a hat, etc. Formerly *Fun on the Pool Table.* 77 figures. 95pp. 21814-7 Paperbound $1.00

HAND SHADOWS TO BE THROWN UPON THE WALL: A SERIES OF NOVEL AND AMUSING FIGURES FORMED BY THE HAND, Henry Bursill. Delightful picturebook from great-grandfather's day shows how to make 18 different hand shadows: a bird that flies, duck that quacks, dog that wags his tail, camel, goose, deer, boy, turtle, etc. Only book of its sort. vi + 33pp. 6½ x 9¼. 21779-5 Paperbound $1.00

WHITTLING AND WOODCARVING, E. J. Tangerman. 18th printing of best book on market. "If you can cut a potato you can carve" toys and puzzles, chains, chessmen, caricatures, masks, frames, woodcut blocks, surface patterns, much more. Information on tools, woods, techniques. Also goes into serious wood sculpture from Middle Ages to present, East and West. 464 photos, figures. x + 293pp.
20965-2 Paperbound $2.00

HISTORY OF PHILOSOPHY, Julián Marias. Possibly the clearest, most easily followed, best planned, most useful one-volume history of philosophy on the market; neither skimpy nor overfull. Full details on system of every major philosopher and dozens of less important thinkers from pre-Socratics up to Existentialism and later. Strong on many European figures usually omitted. Has gone through dozens of editions in Europe. 1966 edition, translated by Stanley Appelbaum and Clarence Strowbridge. xviii + 505pp. 21739-6 Paperbound $3.00

YOGA: A SCIENTIFIC EVALUATION, Kovoor T. Behanan. Scientific but non-technical study of physiological results of yoga exercises; done under auspices of Yale U. Relations to Indian thought, to psychoanalysis, etc. 16 photos. xxiii + 270pp.
20505-3 Paperbound $2.50

Prices subject to change without notice.
Available at your book dealer or write for free catalogue to Dept. GI, Dover Publications, Inc., 180 Varick St., N. Y., N. Y. 10014. Dover publishes more than 150 books each year on science, elementary and advanced mathematics, biology, music, art, literary history, social sciences and other areas.